Landscape Design, Evaluation and Management Created by Novel Technologies

Landscape Design, Evaluation and Management Created by Novel Technologies

Editors

Xiwei Shen
Mary G. Padua
Bo Zhang
Niall Kirkwood
Yang Song
Rosalea Monacella

MDPI

Basel • Beijing • Wuhan • Barcelona • Belgrade • Novi Sad • Cluj • Manchester

Editors
Xiwei Shen
School of Architecture
University of Nevada
Las Vegas
USA

Mary G. Padua
Department of Horticulture and Landscape Architecture
Oklahoma State University
Stillwater
USA

Bo Zhang
School of Architecture
Clemson University
Clemson
USA

Niall Kirkwood
Harvard Graduate School of Design
Harvard University
Cambridge
USA

Yang Song
Department of Landscape Architecture & Urban Planning
Texas A&M University
College Station
USA

Rosalea Monacella
Department of Landscape Architecture
Harvard University
Cambridge
USA

Editorial Office
MDPI AG
Grosspeteranlage 5
4052 Basel, Switzerland

This is a reprint of articles from the Special Issue published online in the open access journal *Land* (ISSN 2073-445X) (available at: https://www.mdpi.com/journal/land/special_issues/C745I5H64M).

For citation purposes, cite each article independently as indicated on the article page online and as indicated below:

Lastname, A.A.; Lastname, B.B. Article Title. *Journal Name* **Year**, *Volume Number*, Page Range.

ISBN 978-3-7258-2251-5 (Hbk)
ISBN 978-3-7258-2252-2 (PDF)
doi.org/10.3390/books978-3-7258-2252-2

Cover image courtesy of Bo Zhang

Contents

About the Editors

Xiwei Shen

Xiwei Shen is interested in the design theory of landscape architecture and the technology of environmental/ecological restoration in landscape design. As a landscape practitioner, Shen has received numerous design awards, including the National ASLA Awards, the Boston Society of Landscape Architects ASLA Awards, the South Carolina Chapter ASLA Awards, and the Landscape Institute (LI) Awards. As a scholar, his articles have been published in numerous peer-reviewed journals. He has served as an editor and reviewer for multiple journals. Shen holds a Ph.D. in planning design and built environment from Clemson University, a Master of Landscape Architecture from Harvard Graduate School of Design, and a Bachelor of Landscape Architecture from Louisiana State University.

Mary G. Padua

Mary G. Padua is Professor of Landscape Architecture at Clemson University's School of Architecture at the College of Architecture, Art and Construction in Clemson, South Carolina, USA. She is an award-winning writer and contemporary theorist, licensed landscape architect, and visual artist, whose published research focuses on adaptive urbanism and the meaning of public space. Dr. Padua served four years as the inaugural Department Chair for the Department of Landscape Architecture at Clemson University's College of Architecture, Arts and Humanities. She is an expert on China's vast urban experiment through the lens of post-Mao cultural landscapes, as illustrated in her award-winning book, *Hybrid Modernity: The Public Park in Late 20th Century China* (Routledge 2020). Dr. Padua is regularly invited to lecture and conduct workshops, and has engaged at international symposia and universities on four continents. She has over twenty years of professional experience as a landscape architect and urban designer working in multidisciplinary settings, involving collaborations with Pamela Burton, FASLA, Cesar Pelli & Associates, the late Lawrence Halprin, and the late Charles W. Moore, among others. She maintains MGP Studio: art design research, a critically minded practice focused on human-centered design and social justice. Additionally, she is a fine art photographer with original works held in public and private collections. She holds degrees from the University of California, Berkeley (BA Landscape Architecture with a minor in Visual Design), UCLA (MA Architecture with the Certificate in Urban Design), and Edinburgh College of Art, University of Edinburgh, Scotland (PhD in Landscape Architecture).

Bo Zhang

Bo Zhang is an award-winning landscape architect, urban designer, architect, LEED-certified professional, and associate professor at Oklahoma State University. Dr. Zhang currently serves as Vice President of Research (VPR) at the Council of Educators in Landscape Architecture (CELA) and Recognition Committee Chair at the Environment Design Research Association (EDRA). Dr. Zhang's research interests include crowdsourcing data methods in design disciplines, ecological design and planning methodology, and intercultural communication in designed landscapes in the US in the 19th century. He has received more than 20 awards for his research and design, including the ASLA National Professional Honor Award of Research (2019), the CELA National Award of Excellence in Teaching (2022), the EDRA Place Award in Research (2023), and the Landscape Institute's Landscape Research and Innovation Finalist Award (2023). He is also the author of the book *Research Methods in Architecture, Urban Planning, and Landscape Architecture*, published in March 2023.

Niall Kirkwood

Niall Kirkwood FASLA is a Charles Eliot Professor of Landscape Architecture at the Harvard Graduate School of Design, where he has taught and carried out research, publishing and consulting since joining the faculty in 1992. He was educated and licensed as a professional landscape architect and architect in the United Kingdom and as a professional landscape architect in the United States. From 2003 to 2009, he was the thirteenth Chair of the Department of Landscape Architecture, the oldest such program in North America, founded in 1901 by Frederick Law Olmsted Jr. and Arthur Shurcliff. From 1999 to 2003 and 2005 to 2007, he was Director of the Master's in Landscape Architecture Degree Programs (MLA), and from 1999 to 2003, he was the coordinator of the "Design and Environment" track of the Master's in Design Studies Program (MDes). Kirkwood serves as Chairman of the GSD Faculty Review Board and Academic Misconduct Panel and has served as a faculty member of the Harvard Medical School Center for Health and the Global Environment, the Harvard University Center for the Environment and a member of the faculty steering committee of The Harvard Global Health Institute. He currently serves as the GSD representative on Harvard University's Title IX Policy Review Advisory Committee and the Vice Provost for Advanced Learning's (VPAL) Planning Council.

Yang Song

Yang Song completed both his Ph.D. and his MLA graduate studies at Clemson University. He works at the intersection between landscape architecture, community planning, and urban design. His academic activities have a strong focus on the role of public placemaking in community health and resiliency. He has a long-standing interest in applying digital technology and data science in landscape research and design. He studies the usage of urban public spaces and develops human-based design theories that enhance active living and economic resilience. His design and research works have been recognized by multiple national and international awards, including the Honor of Research Award from the American Society of Landscape Architects and the Excellence of Research certificate from the Environmental Design and Research Association.

Rosalea Monacella

Rosalea Monacella is a registered landscape architect and has undertaken research on a number of cities around the world, generated urban masterplans for cities in China, USA, South America, Europe and Australia that explore design at the nexus of the urban and natural environments, and has been the recipient of many national and international awards and grants related to her practice based research as co-founder of the OUTR Research Lab at RMIT University Melbourne, Australia. Rosalea's expertise is in the transitioning of the urban environment through a careful indexing and shifting of dynamic resource flows that inform the landscape of contemporary cities. Her research brings together complex urban issues and advanced digital modeling techniques for the generation of sustainable urban futures. Her design approach is one that simultaneously considers forces from the 'ground-up + top-down' through a careful and rigorous exploration of complex economic, ecological, and social systems that shape an ever-changing city. For ten years, she has acted as chief editor leading the development of Kerb Journal to become a significant publication in the discipline that engages and challenges the discourse of landscape architecture. She holds a PhD from RMIT University, a Masters in Landscape Urbanism from the AA School London, UK, and a Bachelor of Architecture RMIT University.

Preface

The landscape, once a passive canvas shaped solely by nature, has for centuries been a dynamic interplay between humanity and the environment. Today, this conversation is undergoing a radical transformation. With the advent of advanced technologies, landscape architecture is no longer confined to the realm of esthetic and functional design; it has become a dynamic discipline at the forefront of innovation, sustainability, and human well-being. This Special Issue represents innovative, experimental, and compelling research from leading experts and offers a glimpse into this exciting new frontier. By showcasing the latest advancements in technology and their application to landscape design, evaluation, and management, these articles illuminate a path toward a future where landscapes are not merely crafted but meticulously engineered for optimal human and ecological benefit.

This collection of research delves into the heart of a rapidly evolving field, where the boundaries between art, science, and technology are increasingly blurred. The articles presented here offer a comprehensive overview of the latest trends, challenges, and opportunities in landscape architecture, demonstrating how technological innovation can be harnessed to create more sustainable, resilient, and equitable environments.

Part 1: The Dawning of Technological Landscape Architecture (TLA)

The opening article by Shen, Padua, and Kirkwood delves into the burgeoning field of TLA. They trace its evolution from its 19th-century roots to its present-day integration with sophisticated digital tools. The authors highlight a critical gap in the current literature—a lack of a comprehensive understanding of TLA's impact on core domains and research themes within landscape architecture. Their study explores how TLA is influencing the generation of new knowledge and its potential to reshape design and sustainability practices. They emphasize the need to differentiate TLA from generic digital tools and identify emerging specializations within this exciting field.

Part 2: Revolutionizing Design Practices with Cutting-Edge Technologies

The subsequent articles showcase the transformative potential of TLA in design workflows. Chen et al. introduce a novel system utilizing Generative Adversarial Networks (GANs) to rapidly convert black-and-white sketches into color renderings for park design. This innovative approach streamlines the iteration process, allowing designers to explore a wider range of possibilities and expedite project development.

Part 3: Optimizing Landscapes for User Experience and Public Health

Public spaces play a vital role in fostering healthy communities. Chen et al. investigate the influence of park pathway features on walking and running activities. Their research, based on extensive field observations, identifies key factors that influence activity density and intensity. This knowledge empowers landscape architects to design parks that promote physical well-being and encourage active lifestyles.

He et al. delve deeper into the realm of urban planning, exploring the impact of multimodal access on property values and land economies. Their research utilizes advanced spatial analysis techniques to illuminate the complex interplay between metro accessibility and urban development. This knowledge empowers policymakers to develop targeted land-use strategies that promote sustainable urban growth.

Zeng et al. propose a groundbreaking method for evaluating public spaces using big data.

Their approach analyzes visitor reviews on travel websites, enabling a data-driven assessment of functionality, design, and landscape elements. This methodology provides valuable insights for optimizing public spaces and ensuring they cater to the needs of diverse user groups.

Hou et al. address the critical issue of green space inequality within urban areas. Their study, employing advanced data collection techniques such as drone imagery, assesses the accessibility of different types of UGSs, encompassing factors related to physical activity. This nuanced analysis reveals disparities in UGS provision and highlights the need for planning strategies that promote equitable access to green spaces for all community members.

This Special Issue extends beyond technical advancements, exploring the social and cultural dimensions of landscape architecture. Ding et al. employ content mining techniques to compare tourist perceptions with official promotional materials of two ancient canal towns. Their findings reveal a significant disconnect between how tourists experience these spaces and how they are presented in promotional narratives. This knowledge underscores the importance of engaging diverse user groups in the design and development of public spaces.

Wang et al. assess public satisfaction with China's innovative "sponge city" initiative—a stormwater management program that prioritizes natural solutions. Their research highlights the importance of social perception and public engagement in the success of environmental projects. By understanding these factors, policymakers can develop strategies that foster long-term community buy-in and ensure the sustainability of urban development initiatives.

Part 4: Redefining Public Space Evaluation and Fostering Inclusive Management

The final two articles address the crucial role of landscape architecture in safeguarding ecological integrity in a rapidly urbanizing world. Zheng et al. utilize GIS and Machine Learning to model the impact of urbanization on bobcat habitats. Their research identifies key factors influencing habitat suitability and outlines conservation strategies such as wildlife corridors and strategic fencing to mitigate road-related threats. This study offers valuable insights for landscape architects and urban planners seeking to integrate wildlife conservation into urban design projects.

Qiu et al. explore the concept of "professional knowledge abstraction" within landscape architecture. They argue that rigorous knowledge acquisition and dissemination are crucial for the continued evolution and sustainability of the profession. Their research highlights the need for landscape architects to embrace higher-level, research-based knowledge and integrate it into design practices.

The articles within this Special Issue collectively paint a compelling portrait of a discipline in flux. By embracing the power of technology, landscape architecture is poised to address some of the most pressing challenges of our time, from climate change and urbanization to public health and social equity. The research presented here offers a roadmap for practitioners, policymakers, and researchers seeking to harness the full potential of this dynamic field.

As we look to the future, it is imperative that we continue to invest in research and development, fostering interdisciplinary collaboration and cultivating a new generation of landscape architects equipped with the skills and knowledge to shape a sustainable and resilient world. This Special Issue serves as a catalyst for innovation, inspiring us to think beyond traditional boundaries and explore uncharted territories in the pursuit of a more harmonious relationship between humans and the natural world.

Xiwei Shen, Mary G. Padua, Bo Zhang, Niall Kirkwood, Yang Song, and Rosalea Monacella

Editors

land

Article

Transformative Impact of Technology in Landscape Architecture on Landscape Research: Trends, Concepts and Roles

Xiwei Shen [1,*], Mary G. Padua [2] and Niall G. Kirkwood [3]

1 School of Architecture, University of Nevada, Las Vegas, NV 89154, USA
2 School of Architecture, Clemson University, Clemson, SC 29631, USA; mgpadua@clemson.edu
3 Graduate School of Design, Harvard University, Cambridge, MA 02138, USA; kirkwood@gsd.harvard.edu
* Correspondence: xiwei.shen@unlv.edu

Abstract: The role of technology in landscape architecture (TLA) has significantly evolved since the 19th century, increasingly integrating with digital tools and technologies in the 21st century. Despite its growing importance, there is a notable deficiency in the scholarly literature regarding the progression of TLA trends and their interplay with the core domains and research themes within landscape research. The influence of TLA on landscape research remains ambiguous, especially concerning its ability to generate new knowledge and impact design and sustainability practices. Furthermore, there is a critical need to delineate how TLA differs from allied general digital technology tools and to identify specific specializations that are emerging within the TLA field. To explore the above gaps, this study utilized a mixed methods approach involving secondary data from peer-reviewed publications, primary data from the archival research of winning projects, and expert interviews based on the two major research types of "Research through Design (RTD)" and "Research for Design (RFD)" to explore the TLA's contribution. This research is significant as it: (1) identified the trend of TLA; (2) conceptualized the TLA, and (3) identified its role in relation to the core domains and research themes of landscape research.

Keywords: technology in landscape architecture; theory-building; digital tools and technology; core domains; trends and theme

Citation: Shen, X.; Padua, M.G.; Kirkwood, N.G. Transformative Impact of Technology in Landscape Architecture on Landscape Research: Trends, Concepts and Roles. *Land* **2024**, *13*, 630. https://doi.org/10.3390/land13050630

Academic Editor: Diane Pearson

Received: 7 April 2024
Revised: 4 May 2024
Accepted: 6 May 2024
Published: 8 May 2024

1. Introduction

Landscape architecture, recognized as an interdisciplinary field, continuously expands its boundaries to incorporate related disciplines like ecology, geology, and architecture [1–3]. Despite this broad scope, the discipline has yet to clearly define its core technologies. Since the late 20th century, numerous scholars in landscape architecture have highlighted digital tools and technologies as pivotal for conducting landscape research [4,5]. However, this perspective is not universally accepted among scholars from allied disciplines. This research aims to bridge these differing views by exploring the relationship between applied digital tools and technologies and the core domains of landscape architecture. It will investigate research themes, analyze trends, and classify outcomes to establish a foundational understanding of the conceptualization of technology in landscape architecture. This study is intended to be a benchmark in landscape research, contributing to the development of a grounded theory and refining the definition of the discipline for the 21st century.

1.1. Trends of Technology in Landscape Research since the Late 20th Century

In the late 20th and 21st centuries, the emergence of technology in landscape research was dominated by the application of digital tools and technology (DTT) [6–11]. Digital tools and technology within landscape architecture research have been observed within topics or specializations such as visualization, geodesign, big data/data analytics, green-blue infrastructure, techniques, and research instruments, among others [12]. This suggests that digital tools and technology have been assisting in expanding the scope of research within the landscape discipline. In this context, digital tools and technology appear to act as a medium and a path to explore the emergence of technology in landscape architecture (TLA). According to several studies on research trends and the theoretical development of the landscape architecture discipline, computers and other types of technological hardware (3D printers, robotics) and software (rendering and 2D representational programs (such as Adobe Suites) ArcGIS Pro, and satellite mapping, among others) have evolved in landscape research as digital tools and technologies, and as a major topic and category for TLA in the 21st century [6–11,13–18].

To be specific, Powers and Walker [6] highlighted the integration of construction technologies, with a particular focus on performance and materials. Also, the sophisticated use of materials, operational methodologies, and system construction have been identified by Margolis and Robinson [13]. Techniques and the utilization of Geographic Information Systems (GIS) for computer graphics and visualization have been discussed by Gobster et al. [7]. Steinitz [19], in particular, concentrated on the technique aspect, and on enhancing GIS's capabilities in relation to landscape research. Also, a notable trend in the literature is the pivot towards digital software and technologies, with a particular focus on advanced techniques, as indicated by authors such as Amoroso [20,21], Amoroso and Hargreaves [22], and Cushing and Renata [8]. Meijering et al. [9] continued this trend, reinforcing the importance of digital software and technologies in shaping the field. There is a growing emphasis on computational design, artificial intelligence, and machine learning among numerous scholars [14–16,23,24], who have collectively underlined the convergence of representation, modeling, animation, responsive environments, data sensing, and the use of artificial intelligence in landscape research. Vicenzotti et al. [10], on the other hand, shed light on the relevance of construction technologies alongside the selection of planting and materials, whereas Langley et al. [17] concentrated on digital software. Newman et al. [11] broadened the scope by integrating tools for virtual reality and geodesign, applying them to applications such as visualization and social media, which opened up new avenues for data analytics and evaluation research within landscape architecture.

The above literature indicates a close relationship between digital tools, technology and TLA. Thus, it represents a breakthrough in exploring the trends, conceptualization, and roles of TLA in landscape research in the 21st century.

1.2. Role of Technology and Core Domains for Landscape Research since the Late 20th Century

To continue, understanding the role of digital tools and technology necessitates an analysis of the core body of knowledge of landscape architecture so as to help reach the goal of this research. This knowledge core has been defined in terms of "domains", "core domains", or in terms of categories and emerging topics of knowledge within the landscape architecture discipline. Key scholarly works have explored the landscape core domains and are summarized chronologically. The Landscape Architecture Body of Knowledge Study Report [25] was a foundational report based on survey research designed to answer questions about the landscape profession's core competencies and the fundamental body of knowledge expected of all graduates of accredited landscape architecture programs. Other studies examine current landscape core domains, and involve analyses of the content of the primary peer-reviewed landscape journals and expert interviews [6–11,17]. Analyzing the above literature, nine landscape core domains were formulated and ranked. They are Landscape Planning and Ecology (C1), Human and Environment Relationships (C2), History, Culture and Theory (C3), Design Education and Pedagogy (C4), Sustainability (C5), Urban Design (C6), Landscape Performance (C7), Landscape Design and Implementation (C8), and Digital Tools and Technology (C9). The formulation of these nine domains, noted below, was achieved through an analysis of the number of times or frequency at which a particular core domain was indicated, as below (Figure 1).

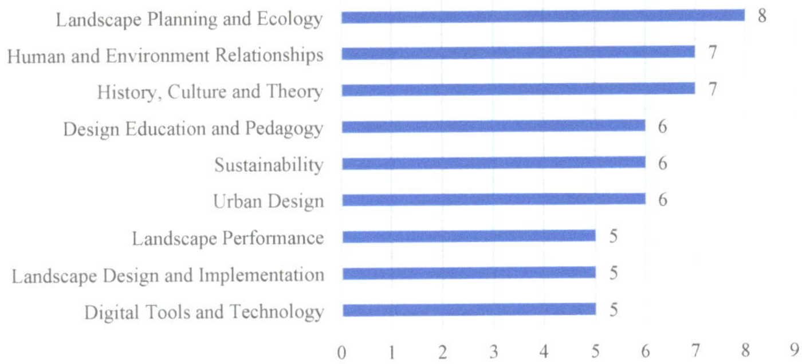

Figure 1. Nine primarily considered core domains in landscape research.

Figure 1 identifies digital tools and technology as a core domain, supporting the initial claim that landscape architecture scholars consider digital tools and technology essential for research in the field. However, Figure 1 does not adequately discuss the core technologies specific to landscape research, leading to potential confusion between digital tools and TLA. It is worth questioning whether these concepts are conflated by scholars or if TLA has been insufficiently defined. Previous studies [6,11] have noted similar concerns.

1.3. TLA from Digital Tools and Technology and Priorities Research Approaches

Carl Steinitz emphasizes the importance of digital tools and technology in exploring various subjects such as climate, geology, hydrology, ecology, and perception. He highlights its role in generating new research outcomes, which he categorizes as aspects of TLA [26]. Frayling [27] notes two priority research approaches within landscape architecture that contribute to generating it: (1) Research through Design (RTD) and (2) Research for Design (RFD). RTD refers to researchers who generate new technology by understanding the current state and then suggesting an improved future state in the form of a design [28]. RFD is another research approach that landscape scholars utilize to generate new technology within the landscape discipline. It refers to scholarly research that informs design as a way to improve the quality of the designed artifact and to increase its reliability. The designer then translates such knowledge to substantiate the design, with examples illustrated by Deming and Swaffield [29] (pp. 90–100). In Frayling's words [27], this "gathering of reference materials" culminates as a product. The major sources for RTD are professional projects, and the major sources for RFD are based on scientific research from disciplines related to the field of landscape architecture. However, works from both RTD and RFD are rarely discussed in terms of their contributions to new knowledge in 21st century TLA, trends and themes of landscape architecture research, or ways these emergent technology-based strategies could be framed now, or applied in the future [30,31].

The literature discussed above indicates that TLA has not been investigated meaningfully. This suggests a gap that has been exasperated by rapid changes and innovations in technology, particularly as regards the common deployment of personal computers and the use of computer workstations among landscape practitioners and researchers. This research also suggests that digital tools and technology, landscape core domains, and trends and themes in landscape architecture research appear to be interconnected. Yet, few studies have examined the phenomenon of TLA, its related specializations, and the ways that digital tools cross and intersect the various landscape core domains.

1.4. Research Goals and Questions

Between Sections 1.1 and 1.3, the text identifies three significant gaps that obscure the role of TLA in landscape research:

- The ambiguous relationship between digital tools and technology and TLA;
- The unclear identification of TLA within the core domain of landscape architecture;

- The lack of conceptualization and categorization for TLA.

To fill the above gaps, understanding the shift in the relationship among TLA, landscape core domains, digital tools and technology and research themes helps us to better explore the role of TLA. Hence, the research will focus on the following research questions:

- RQ1—How can the relationship between TLA, the core domains of landscape architecture, digital tools and technology, and research themes be interpreted?
- RQ2—How should TLA be conceptualized based on these explored relationships?

This work is instrumental in laying the foundation for defining core domains and related emerging topics. It also establishes a theoretical framework for investigating TLA, along with potential correlations or interrelationships among the emerging topics within these core domains of the landscape discipline. The intent of this research is to bridge this knowledge gap by systematically exploring the role of TLA in the 21st century. Based on the findings from the literature review, the research questions have been refined into a series of primary and secondary questions that reflect these identified gaps.

2. Methodology

Figure 2 provides an overview of the mixed methods (both for data collection and data analysis) utilized for this exploratory study. Secondary data were drawn from the major landscape architecture peer-reviewed publications from certain periods that have yet to be analyzed. Primary data were drawn from archival research on projects that received awards from ASLA from 2005 to 2021, as well as expert interviews. This research also incorporated grounded theory, whereby content analysis of the primary data generated from responses to the semi-structured interviews of "experts" (practitioners and scholars) helped determine patterns, categories, and codes.

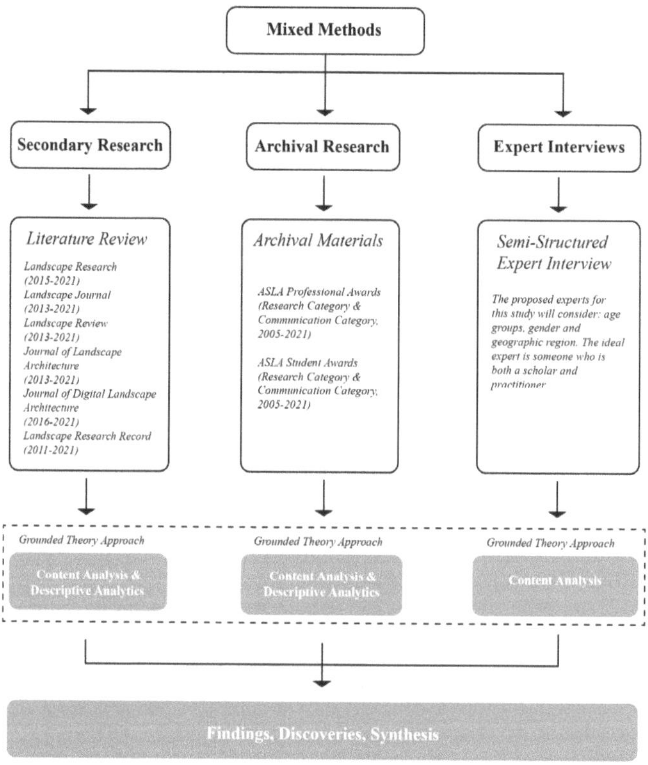

Figure 2. Analytical framework.

2.1. Data Collection

2.1.1. Stage One: Secondary Research

The initial phase of data collection focused on secondary research, specifically tailored towards RFD within the realm of landscape architecture. This comprehensive review targeted peer-reviewed literature from a select set of journals between 2011 and 2021, including the Journal of Digital Landscape Architecture, Landscape Research, Landscape Journal, Landscape Review, Journal of Landscape Architecture, and Landscape Research Record. The primary goal was to unearth and analyze studies specifically addressing the integration and application of digital tools and technology within RFD. This methodical scrutiny was designed to collate data on the research objectives, methodologies, tools utilized, and the significant outcomes from each study, providing a solid foundation for understanding how digital tools and technology contribute to and enhance the design process in landscape architecture.

2.1.2. Stage Two: Archival Research

The second data collection stage involved primary data drawn from archival research and the ASLA archive of award-winning annual projects. As stated earlier, Zimmerman and Forlizzi [28] noted that the major source of RTD is derived from design research projects. Primary data were collected through a systematic review of the ASLA archive of award-winning projects by professionals and students in the "Research" and "Communication" categories from 2005 to 2021. These two categories represent RTD's standard of excellence. Key data were drawn from the project statement, narrative, and captions from award-winning presentation boards and images.

2.1.3. Stage Three: Expert Interviews

The third stage of data collection consisted of expert interviews. Primary data were gathered from the responses to semi-structured (open-ended) survey questions. This interview process consisted of a one-on-one dialogue between the researcher and participant, guided by a flexible interview protocol that was supplemented with follow-up questions, probes, and interactive discussions. This complied with social science interview standards and allowed the researcher to collect open-ended data from the interviewees' responses in order to explore their thoughts, feelings, and beliefs about a particular topic [32]. These responses provided primary data. To achieve reliability and validity in the data, the experts represented a range of ages, genders, and geographic regions. All invited experts had been scholars and design practitioners for at least ten years. Audio recordings of the expert interviews allowed for effective and fluid communication; the recordings were later transcribed. According to Creswell [33], 15–20 experts represents an appropriate sample population for expert interviews. Thus, 18 experts served as the target number for this study. For the semi-structured interviews with open-ended questions, the protocol and procedure began by describing the context of the research and then asking the fundamental questions as follows:

1. What are your thoughts on conceptualizing TLA in the 21st century?
2. What are your thoughts on the relationships between TLA and landscape core domains?
3. What are your thoughts on the relationships of research themes and associated topics with TLA?
4. What are the differences between digital tools and technology and TLA?
5. What methods of inquiry do you apply in your research, and what is the meaning of TLA to your research?

2.2. Data Analysis

2.2.1. Data from Secondary Research

Content analysis and descriptive analytics were applied to the data gathered in the first stage of Secondary Research. The major components considered for analysis included landscape core domains, research topics, methods of inquiry, research strategies, digital tools and technology, and TLA specializations. Table 1 illustrates how the data from secondary research were sorted and organized for the analysis. The reference number, notably, represents the nine landscape core domains summarized in Figure 1, Section 1.2. Appendix A shows the full list of datasets [34].

Table 1. Data collection sample.

Reference Number	*C1*
Article Title	*Digital Age for Observations: The Use of GIS for Analyzing Observations and Behavior Mapping*
Core Domain	*Landscape design and implementation*
Research Topic	*Landscape assessment*
Method of Inquiry	*Secondary research—RFD*
Digital Tools and Technology Incorporated	*GIS*
TLA Specializations	*A tool for observing the water experiences of children (TOWEC)*

To further illustrate the Table 1, the data from secondary research were sorted, and the majority claimed Research for Design (RFD) as their dominant "Method of Inquiry". In addition, based on the data analysis strategies in grounded theory, content analysis comprised two steps to classify the data taken from the "Abstract", "Research Method", and "Conclusions" of each sample to "Digital Tools and Technology Incorporated", "Research Topic" and "TLA Specializations". The categories in the "Core Domain" were drawn from Figure 1. After the data were converted from samples to Table 1, descriptive analytics was applied to identify and determine specializations, trends, and themes related to digital tools and technology and landscape core domains.

2.2.2. Data from Archival Research

Data gathered in this second stage of archival research and reviews of ASLA Award-winning projects utilized content analysis and descriptive analytics. In this archival review, the content analysis involved extracting keywords from the "Project Statement", "Method", and "Conclusions" for each ASLA Award-winning project, which were then categorized by applying the same data classification method as used in Stage Two in Section 2.1.1 to landscape core domains, research themes, and associated topics and digital tools and technology (Table 1). Descriptive analytics helped with pattern recognition and with determining themes and trends in relation to digital tools and technology, landscape core domains, and landscape research topics.

2.2.3. Data from Expert Interviews

The content analysis of the responses to the expert interviews (primary data) was performed via descriptive analytics. Similar to the two datasets from the secondary research (select literature) and archival research of ASLA Award-winning projects, the objective was to discover the relationship between digital tools and technology, specializations, trends and themes, and landscape core domains for TLA in the 21st century. The data from expert interviews were sorted into two groups based on their major research approaches (RTD or RFD). This assisted with the subsequent analytical and synthetic stages, which will be discussed later.

3. Results

A total of 298 peer-reviewed articles were reviewed for the secondary research phase and represent RFD. A total of 42 ASLA Award-winning projects from the "Research" and "Communication" categories were reviewed during the archival research phase and represent RTD. Eighteen experts were interviewed for the Expert Interview phase. The data from the Expert Interviews were classified as either RTD or RFD according to each interviewee's expertise.

3.1. Findings from Secondary Research (RFD)

3.1.1. RFD Trends of the Landscape Core Domains Intersecting with TLA

This section discusses the analysis of landscape core domains intersecting with TLA in Secondary Research. It examines trends based on patterns of annual changes and the frequency of TLA-related research intersecting with each core domain (Figure 3). This research explored the TLA in the 21st century based on research incorporating digital tools and technology, meaning that digital tools and technology was not considered with the other landscape core domains. Thus, they were excluded from this analysis. The findings indicate a trend for research based on RFD emerging for TLA, intersecting with "Landscape Performance" and "Landscape Planning and Ecology" from 2013 to 2018. From 2018 to 2021, TLA's intersection with "Landscape Planning and Ecology", "Human and Environment Relationships", and "Sustainability" appeared to increase more rapidly than in the other landscape core domains.

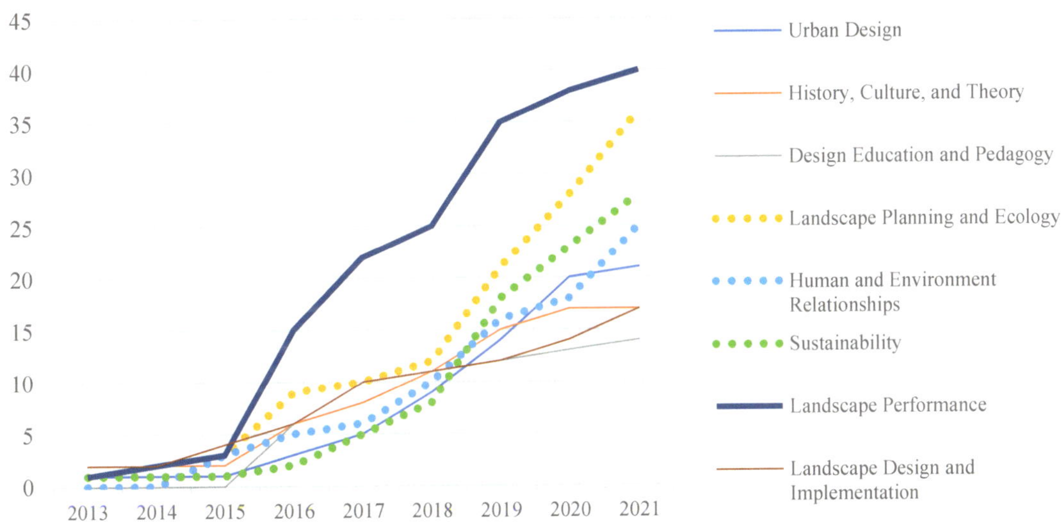

Figure 3. Secondary research—finding: yearly annual change of intersected landscape core domains with TLA (2013–2021).

3.1.2. RFD Trends of the Research Themes Intersecting with TLA

Based on the analysis of data collected through secondary research, 14 themes were identified, referring to the 2022 CELA research tracks [12], and organized according to the landscape core domains. The frequency of research themes emerging in the database is shown in Table 2. As noted in Figure 1, nine core domains were utilized for this study: Human and Environment Relationships (HER); History, Culture and Theory (HCT); Design Education and Pedagogy (DEP); Landscape Planning and Ecology (LPE); Urban Design (UD); Sustainability; Landscape Performance (LP); Landscape Design and Implementation (LDI), and Digital Tools and Technology (DTT).

Table 2. Secondary Research—Analysis and Frequency: Research Themes for TLA.

Research Themes	Freq.	LPE	HCT	HER	UD	SS	DEP	LDI	LP
1 Environmental Monitor and Assessment	112	x			x	x		x	x
2 Coastal Resilience, Hazard and Water Management	91	x				x		x	x
3 Urban Public Space Design and Management	85	x		x	x	x		x	x
4 Ecology Restoration Enhancement	72	x		x	x			x	x
5 Construction Materials and Infrastructure	51	x			x	x	x		x
6 Climate Change and Adaptive Design	51	x		x	x	x		x	x
7 Human Behavior and Environment.	44	x	x	x				x	
8 Community Engagement and Participatory	32	x		x	x	x		x	
9 LA Education Strategies	28	x	x		x	x	x	x	
10 LA Research and Experiments	28	x			x	x	x	x	
11 Visual Communication	17	x					x	x	
12 Social Equity and Landscape Identity	14		x	x			x		x
13 Historical Preservation and Management	11		x		x	x	x	x	
14 Wildlife Habitat and Biodiversity	10	x			x	x		x	

To explore the trends of these research themes, this research examined the yearly patterns of their frequency (Figure 4).

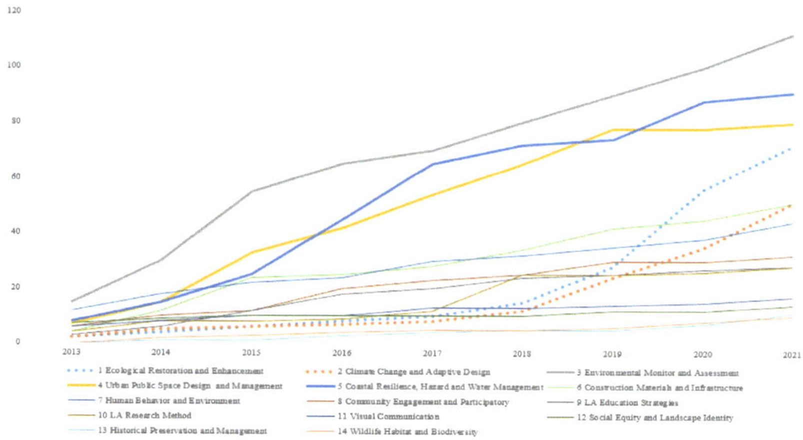

Figure 4. Secondary research—findings: yearly patterns of research themes (2013–2021).

The findings indicate that, in the five years from 2013 to 2018, "Environmental Monitor and Assessment", "Urban Public Space Design and Management" and "Coastal Resilience, Hazard, and Water Management" were the three dominant research themes. However, from 2018 to 2021, "Ecological Restoration and Enhancement" and "Climate Change and Adaptive Design" appear to have increased more rapidly than the other research themes. In the analysis of the dominant research themes of the topics, it appears that studies based on RFD shifted their themes from "Environmental Monitor and Assessment" and "Coastal Resilience, Hazard, and Water Management" to "Ecological Restoration and Enhancement" and "Climate Change and Adaptive Design".

3.1.3. RFD Categories and Trends of Digital Tools and Technology Intersecting with TLA

Based on the analysis of the data derived from Secondary Research, nine categories of digital tools and technology have been identified and ranked by their frequency of application in the literature (Table 3).

Table 3. Secondary research—findings: categories of digital tools and technology for TLA.

Research Themes	Freq.	LPE	HCT	HER	UD	SS	DEP	LDI	LP
1 Geographic Information System (GIS)	151	x	x	x	x	x	x	x	x
2 Parametric and Computational Algorithms	62	x	x	x	x	x	x	x	x
3 3D Modeling	60	x	x	x	x	x	x	x	x
4 Photography-Based Digital Visualization	47	x	x	x	x	x	x	x	x
5 Statistical Modelling	41	x		x	x				
6 Virtual Reality	30	x	x	x	x		x	x	
7 Remote Sensing	29	x	x	x	x	x	x	x	x
8 Mobile Technologies	23		x	x	x	x	x	x	
9 Computational Analysis and Evaluation	21	x		x		x			x
10 Social Media Data Mining Processes	19	x	x	x		x	x	x	x
11 Global Positioning System (GPS)	16	x	x	x	x	x	x	x	x
12 Unmanned Aerial Vehicles	16	x	x		x	x	x	x	x
13 Web-Based Interactive Map	15		x	x	x	x	x	x	x
14 Augmented Reality	14	x	x	x		x	x	x	x
15 Crowdsourcing	14		x	x				x	x
16 Physical Sensors	12	x	x			x	x	x	x
17 Gaming Engine	11	x	x	x	x		x	x	
18 3D Point Cloud	11		x	x	x	x	x	x	
19 Building Information Modeling (BIM)	11	x						x	
20 Topology Evaluation	11	x			x	x			x
21 Web-based App	10					x	x		x
22 Open Source and Programming	9		x		x		x	x	x
23 Video, Audio and Visual Technology	8				x	x	x	x	
24 Planning Support System	5	x	x	x		x			x
25 Artificial Intelligence	5	x	x	x				x	x
26 Tangible User Interface (UI)	4	x					x	x	
27 Computational Fluid Dynamics (CFD)	4	x				x		x	x
28 Robotic Processes	4					x		x	
29 Digital Twin	4	x						x	
30 Agent-Based Modeling	3			x	x			x	
31 Digital Surface Models (DSM)	3			x	x			x	
32 Fractal Dimension	3				x				x
33 Image Segmentation	3				x				x
34 Decision Support System (DSS)	2		x		x				
35 Emotion-Detection Software	2			x					

To determine trends, the annual changes in frequency for each category of digital tools and technology were analyzed (Figure 5).

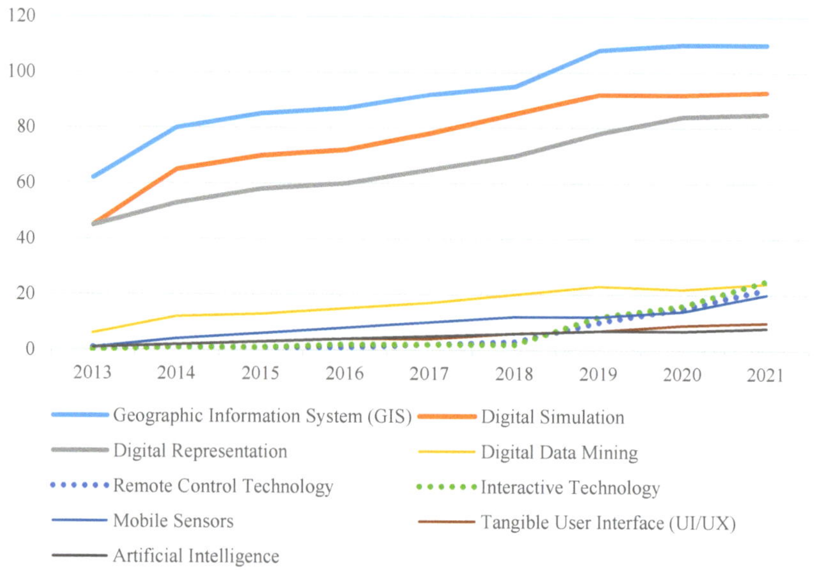

Figure 5. Secondary research—findings: annual change for categories of digital tools and technology (2013–2021).

From 2013 to 2018, the frequency of GIS and "Digital Simulation" increased and appeared to dominate the category of digital tools and technology. However, after 2018, the trend for research based on RFD appeared to incorporate more "Remote Control Technology" and "Interactive Technology". This suggests a change in digital tools and technology for TLA.

The specific modes of implementation of different types of digital technologies in the eight landscape core domains (except DTT) are illustrated in Figure 6.

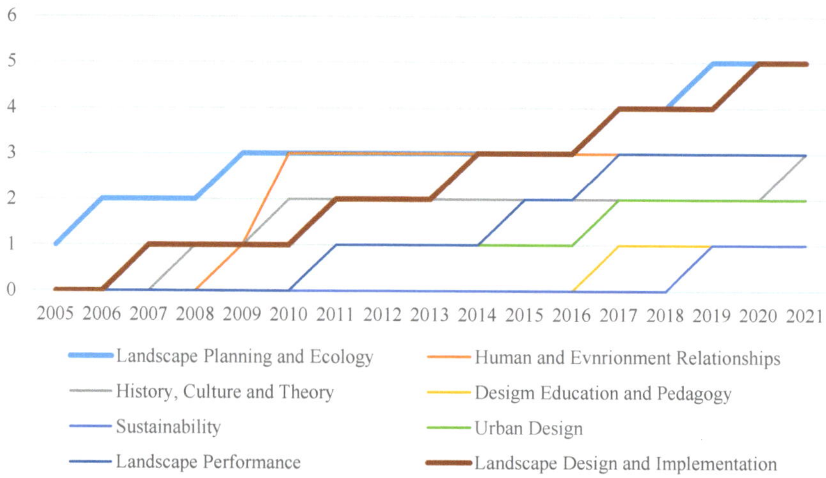

Figure 6. Archival research—finding: annual change of intersected landscape core domains with TLA for the professional projects (2005–2021).

According to the analysis of 298 peer-reviewed articles from Secondary Research, TLA was mentioned in 195 articles. Based on the analysis and findings, discoveries of dominant themes and emerging specializations were made. These are summarized in Table 4.

Table 4. Secondary research—summary of findings.

Trends	2013–2018	2018–2021
Dominant Landscape Core Domains	• Landscape Planning and Ecology • Landscape Performance	• Landscape Planning and Ecology • Sustainability • Human and Environment Relationships
Dominant Research Themes	• Environmental Monitor and Assessment • Coastal Resilience, Hazard and Water Management • Urban Public Space Design and Management	• Ecological Restoration and Enhancement • Climate Change and Adaptive Design
Categories of DTT	• Geographic Information System (GIS) • Digital Simulation	• Remote Control Technology • Interactive Technology

3.2. *Fidnings from Archival Research (RTD)*

Data analysis strategies in Secondary Research were applied in the second stage of data collection for archival research. As noted earlier, archival research was analyzed using the RTD approach. In total, 18 student awards and 26 professional awards were selected for analysis in this research.

3.2.1. RTD Trends of the Landscape Core Domains Intersecting with TLA

This section discusses the analysis of landscape core domains intersecting with TLA in archival research. It examines trends based on patterns of annual changes and frequency of TLA-related research intersecting with each core domain. The findings indicate a trend for research (both professional and student projects) based on the RTD approach that emerged for TLA, where it intersected with "Landscape Design and Implementation" and "Landscape Planning and Ecology" from 2005 to 2021 (Figures 6 and 7). These two core domains increased more rapidly than the others.

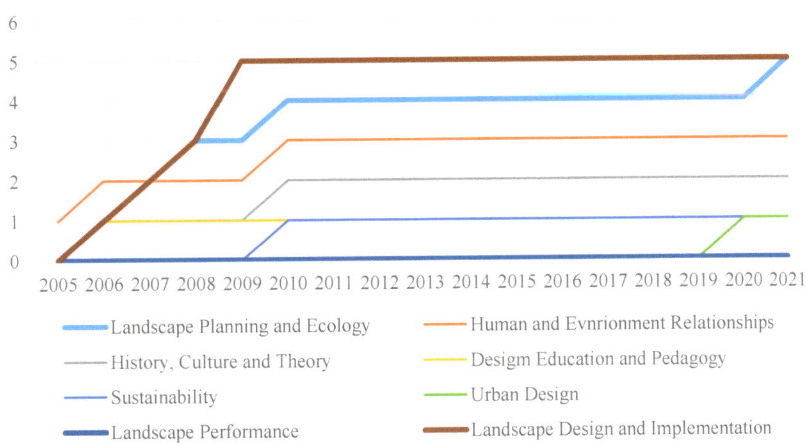

Figure 7. Archival research—finding: annual change of landscape core domains intersecting with TLA for the student projects (2005–2021).

3.2.2. RTD Trends of the Research Themes Intersecting with TLA

Based on the analysis of the data generated through archival research, 12 research themes were sorted into the landscape core domains according to the CELA tack [12]. This involved frequency analyses of the archival database (Table 5).

To determine trends in professional projects, the frequency analysis found that "Visual Communication" and "Urban Public Space Design and Management" were two of the dominant research themes from 2005 to 2018. "Coastal Resilience, Hazard and Water Management" and "Climate Change and Adaptive Design" emerged as two of the dominant research themes in terms of frequency from 2018 to 2021 (Figure 8).

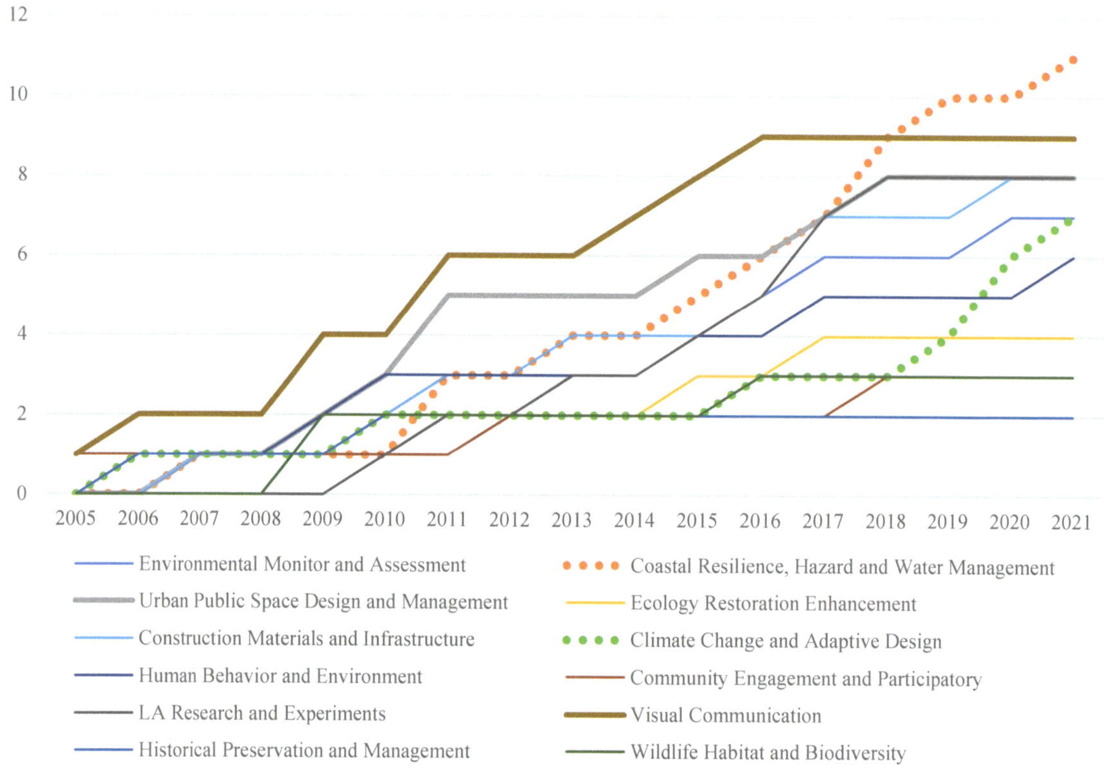

Figure 8. Archival research—findings: yearly patterns of research themes in the professional projects (2005–2021).

To determine trends in student projects, we found through frequency analysis that "Visual Communication" and "Urban Public Space Design and Management" were two of the dominant research themes from 2005 to 2018. "Coastal Resilience, Hazard and Water Management", "Ecological Restoration and Enhancement" and "Climate Change and Adaptive Design" emerged as three of the dominant research themes in terms of frequency from 2018 to 2021 (Figure 9).

Table 5. Archival research—analysis and frequency: research themes.

Research Themes	Freq.	LPE	HCT	HER	UD	SS	DEP	LDI	LP
1 Construction Materials and Infrastructure	30				x	x	x	x	x
2 Coastal Resilience, Hazard and Water Management	25	x				x		x	x
3 Visual Communication	18	x					x	x	
4 Urban Public Space Design and Management	18	x			x	x		x	
5 Environmental Monitor and Assessment	15	x			x	x		x	x
6 Human Behavior and Environment	14		x	x		x		x	
7 Community Engagement and Participatory	12			x	x	x		x	
8 Ecology Restoration Enhancement	10	x						x	x
9 Climate Change and Adaptive Design	10	x			x	x		x	x
10 LA Research and Experiments	8					x	x	x	
11 Wildlife Habitat and Biodiversity	8					x		x	
12 Historical Preservation and Management	5		x				x		

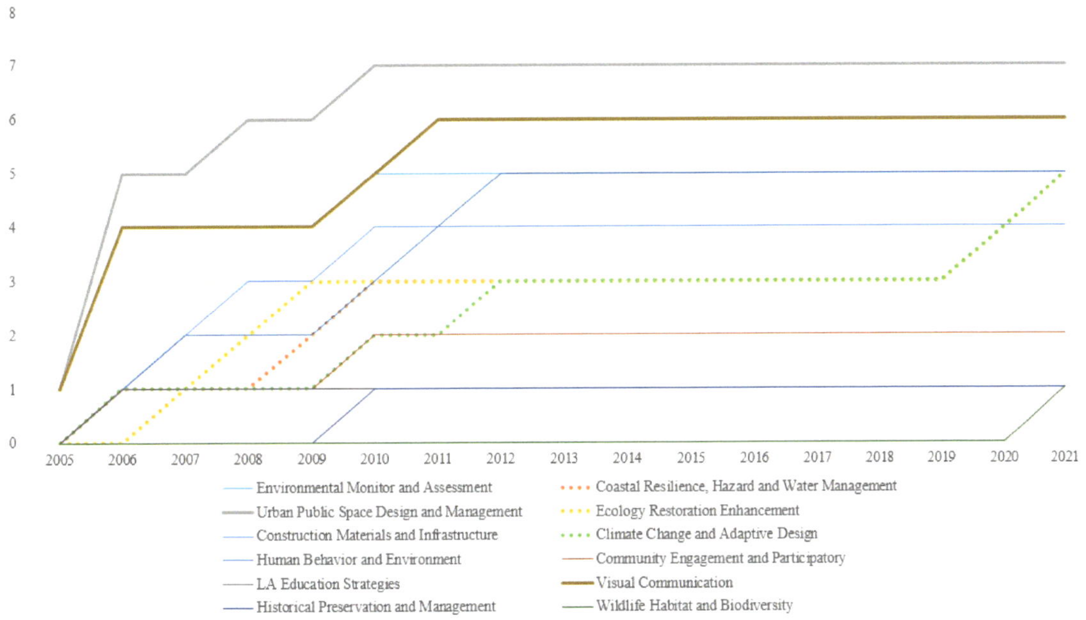

Figure 9. Archival research—findings: yearly patterns of research themes in the student projects (2005–2021).

3.2.3. RTD Categories and Trends of Digital Tools and Technology Intersecting with TLA

Based on the analysis of the primary data generated through Archival Research, seven categories were identified according to the CELA research track in 2022 [12] (Table 6).

Table 6. Archival research—findings: categories of digital tools and technology for TLA.

Research Themes	Freq.	LPE	HCT	HER	UD	SS	DEP	LDI	LP
1 Geographic Information System (GIS)	151	x	x	x	x			x	x
2 3D Modelling	60							x	x
3 Photography-Based Digital Visualization	47	x	x	x	x		x	x	
4 Statistical Modeling	41		x		x			x	x
5 Remote Sensing	29		x						
6 Virtual Reality	24						x		
7 Mobile Technologies	23						x		
8 Computational Analysis and Evaluation	21	x		x				x	x
9 Global Positioning System (GPS), Google Earth	16	x	x						
10 Web-based Interactive Map	15	x		x	x				
11 Crowdsourcing	14							x	
12 Video, Audio and Visual Technology	12	x		x			x		
13 Physical Sensors	12					x			x
14 Web-based App	10				x		x		x
15 Social Media Data Mining Processes	10				x			x	
16 Parametric and Computational Algorithms	5							x	

From 2005 to 2018, the frequencies of the "GIS" and "Digital Representation" categories of digital tools and technology appeared to dominate both student and professional projects. From 2018 to 2021, the frequencies of the "Interactive Technology" and "Digital Simulation" categories appeared to dominate both the student and professional projects (Figures 10 and 11).

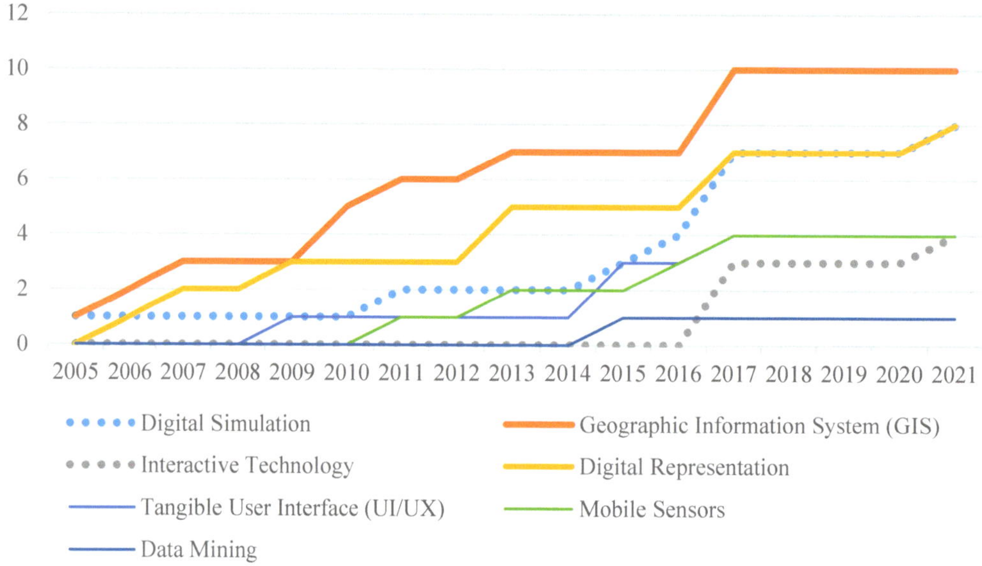

Figure 10. Archival research—findings: annual change for categories of digital tools and technology for the professional projects (2005–2021).

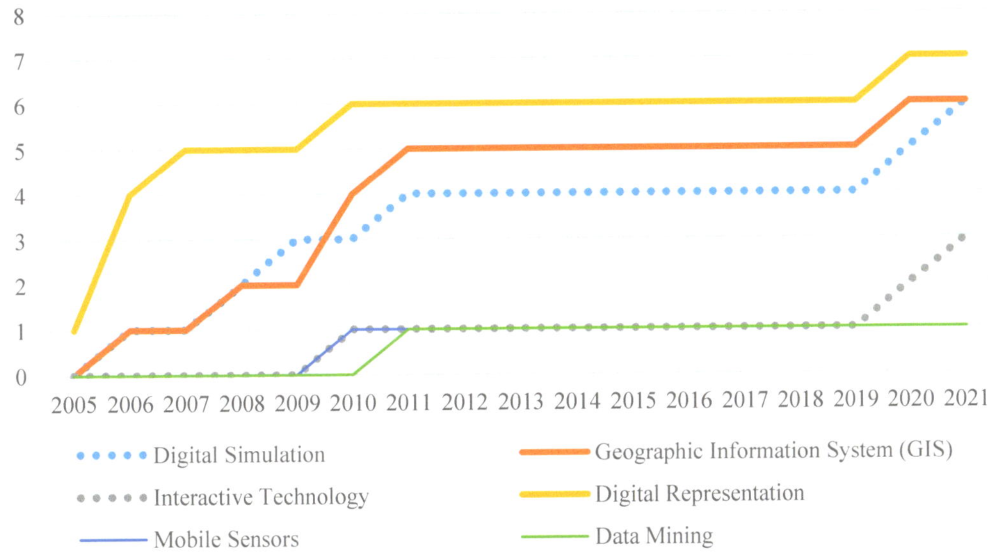

Figure 11. Archival research—findings: annual change for categories of digital tools and technology for the student projects (2005–2021).

According to the analysis, 43 projects (18 student awards and 25 professional awards) appear to represent TLA knowledge. Based on the above analysis, the findings are summarized below (Table 7).

Table 7. Archival research—summary of findings.

Trends	2005–2018	2018–2021
Dominant landscape core domains (student and professional projects)	• Landscape Design and Implementation and Landscape Planning and Ecology	
Dominant research themes (student and professional projects)	• Visual Communication • Urban Public Space Design and Management	• Coastal Resilience, Hazard and Water Management • Climate Change and Adaptive Design • Ecological Restoration and Enhancement (trend for student projects only)

3.3. *Findings from Expert Interviews*

Among the 18 experts, 11 stated their research was significantly based on the RFD, while the remaining 7 described their work as significantly based on RTD. This research was categorized into two groups (Group A—RFD and Group B—RTD) and utilized axial codes.

3.3.1. Expert Interviews Trends of the Landscape Core Domains Intersecting with TLA

Finding 1: Research themes intersected across different landscape core domains. In the code analysis, Group A and Group B experts pointed out different core domains incorporating similar research themes and topics. Their intersection provides ways to understand the trends for TLA.

Finding 2: TLA should be considered one of the landscape core domains. In the analysis of responses from both Group A and Group B, TLA should be considered one of the landscape core domains. In the responses, "Technology" was shown to be integral to the landscape architecture discipline. For example, experts noted that technology in architecture refers to fields like structure engineering, plumbing design, and mechanical engineering. These were the exclusive technologies for architecture. However, landscape architecture appears to offer no definition of its exclusive core technology. The concept of TLA has the potential to determine the technology used exclusively in the landscape architecture discipline. Hence, there is a need to designate TLA as one of the core domains and expand upon landscape architectural knowledge.

3.3.2. Expert Interviews—Trends of the Research Themes Intersecting with TLA

"Climate Change and Mitigation" was a trend in the research based on both RFD and RTD, and contributed to TLA. The experts from both Group A and Group B indicated a trend whereby current research based on the RFD and RTD approaches was focused on environmental issues such as climate actions, sustainability, and ecological planning. The responses noted that landscape discipline was helping to realize a carbon-neutral future in the built environment. For example, the practitioners planned and designed dense, walkable communities that reduced emissions from transportation and sprawl.

For site-scale research based on RTD, the experts' responses indicated landscape architecture's contribution to the built environment regarding energy and carbon efficiencies with strategies such as green roofs, water-efficient design, and the use of sustainable materials and construction practices. For regional-scale research based on RFD, the landscape experts discussed defense and expanded their knowledge on carbon-sequestering landscapes such as forests, wetlands, and grasslands, which help to draw down atmospheric carbon dioxide. These can be interpreted as community-oriented strategies whereby communities are enabled to better adapt to climate change and improve their resilience.

3.3.3. Expert Interviews—Categories and Trends of Digital Tools and Technology to TLA

Finding 1: Clarification between digital tools and technology and TLA. In the code analysis, both Group A and Group B responded that digital tools and technology and TLA were different. The experts' responses indicated that digital tools and technology evolved in three stages. In the first stage, digital tools and technology was a digital simulation toolkit focused on visual communication, such as Adobe products and AutoCAD by Autodesk. In the second stage, digital tools and technology represented a digital analytics tool that focused on simulation and modeling research, such as GIS, Landscape Information Model (LIM), Building Information Model (BIM), Stormwater Management Model (SWMM), and Grasshopper. These tools also inform data-driven landscape design. The third stage comprised a digital data communication toolkit used to request data or to send a result of the action back to the user, such as Augmented Reality (AR), Virtual Reality (VR), and An Unmanned Vehicle (UAV). In addition, the experts identified that all of this technology was from other disciplines, not landscape architecture. These digital tools and technologies were described as tools that helped research and could be interpreted as contributions to TLA, like the Land F/X software program. The technical product may be a digital-based application, a method or a process. It appears that landscape research lacks the use of 21st-century technology. This may imply a "blurring" of the core of landscape and its boundary with allied disciplines. The evolution of digital tools and technology appears to have expanded the TLA. Some digital-based TLA appears to be a part of digital tools and technology. However, digital tools and technology from other disciplines are not defined as TLA.

Finding 2: The trend of incorporating "Interactive Technology" based on RTD and RFD contributes to TLA. The codes in the analysis indicate the experts' responses to applied RTD and RFD, representing a growing preference for "Interactive Technology" such as AR, VR, and wearable devices in landscape research. The analysis of the experts' responses illustrates that digital analog software such as GIS, Grasshopper, and Rhino were the dominant forms of technology utilized in landscape research 8–10 years ago. Within the last 5–8 years, the experts' responses indicate that more scholars tend to use interactive equipment to explore the human perception of the built environment. For example, one expert cited a study that generated a framework to use spontaneous invasive plants in reconstructing the landscape of contaminated local sites through the use of interactive equipment and analyses of the plants' ecological performance. Table 8 summaries the findings and compare the differences between RFD approach and RTD approach.

Table 8. Expert interviews—summary of findings.

<table>
<tr><th></th><th>RFD Approach</th><th>RTD Approach</th></tr>
<tr><td>Relationship between Research Themes and Associated Topics with TLA</td><td colspan="2">Finding 1: Climate change and mitigation are trends in research based on both RFD and RTD, and contribute to TLA.</td></tr>
<tr><td rowspan="2">Relationship Between Landscape Core Domains and TLA</td><td colspan="2">Finding 1: Research themes intersected different landscape core domains.</td></tr>
<tr><td colspan="2">Finding 2: TLA should be considered as one of landscape core domains.</td></tr>
<tr><td rowspan="2">Relationship Between Digital Tools and Technology and TLA</td><td colspan="2">Finding 1: Clarification between digital tools and technology and TLA.</td></tr>
<tr><td colspan="2">Finding 2: There is a trend of using "Interactive Technology" based on RTD and RFD approaches to contribute to TLA.</td></tr>
</table>

4. Discussion

This chapter involves interpreting the result of findings discussed in Sections 3.1–3.3, which concerned the (1) relationship between TLA, the core domains of landscape architecture, digital tools and technology, and research themes, and the (2) conceptualization of TLA in the 21st century.

4.1. Relationship between TLA, the Core Domains of Landscape Architecture, Digital Tools and Technology, and Research Themes

4.1.1. Interrelationship of Landscape Core Domains within the Context of TLA

One key finding of the analysis of the interviewees' responses indicates that TLA should be considered as its own landscape core domain. The experts stated that TLA supports the development of other core domains. Additionally, digital tools and technology have been designated as a landscape core domain. However, digital tools and technology from other disciplines (DTT-od) should not be considered part of the landscape core domain. The analysis of the interviewees' responses indicates that digital tools and technology can be considered one of the landscape core domains. This analysis indicates the experts' preference for TLA, essentially disregarding DTT-od and promoting TLA as its replacement. This signifies a growing recognition that existing, externally developed technologies may not fully address the unique challenges and opportunities faced by landscape professionals. By fostering a deeper understanding of technological capabilities, landscape professionals can seamlessly integrate technology into the design process, leading to groundbreaking solutions. Examples include designing smart green spaces that dynamically adapt to environmental conditions or utilizing virtual reality to create immersive experiences within planned environments. The landscape architecture field grapples with issues such as climate change, urbanization, and resource scarcity. A TLA focus could equip professionals with the tools to tackle these challenges head-on. For instance, developing bioremediation techniques or designing sustainable infrastructure solutions might necessitate expertise in specific technological areas.

In addition, to interpret the shift in the intersection of TLA with other landscape core domains, the research findings illustrate the impacts of Research for Development (RFD) and Research through Design (RTD) on TLA over different periods. From 2013 to 2018, RFD helped TLA gain prominence through its integration with "Landscape Ecology and Planning" and "Landscape Performance", highlighting a focus on ecological systems and the effectiveness of landscape interventions. The trend shifted from 2018 to 2021, with TLA increasingly intersecting with "Sustainability", reflecting a global shift towards sustainable development practices. Simultaneously, from 2005 to 2021, RTD influenced TLA predominantly in "Landscape Design and Implementation" and "Landscape Planning and Ecology", indicating a strong emphasis on practical, ecologically informed landscape solutions. The shift in research focus over time reflects broader trends in landscape architecture towards more sustainable and ecologically integrated practices. The prominence of ecology and sustainability in recent years highlights the profession's response to global environmental challenges, such as climate change and biodiversity loss, emphasizing the role of landscape architects in creating resilient and sustainable environments.

4.1.2. Interrelationship of Research Themes within the Context of TLA

To interpret the findings from the research across three different datasets—secondary research, archival research, and expert interviews—a structured approach can be helpful. Both Research through Design (RTD) and Research from Design (RFD) have significantly contributed to the growth of TLA across multiple themes. Prior to 2018, RTD was primarily influential in themes like "Visual Communication" and "Urban Public Space Design and Management", while RFD was dominant in "Environmental Monitoring and Assessment", "Coastal Resilience, Hazard and Water Management", and "Urban Public Space Design and Management". From 2018 to 2021, both RTD and RFD played dominant roles in "Ecological Restoration and Enhancement" and "Climate Change and Adaptive Design". The emergence of "climate actions" as a research trend since 2018 indicates a shared direction in both RTD and RFD approaches, emphasizing their importance in addressing contemporary issues like climate change.

A key difference observed is that RTD, unlike RFD, continued to strongly influence the theme of "Coastal Resilience, Hazard and Water Management" even after 2018. RTD-based research tends to focus more on practical applications that expand knowledge in TLA through solutions like green roofs and walls, which are direct interventions for climate mitigation and habitat creation. In contrast, RFD-based research has expanded the knowledge base by developing ecological strategies, urban forestry policies, wetlands management, and human-centered design approaches, suggesting a more policy- and strategy-oriented approach compared to the more direct design interventions of RTD.

The findings suggest that RTD is more closely aligned with direct design and physical interventions in landscape architecture, contributing both aesthetically and functionally to urban and environmental challenges. RFD, however, leans towards a broader strategic and policy-making role, influencing landscape architecture through research that informs guidelines, policies, and broader management strategies. The convergence of RTD and RFD in newer research themes like "Ecological Restoration and Enhancement" and "Climate Change and Adaptive Design" post-2018 reflects a collaborative and integrated approach towards solving modern challenges through both design and strategic research. The insights from expert interviews indicate a recognition and validation of these varied approaches, emphasizing the need for a diverse yet cohesive research strategy within TLA to effectively address emerging global concerns such as climate change. This exposes a knowledge gap—traditional experience may not suffice. These topics necessitate a more comprehensive TLA. Ecological strategies require environmental modeling expertise. Urban forestry/wetland policies demand knowledge of urban planning frameworks. Human-centered design necessitates a familiarity with UX research. A broader TLA interpretation is needed, encompassing skills like parametric design software, 3D modeling, and UX research, alongside traditional design. This shift brings opportunities and challenges. Landscape architecture can become key in environmental issues and human well-being design. However, overcoming resource limitations and refining the definition of TLA are crucial for the successful navigation of this paradigm shift.

4.1.3. Interrelationship of Digital Tools and Technology within the Context of TLA

The differentiation between digital tools and technology and TLA is pivotal for advancing the field. Digital tools and technology encompass a wide array, such as Geographic Information Systems (GIS), Digital Simulation, Interactive Technology, and Remote Control Technology. The application of these tools has evolved significantly, marking the transition from simple visualization to complex interactive design processes.

Evolution of digital tools and technology in landscape architecture:

1. Visualization and communication—Initially, digital tools and technology were used primarily for digital simulations that focused on visualizing and communicating landscape designs;
2. Modeling and simulation—Over time, digital tools and technology expanded to include sophisticated modeling and simulation tools, enhancing data-driven design processes;
3. Interactivity and data generation—The latest developments in digital tools and technology incorporate interactive technologies such as AR, VR, and UAVs. These tools are pivotal for generating new data and fostering original research, representing a shift towards dynamic, user-centric design experiences.

TLA, in contrast, refers to the application of these digital tools within landscape architecture, aiming to foster innovative design practices and research methodologies. TLA encompasses a methodological and conceptual shift towards technology-driven design and research, transforming traditional landscape architecture practices. The evolving role of digital tools and technology in landscape architecture signifies a significant shift. Once primarily a tool for visual communication, digital tools and technology is expanding its functionalities to encompass simulation, modeling research, data-driven design, and interactive data exploration via AR/VR/UAV technologies. This empowers landscape architects to not only analyze data, but also generate their own through these interactive toolkits, fostering a deeper understanding of design challenges and contributing to the growing body of knowledge in TLA. Furthermore, digital tools and technological transformation open doors for collaboration with other disciplines, and potentially lead to more user-centric design outcomes. However, this evolution also necessitates the development of new skillsets in data analysis, as well as modeling techniques, and interactive technologies for landscape architects. In general, the field needs to be aware that digital tools and technology, as well as TLA, may generate unsolved conceptual and value conflicts, which need to be resolved in a timely fashion in the process.

To further interpret the trends of intersection of digital tools and technology with TLA within landscape research, the findings indicate that research using the Research through Design (RTD) approach identified Geographic Information Systems (GIS) and Digital Representation as the primary digital tools used until 2018. After this, there was a noticeable shift towards employing Digital Simulation and Interactive Technology. Similarly, the research based on Research for Design (RFD) findings confirmed a strong emphasis on GIS and Digital Simulation from 2013 to 2018. Post-2018, there was a significant increase in the use of "Remote Control Technology" and "Interactive Technology", indicating a trend towards more advanced digital tools and highlighting the ongoing development of knowledge within TLA.

4.2. Conceptualizations of TLA in the 21st Century

The expert interviews reveal a need to clarify, define and potentially categorize TLA. This next section suggests ways to clarify TLA. Based on the analysis of the three datasets, TLA is framed in terms of categories and specializations. As an interpretive approach, several terminologies were used to understand TLA. Through analysis of three datasets, several specializations emerged: "Method", "Approach", "Guideline", "Strategy", "Framework", and "Software". However, most of these terms appeared to yield similar interpretations, and were difficult to distinguish. This necessitates future analysis and a deeper understanding, which can be achieved by reviewing the literature and the experts' responses.

According to the code analysis of interviewees' responses, "Method" is a particular design or research process. In other words, it is the way in which something is done. In this sense, it is similar to "Approach", which is the way something is proposed. "Approaches" have to be decided before selecting the "Method". "Approaches" and "Methods" are approaches that deal with general philosophies of implementation [35]. However, a "Method" practically implements an "Approach".

In the code analysis of interviewees' responses, instructional TLA is included, which includes "Guideline" and "Strategy". "Strategy" refers to a plan of action designed to achieve an aim [36]. Based on that, a "Guideline" establishes the LA design and research criteria. It could be a new standard, metric, or principle that could affect design considerations that designers apply with discretion.

The code analysis from interviewees' responses shows that TLA is expected to be a technical product/application. The responses note that it could be the "Framework" of the application and the "Software". Pragmatically, a "Framework" is a platform for developing specific software and applications. It supports logical, functional, computational, interaction, and application aspects [37]. "Software" is a set of instructions, data, or programs based on the implementation of a "Framework", and used to operate computers and execute specific tasks [38].

Based on the analysis of the above terms, this research has identified three specializations of TLA:

- Instruction (Strategy and Guideline);
- Process (Approach and Method);
- Application (Framework and Software).

To explain the three categories, "Process" describes "what" is done, and "Instruction" describes "how" it is done [39]. "Application" is a digital program that computers can use directly. Figure 12 illustrates their relationships.

Figure 12. Relationships of categories and specializations of TLA in the 21st century.

In addition, according to the analysis of the data generated from Archival Research and Secondary Research, this research found that both RTD and RFD contribute to "Approach" and "Method" as two dominant specializations of TLA. In addition, these two specializations of TLA appeared to increase more rapidly than others based on the patterns of annual change and frequency prior to 2018. From 2018 to 2021, both the patterns of annual change and frequency and the analysis of codes based on interviewees' responses indicated a trend towards utilizing TLA as an "Application". Overall, the analysis of the experts' responses indicates that scholars became increasingly aware that TLA had the potential to become a technical product for future landscape research. Figure 13 indicates the structure of TLA in the 21st century.

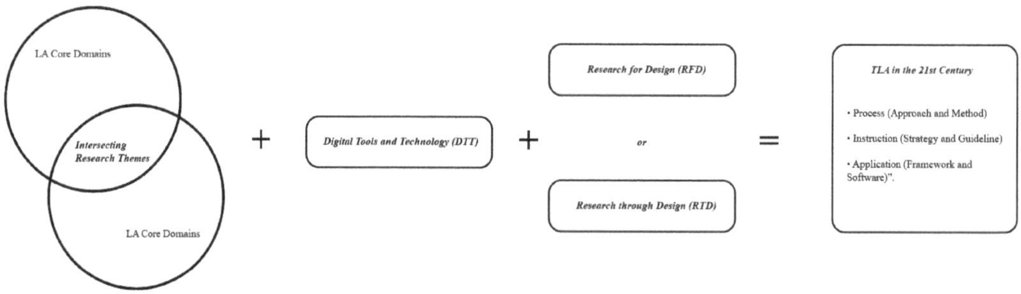

Figure 13. Structure of TLA in the 21st century.

4.3. Impact on Future Landscape Research

The proposed conceptualization could potentially impact on three future key areas of landscape architecture: "Digital Practice", "Design Computation", and "Design Representation".

(1) Digital Practice—It is crucial to integrate TLA as a data source and communication medium. However, the research suggests that TLA's impact on the design process might be less significant in areas like sustainability and human-environment relationships. Here, the focus should be on equipping students with a stronger understanding of TLA terminology and mechanisms through relevant coursework. The core objective should be expanding the scope of TLA knowledge. The scholarships should also encourage leveraging TLA to generate practical knowledge for landscape practice;
(2) Design Computation—Design computation aims to optimize design efficiency and accuracy through TLA. The research suggests that certain design tasks, especially in landscape planning and ecology, and potentially in design and implementation, could be automated with new software, platforms, and methods. The focus of TLA education in this aspect should equip students with the necessary software and platform proficiency;
(3) Design Representation—While the study found the limited integration of TLA in RTD and RFD, the value of TLA in communicating final design solutions to an audience with various backgrounds is undeniable. The integration process is likely smoother, as it requires fewer theoretical debates than other areas. Therefore, we should enhance the training on visual representation, integrating TLA across diverse landscape architecture courses.

5. Conclusions

The objective of this study is to examine the roles and conceptual frameworks within landscape research, interpreting trends and relationships across landscape core domains, research themes, digital tools and technologies and TLA. This exploration was conducted through Archival Research, Secondary Analysis, and expert interviews. The findings reveal that digital tools and technology categories have transitioned from simple digital simulations to complex interactive technologies that have played a crucial role in environmental design and planning over recent years. This shift is marked by the increased utilization of Geographic Information Systems (GIS), Digital Simulation, and newer technologies such as Remote Control and Interactive Technologies, which are now pivotal in fostering innovative, data-driven design solutions. Further, TLA has emerged as a pivotal domain within landscape research, markedly enhancing traditional practices and spearheading the development of sophisticated methodologies by integrating digital tools and technologies. Notably, there is a pronounced trend towards utilizing digital tools and technology in aspects ranging from communication and visualization to interactive technology. This integration substantially intersects with key research themes, such as landscape planning, ecology, and sustainable urban design—all vital for tackling issues like ecological resilience, climate change and enhanced human-environment interactions.

The relationship between digital tools and technology categories and TLA within landscape architecture reveals a dynamic and evolving connection. Digital tools and technology have been instrumental in expanding the research and practical capabilities of landscape architecture, pushing the field towards more data-driven and interactive approaches, and enabling landscape architects to undertake complex planning and design tasks with greater precision and effectiveness. On the other hand, TLA represents a broader conceptualization of technology within landscape architecture, embodying the integration of digital tools into the core practices of the field. It suggests a shift towards a more integrated approach, whereby technology is not just a tool but a fundamental aspect of landscape architectural practice. TLA includes the application of digital tools and technology in specific projects and the broader methodological shifts these tools enable, such as improved environmental modeling, user interaction through immersive technologies, and enhanced data analysis capabilities.

Furthermore, the study underscores the significance of categorization and specialization within TLA, delineating three main areas: "Instruction" (encompassing strategies and guidelines), "Process" (covering approaches and methods), and "Application" (involving frameworks and software). The research identifies two major research types—RTD and RFD—that leverage digital tools and technologies, intersecting with various research themes linked to different core domains to effectively cultivate TLA.

6. Limitation

In the data collection process, particularly during the archival research phase, while ASLA provides a comprehensive collection of award-winning projects with detailed descriptions that facilitate thorough analysis, it predominantly reflects trends and practices within the United States. Recognizing the need for a more diverse understanding of landscape architecture practices globally, other significant sources were considered, including awards from the International Federation of Landscape Architects (IFLA) and the European Foundation for Landscape Architecture (EFLA). However, these additional sources often lacked sufficient project descriptions, which are crucial for an in-depth analysis. The limited availability of detailed project information from these international sources constrained the ability to incorporate a wider perspective into our research. Consequently, while this study continues to rely on the ASLA Awards to yield a substantive dataset, we acknowledge this reliance as a limitation. This restriction highlights the challenge of accessing detailed, comparable data across different global landscape architecture awards, and underscores the necessity for future research to explore more varied sources, potentially encouraging international bodies to standardize the documentation of award-winning projects to support academic investigation.

Author Contributions: Conceptualization, X.S. and M.G.P.; methodology, X.S., M.G.P. and N.G.K.; software, X.S.; validation, N.G.K.; formal analysis, X.S.; investigation, X.S.; resources, X.S.; data curation, X.S.; writing—original draft preparation, X.S.; writing—review and editing, X.S. and M.G.P.; visualization, X.S.; supervision, M.G.P. and N.G.K.; project administration, X.S.; funding acquisition, X.S. All authors have read and agreed to the published version of the manuscript.

Funding: This research received no external funding.

Data Availability Statement: The original contributions presented in the study are included in the article, further inquiries can be directed to the corresponding author.

Conflicts of Interest: The authors declare no conflicts of interest.

Appendix A. Datasets of Secondary Research and Archival Research

Due to the length of Appendix A, we have uploaded it to the google sheet. The link is: https://docs.google.com/spreadsheets/d/14ddDEtjNZ7lHwzbx_G_sVEhgsmhJK6dekoyhtgDfuRs/edit?usp=sharing (accessed on 1 May 2023)

Secondary Research:

* A = Journal of Digital Landscape Architecture; B = Journal of Landscape Architecture; C = Landscape Research Record; D = Landscape Research; E = Landscape Review; F = Landscape Journal.

Archival Research:

G = ASLA Professional Award-Winning Projects; H = ASLA Student Award-Winning Projects.

* Red text refers to the articles or projects that do not generate specific TLA knowledge.

References

1. Newton, N.T. *Design on the Land: The Development of Landscape Architecture*, 1st ed.; Belknap Press (An Imprint of Harvard University Press): Cambridge, MA, USA, 1971.
2. Beardsley, J. A Word for Landscape Architecture. Harvard Design Magazine. Available online: https://www.harvarddesignmagazin org/articles/a-word-for-landscape-architecture/ (accessed on 5 April 2024).

3. Kempenaar, A.; Brink, A. van den. Regional Designing: A Strategic Design Approach in Landscape Architecture. *Des. Stud.* **2018**, *54*, 80–95. [CrossRef]
4. Weilacher, U. Syntax of Landscape: The Landscape Architecture of Peter Latz and Partners. In *Syntax of Landscape*; Birkhäuser: Basel, Switzerland, 2007. [CrossRef]
5. Senbel, M.; Girling, C.; White, J.; Kellett, R.; Chan, P.F. Precedents Reconceived: Urban Design Learning Catalysed through Data Rich 3-D Digital Models. *Des. Stud.* **2013**, *34*, 74–92. [CrossRef]
6. Powers, M.; Walker, J. Twenty-Five Years of Landscape Journal: An Analysis of Authorship and Article Content. *Landsc. J.* **2009**, *28*, 96–110. [CrossRef]
7. Gobster, P.; Nassauer, J.; Nadenicek, D. Landscape Journal and Scholarship in Landscape Architecture: The Next 25 Years. *Landsc. J.* **2010**, *29*, 52–70. [CrossRef]
8. Cushing, D.F.; Renata, A. Themes in Landscape Architecture Publishing: Past Trends, Future Needs. *Landsc. J.* **2015**, *34*, 15–36. [CrossRef]
9. Meijering, J.V.; Tobi, H.; van den Brink, A.; Morris, F.A.; Bruns, D. Exploring Research Priorities in Landscape Architecture: An International Delphi Study. *Landsc. Urban Plan.* **2015**, *137*, 85–94. [CrossRef]
10. Vicenzotti, V.; Jorgensen, A.; Qvistrom, M.; Swaffield, S. Forty Years of Landscape Research. *Landsc. Res.* **2016**, *41*, 388–407. [CrossRef]
11. Newman, G.; Li, D.; Tao, Z.; Zhu, R. Recent Trends in LA-Based Research: A Topic Analysis of CELA Abstract Content. *Landsc. J.* **2021**, *39*, 51–73. [CrossRef]
12. 2022 Conference Tracks. CELA (Council of Educators in Landscape Architecture). Available online: https://thecela.org/conference-tracks/ (accessed on 6 April 2024).
13. Margolis, L.; Robinson, A. *Living Systems: Innovative Materials and Technologies for Landscape Architecture*, 1st ed.; Birkhäuser Architecture: Basel, Switzerland; Boston, MA, USA, 2007.
14. Cantrell, B.; Michaels, W. *Digital Drawing for Landscape Architecture: Contemporary Techniques and Tools for Digital Representation in Site Design*, 1st ed.; Wiley: Hoboken, NJ, USA, 2010.
15. Cantrell, B.; Yates, N. *Modeling the Environment: Techniques and Tools for the 3D Illustration of Dynamic Landscapes*, 1st ed.; Wiley: Hoboken, NJ, USA, 2012.
16. Cantrell, B.E.; Holzman, J. *Responsive Landscapes: Strategies for Responsive Technologies in Landscape Architecture*, 1st ed.; Routledge: New York, NY, USA, 2015.
17. Langley, W.; Corry, R.; Brown, R. Core Knowledge Domains of Landscape Architecture. *Landsc. J.* **2018**, *37*, 9–21. [CrossRef]
18. Nitavska, N.; Mengots, A. Digital Tools in Landscape Architecture. *Landsc. Archit. Art* **2018**, *11*, 42–50. [CrossRef]
19. Steinitz, C. *A Framework for Geodesign: Changing Geography by Design*; Esri: Redlands, CA, USA, 2012.
20. Amoroso, N. (Ed.) *Representing Landscapes: A Visual Collection of Landscape Architectural Drawings*, 1st ed.; Routledge: London, UK; New York, NY, USA, 2012.
21. Amoroso, N. (Ed.) *Representing Landscapes: Digital*, 1st ed.; Routledge: Abingdon, VA, USA, 2015.
22. Amoroso, N.H. *Digital Landscape Architecture Now*; Thames & Hudson: London, UK, 2012.
23. Cantrell, B.; Mekies, A. (Eds.) *Codify: Parametric and Computational Design in Landscape Architecture*, 1st ed.; Routledge: London, UK; New York, NY, USA, 2018.
24. Cantrell, B.; Zhang, Z.; Liu, X. Artificial Intelligence and Machine Learning in Landscape Architecture. In *The Routledge Companion to Artificial Intelligence in Architecture*; Routledge: New York, NY, USA, 2021.
25. American Society of Landscape Architects. Landscape Architecture Body of Knowledge Study Report. Available online: https://www.asla.org/search.aspx?q=Landscape%20Architecture%20Body%20of%20Knowledge%20Study%20Report (accessed on 6 April 2024).
26. Steinitz, C. On Landscape Architecture Education and Professional Practice and Their Future Challenges. *Land* **2020**, *9*, 228. [CrossRef]
27. Frayling, C. Research in Art and Design (Royal College of Art Research Papers, Volume 1, Number 1, April 1993). Available online: https://researchonline.rca.ac.uk/384/ (accessed on 6 April 2024).
28. Zimmerman, J.; Forlizzi, J. Research Through Design in HCI. In *Ways of Knowing in HCI*; Olson, J.S., Kellogg, W.A., Eds.; Springer: New York, NY, USA, 2014; pp. 167–189. [CrossRef]
29. Deming, M.E.; Swaffield, S. *Landscape Architectural Research: Inquiry, Strategy, Design*; John Wiley & Sons: Hoboken, NJ, USA, 2011.
30. Matthews, B.; Doherty, S.; Johnston, J.; Foth, M. The Publics of Design: Challenges for Design Research and Practice. *Des. Stud.* **2022**, *80*, 101106. [CrossRef]
31. Shen, X. Identifying the Role of Technology within the Discipline of 21st Century Landscape Architecture. *Des. J.* **2023**, *26*, 351–361. [CrossRef]
32. McGowan, M.L.; Prows, C.A.; DeJonckheere, M.; Brinkman, W.B.; Vaughn, L.; Myers, M.F. Adolescent and Parental Attitudes about Return of Genomic Research Results: Focus Group Findings Regarding Decisional Preferences. *J. Empir. Res. Hum. Res. Ethics* **2018**, *13*, 371–382. [CrossRef] [PubMed]
33. Creswell, J.W. *Research Design: Qualitative, Quantitative, and Mixed Methods Approaches*, 3rd ed.; Sage Publications, Inc.: Thousand Oaks, CA, USA, 2009; pp. xxix, 260.

34. Shen, X.; Kong, J.; Padua, M.G.; Kirkwood, N. 21st Century Technology in Landscape Architecture and Its Significance within the Discipline of Landscape Architecture. *Chin. Landsc. Archit.* **2023**, *39*, 34.
35. Nilsen, P. Making Sense of Implementation Theories, Models and Frameworks. *Implement. Sci.* **2015**, *10*, 53–79. [CrossRef] [PubMed]
36. Whalley, J. A Plan of Action Designed to Achieve a Long-Term Goal or Overall Aim. | LinkedIn. Available online: https://www.linkedin.com/pulse/plan-action-designed-achieve-long-term-goal-overall-aim-john-whalley/ (accessed on 6 April 2024).
37. Horváth, I. Connectors of Smart Design and Smart Systems. *AI EDAM* **2021**, *35*, 132–150. [CrossRef]
38. Rosencrance, L. What Is Software? Definition, Types and Examples. App Architecture. Available online: https://www.techtarget.com/searchapparchitecture/definition/software (accessed on 6 April 2024).
39. Walliss, J.; Rahmann, H. *Landscape Architecture and Digital Technologies: Re-Conceptualising Design and Making*; Routledge: London, UK, 2016. [CrossRef]

land

MDPI

Article

Enhancing Urban Landscape Design: A GAN-Based Approach for Rapid Color Rendering of Park Sketches

Ran Chen [1], Jing Zhao [1,*], Xueqi Yao [1], Yueheng He [1], Yuting Li [1], Zeke Lian [2], Zhengqi Han [1], Xingjian Yi [1] and Haoran Li [3]

1 School of Landscape Architecture, Beijing Forestry University, Beijing 100083, China; chenran705367787@bjfu.edu.cn (R.C.); yxq18210187099@bjfu.edu.cn (X.Y.); candyhyh808@bjfu.edu.cn (Y.H.); liyuting1625571259@bjfu.edu.cn (Y.L.); zenkii@bjfu.edu.cn (Z.H.); yixingjian0906@bjfu.edu.cn (X.Y.)
2 Ningbo City College of Vocational Technology, Ningbo 315100, China; lianzeke@nbcc.edu.cn
3 China United Network Communications Group Co., Beijing 100031, China; lihr21@chinaunicom.cn
* Correspondence: zhaojing@bjfu.edu.cn; Tel.: +1-3811-993-788

Abstract: In urban ecological development, the effective planning and design of living spaces are crucial. Traditional color plan rendering methods, mainly using generative adversarial networks (GANs), rely heavily on edge extraction. This often leads to the loss of important details from hand-drawn drafts, significantly affecting the portrayal of the designer's key concepts. This issue is especially critical in complex park planning. To address this, our study introduces a system based on conditional GANs. This system rapidly converts black-and-white park sketches into comprehensive color designs. We also employ a data augmentation strategy to enhance the quality of the output. The research reveals: (1) Our model efficiently produces designs suitable for industrial applications. (2) The GAN-based data augmentation improves the data volume, leading to enhanced rendering effects. (3) Our unique approach of direct rendering from sketches offers a novel method in urban planning and design. This study aims to enhance the rendering aspect of an intelligent workflow for landscape design. More efficient rendering techniques will reduce the iteration time of early design solutions and promote the iterative speed of designers' thinking, thus improving the speed and efficiency of the whole design process.

Keywords: hand-drawn sketch; image color rendering; generative adversarial networks; data augmentation; landscape design

Citation: Chen, R.; Zhao, J.; Yao, X.; He, Y.; Li, Y.; Lian, Z.; Han, Z.; Yi, X.; Li, H. Enhancing Urban Landscape Design: A GAN-Based Approach for Rapid Color Rendering of Park Sketches. *Land* **2024**, *13*, 254. https://doi.org/10.3390/land13020254

Academic Editor: Eckart Lange

Received: 13 December 2023
Revised: 9 February 2024
Accepted: 10 February 2024
Published: 18 February 2024

1. Introduction

The construction of urban living environments is a crucial aspect of realizing urban ecological civilization. It directly impacts residents' quality of life and the sustainable development of cities. Park green spaces, as essential elements for enhancing the ecological environment and city image, are among the city's most vital resources. In the planning process of urban green spaces, designers integrate various design concepts, such as site space layout, landscape nodes, and road traffic, into hand-drawn plans. These plans directly express the designer's thoughts, and the rendered colorful plans illustrate the design effect. This process aids communication with multiple parties and enhances the design's quality and level. However, the current workflow from the research stage to the color flat drawing stage is often time-consuming. Therefore, improving the efficiency of plan rendering is a significant challenge in the current design process and the primary motivation and goal of this study.

To address this challenge, we explored the usefulness of our study based on traditional landscape design methods and their inherent challenges. In the traditional landscape design process, designers typically go through stages like site analysis, concept development, design evolution, and drawing production to complete a project [1,2]. Initially, designers conceive the spatial arrangement of landscape elements based on the site conditions and client requirements. They often sketch designs by hand, as it better captures smooth curves. Eventually, these sketches are turned into colored flat plans using computer software to present to clients. Since clients usually lack professional knowledge, designers need to use these colored plans to explain the overall design. The process requires repeating these steps in communication with clients. The step from sketch to colored plan is time-consuming, posing a major challenge. Therefore, our study aims to integrate intelligent tools into the traditional process, enhancing the efficiency of sketch rendering and accelerating the process of design thinking and scheme iteration.

In recent years, generating images with appropriate colors and textures based on sketches has become a research hotspot. The advancement of machine learning has greatly facilitated designers. With the evolution of artificial intelligence, many painting tasks are now accomplished using deep-learning technology. Generative adversarial networks, an innovative architecture, have found widespread application. However, landscape design drawings often struggle to render satisfactory results due to the irregularity of elements, intricate details, and limited available corresponding data. Currently, designers still need to manually complete it through software such as Photoshop, relying on their perception of color matching and the coordination of different detail textures. If deep learning can assist designers in quickly completing the design, it will significantly improve design efficiency and generate more application value. However, the drawing of landscape design diagrams makes it difficult to render better results due to defects such as the weak regularity of elements, complexity of details, and less corresponding data available. Currently, designers are still required to complete the design manually through Photoshop and other software based on the perception of color matching and the coordination of different details and textures. If deep learning can assist designers to quickly complete the design, the design efficiency will be greatly improved and generate more application value.

To address these issues, this study proposes an automatic plan rendering system based on generative adversarial networks, combining the features of machine learning and landscape planning and design. We use hand-drawn line-draft plans and color plan design drawings for training. Compared to the edge extraction commonly used in other studies, hand-drawn line-draft maps are more likely to reflect the actual semantic structure of images, conform to the aesthetics and habits of designers, and facilitate us in screening out the generation algorithm most suitable for the actual design process. After selecting the appropriate algorithm, we built a data enhancement module to optimize the rendering system. The main contributions of this paper are as follows: (1) Based on Pix2pix and CycleGAN, we built a fully automated park plan rendering system and screened out algorithms suitable for the design process. (2) By employing data augmentation models, we were able to expand our small sample size and optimize the training process. (3) We analyzed the differences and similarities in the results generated by various algorithms and sample data volumes. (4) We utilized a novel approach to visualize urban park green spaces, using hand-drawn line drafts as experimental data. (5) Our proposed automated rendering system enhanced the speed and quality of sketch rendering in landscape design. Integrating this system into the design development and drawing production stages of traditional landscape design processes can improve the overall design efficiency.

The remainder of this paper is organized as follows: Section 2 provides a review of the related literature. Section 3 presents our proposed architecture and training methodologies. Section 4 details the experiments conducted and their respective results. Section 5 discusses the related research. Finally, Section 6 elaborates on the results and potential future work.

2. Related Work

Our work is primarily related to two research areas: image coloring and conditional generative adversarial networks. This section provides a comprehensive review of the pertinent literature in these fields.

2.1. Image Coloring

Image coloring, a technique that infuses grayscale or sketch images with color and texture, holds significant value in the realm of nonphotorealistic rendering [3,4]. However, image coloring presents an ill-posed problem due to the potential for one image to correspond to multiple color schemes. Consequently, conditional information is required to generate vivid and reasonable results. The earliest image coloring methods were grounded in CG methods, employing physical simulation or programming to emulate effects such as watercolor and oil painting, or utilizing feature lines for line rendering. For instance, Chu and Tai leveraged GPU computations to generate watercolor effects [5]. While this method can yield realistic outcomes, it is computationally expensive, complex in design, and challenging to adapt to varying styles and scenes. Traditional methods, such as LazyBrush [6], can only manage simple shape line drafts, and issues like unnatural colors or vacancies arise when dealing with complex patterns. Over the past decade, deep-learning methods have been employed for image coloring tasks. Researchers like Deshpande utilized variational autoencoders (VAEs) to learn the low-dimensional embedding of the color domain [7], thereby constructing a conditional model for generating diversified coloring results. Conversely, researchers like Mouzon integrated variational methods with convolutional neural networks (CNNs) to design a fully automatic image coloring framework [8].

With the advancement of deep learning, researchers have begun to employ generative adversarial networks (GANs) for image coloring across various fields, yielding notable results. In the realm of portrait synthesis, Fang and colleagues concentrated on synthesizing critical identity markers such as eyes and noses [9]. They utilized CycleGAN to generate high-resolution images, transitioning from sketches to facial photos. Simultaneously, in the medical sector, Long and others proposed an enhanced CycleGAN that incorporated perceptual loss [10]. This successfully facilitated the interactive translation of echocardiograms and enabled arbitrary conversions between sketch images and ultrasound images. Subsequently, GAN-based coloring methods were applied to painting tasks. For instance, Peng and others employed the CE-CycleGAN framework to process edge contours and transform landscape photos into Chinese landscape painting styles [11]. Sun and others developed a system capable of fitting edges based on semantic label maps [12], generating exquisite paintings of a specific type. Ren and others used a superpixel segmentation algorithm to optimize the coloring scheme of anime line drafts, addressing issues such as color diffusion and color loss [13]. Wang and others proposed a Thangka color simulation algorithm (SMAC-CGAN), which restored the actual colors of Tibetan paintings based on line drafts, resolving the halo problem [14]. Additionally, GAN technology has been applied to complete the automatic generation process from line sketches to color pictures based on nonspatially corresponding data. For example, Chen and others constructed the SketchyGAN to transform rough sketches of 50 categories, including various animals and daily necessities, into near-real color pictures [15]. Moreover, using similar images as virtual references, Wang and others [16], as well as Lee and others [17], have successfully completed the color transfer of the same type of sketch. This demonstrates the potential of these techniques in various applications.

Existing methods primarily focus on the coloring of grayscale painting images or line drawings, with less emphasis on the rendering of design drawings. In the landscape architecture domain, rendering design sketches involves transforming hand-drawn sketches into images with color, texture, and shadow effects to enhance the expressiveness and appeal of the design. This differs significantly from standard line-drawing coloring, as the lines and colors in design drawings carry landscape element information. These elements not only reflect the designer's creativity and thought process but also the functionality and form of the design. Therefore, rendering landscape design sketches requires a comprehensive consideration of the design's semantics, style, and rules, rather than merely coloring the lines.

In various design fields, researchers utilize different algorithms for rendering design sketches. A common approach is based on generative adversarial networks (GAN), leveraging deep learning to automatically learn design features from data and generate realistic images. In architecture, GAN methods have been used for rendering architectural effect and layout drawings. Qian et al. [18] proposed a GAN-based method to render sparse architectural sketches into textured effect drawings. Concurrently, their team also presented a generation process from user-preferred geometric shapes to sketches to high-rise architectural renderings [19]. However, when generating architectural images, the GAN model tended to distort the surface instead of maintaining the original linear structure. To address this issue, Zhao and others combined the Y-shaped GAN and denoising diffusion implicit models [20]. For rendering flat layouts within given boundaries, the system proposed by Wu and others can automatically and efficiently generate residential layout floor plans [21]. Building upon this, Huang Weixin [22] and colleagues used the GAN variant model Pix2pixHD to generate house-type diagrams based on room layouts. On a larger scale, Yang Liu [23] similarly utilized this approach to generate apartment floor plans based on function partitioning diagrams of youth apartments.

In the field of landscape architecture, GAN methods have also been applied to render design line drawings. Zhou et al. [24] used CycleGAN to render different types of plots in parks from color block diagrams to textured color maps, but faced issues with detail recognition, monotonous planar colors, and poor transitions. Our study attempts to compare different algorithms to determine which is more suitable for rendering landscape design drawings. We employed two distinct GAN models, namely, CycleGAN and Pix2pix, chosen for their respective advantages and applicable scenarios. CycleGAN is suitable for situations without paired data, while Pix2pix handles cases with paired data.

Regarding data selection, current image coloring tasks based on deep learning primarily obtain line-sketch datasets through edge extraction. Some researchers use the Canny algorithm to extract edges from real images to complete the coloring task. For instance, Zou and others used traditional image processing technology, Canny edge detection, to extract the line sketches of the Forbidden City patterns [25]. They completed the color restoration of the details of the Forbidden City murals. In terms of design, Sun and others also used the same method to extract the edge contours of billboards and icons for coloring [26]. In painting, because the line sketches obtained by the Canny algorithm cannot meet professional requirements, Aizawa and others used a model specifically based on the extraction of painting image line structure "LineDistiller" to build a dataset [27]. However, the XDoG algorithm used by Golyadkin and Makarov in comic coloring research could generate images with more visual appeal compared to it, which is more suitable for artistic line coloring [28]. Similarly, Zhao and others used this algorithm to obtain edge data for architectural sketch coloring [20]. When Li and others did the rendering of interior effect line drafts [29], some data were extracted by the Canny algorithm, and other data were nonmatching line drafts. Most researchers choose to obtain line-draft sketches by applying algorithms to images for the edge extraction in coloring tasks, and fewer researchers use corresponding hand-drawn sketches with more details. However, in the coloring of plan design drawings, hand-drawn line sketches have richer structures and light and shadow effects than computer-generated edge contour images, which can better express design effects. We choose hand-drawn line-draft sketches as the original data from the actual design process.

2.2. Conditional Generative Adversarial Networks

2.2.1. GAN Generation

Generative adversarial networks (GANs), proposed by Goodfellow [30] in 2014, represent a neural network framework grounded in game theory. This framework is unique in its inclusion of two competing subnetworks: the generator and the discriminator. The generator's role is to create virtual data that appears real from random noise, while the discriminator is tasked with discerning the authenticity of the input data. During the training process, these two subnetworks operate in an adversarial manner, striving for a Nash equilibrium. Consequently, the generator can produce virtual data capable of deceiving the discriminator.

2.2.2. Pix2pix and CycleGAN

Currently, the application of generative adversarial networks (GANs) in landscape design generation remains somewhat uninterpretable. This paper primarily focuses on Pix2pix and CycleGAN to construct a more scientific and controllable plan rendering design. Pix2pix, through its label set, can control the generation results, making it widely applicable in layout scheme design generation. For instance, contour images have been used as labels to generate robust arrangements for residential areas [31]. Additionally, related design indicators have been introduced to control the generation of urban land distribution schematic diagrams [32]. However, while Pix2pix is a supervised learning model that supports conditional control, it lacks the ability to explain the extraction of hidden features. On the other hand, CycleGAN, an unsupervised learning model, emphasizes adversarial training and cycle consistency loss. It can autonomously extract and explain implicit features. It has been extensively used in image style transfer, image super-resolution, image translation, and other fields. Chen et al. [33] in their research developed a park layout generation system using CycleGAN technology. The system primarily renders color block layouts, but it does not address the rendering of hand-drawn line design drawings, a common aspect in practical engineering design. In rendering color block layouts, the focus is on the overall visual effect of color and spatial element arrangements. However, the rendering of hand-drawn line drawings presents a more complex challenge, requiring the precise expression of line texture and detailed color nuances. Moreover, sketch rendering plays a crucial role in the design process of landscape architects, needing to reflect not just the precision of computer vision but also to aid designers in expressing their creative ideas. In light of these considerations, this study compared the Pix2Pix and CycleGAN algorithms and established an automated hand-drawn line-drawing rendering system based on GANs aimed at better addressing these challenges.

3. Methodology

3.1. Analytical Framework

Figure 1 presents the comprehensive research framework for the line-sketch rendering design scheme generation system proposed in this paper.

During the dataset construction phase, the quality and quantity of the dataset significantly influence the design learning of the algorithm. Given the challenge of obtaining paired park plan layout and design scheme data, we enhanced the black-and-white line-sketch data using CycleGAN. This enhancement expanded the original 652 pairs of park data to a dataset of 1699 pairs, thereby constructing a more diverse and accurate database for subsequent generation experiments.

In the model construction phase, we employed two neural networks, Pix2pix and CycleGAN, for image generation. The corresponding sketch and design scheme datasets were simultaneously input into the neural network for initial training of the small-sample data rendering system. However, the output results were suboptimal, leading us to add a data enhancement module based on CycleGAN. Consequently, we constructed an optimized plan rendering system using the expanded large-sample data.

During the application phase, the black-and-white line sketch is input into the model. The algorithm then automatically generates a rendering design scheme of the corresponding style within seconds.

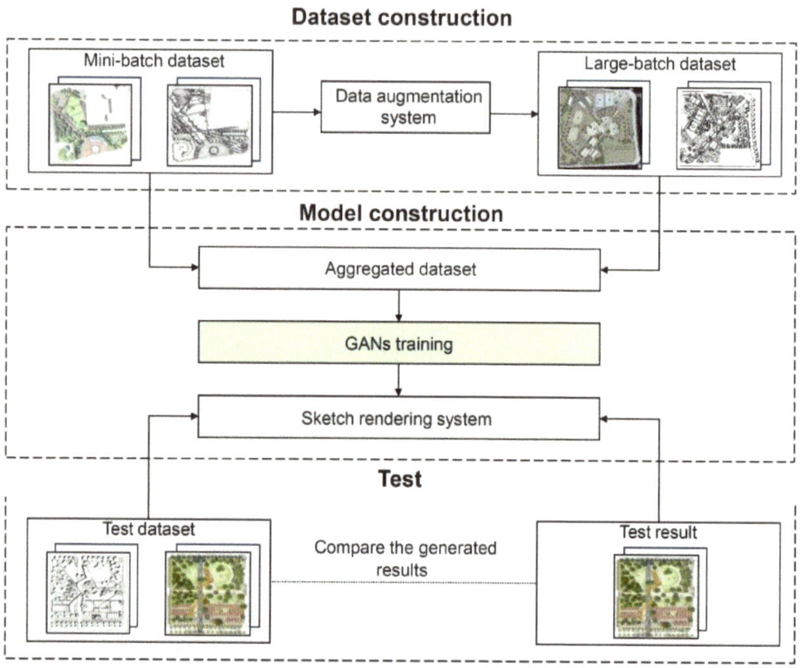

Figure 1. This is the process of building a line-sketch rendering system.

The following sections will provide a detailed discussion of each step.

3.2. Data Collection and Preparation

Due to the difficulty in discerning image details in large-scale design plans, this experiment uses park green space plans as its subject. Since publicly available data on green space design are limited, we collected various types of public training data from design websites and other channels. After initial screening and processing, these were used as our training dataset. The aim of our experiment is to improve the rendering efficiency in landscape design workflows. Therefore, we chose black-and-white hand-drawn sketches, commonly used in design, as our input data instead of colored images. Ultimately, we gathered 152 pairs of black-and-white hand-drawn sketches and design plans as training samples and standardized the dataset into 512 × 512 pixel jpg images. Part of our experimental data is shown in Figure 2.

3.3. Sketch Rendering Model

Our approach employs Pix2pix and CycleGAN to establish an automated plan rendering system. These are prevalent techniques in style transfer-related research, yet they exhibit substantial differences in their learning methodologies.

3.3.1. Supervised Learning Based on Pix2pix: Working Principle and Process

Pix2pix is developed on the foundation of conditional generative adversarial networks (CGANs). It treats the input image as a condition, learns the mapping relationship between the input and output images, and consequently generates the results [34].

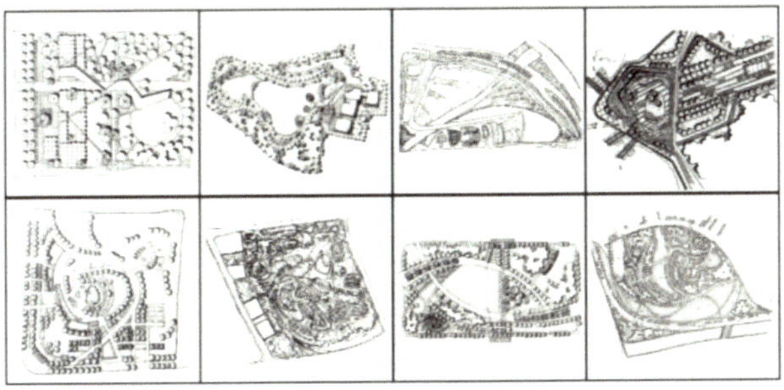

(a)Park sketch dataset

(b)Corresponding design dataset

Figure 2. These are the representative examples of hand-drawn sketches and corresponding design drawings dataset: (**a**) park sketch dataset; (**b**) corresponding design dataset.

The fundamental structure of the Pix2pix algorithm, as depicted in Figure 3, comprises two components: the generator (G) and the discriminator (D). The generator employs the U-net framework for feature extraction and upsampling. The discriminator, on the other hand, utilizes PatchGAN to discern whether the local details in the generated scheme closely resemble reality.

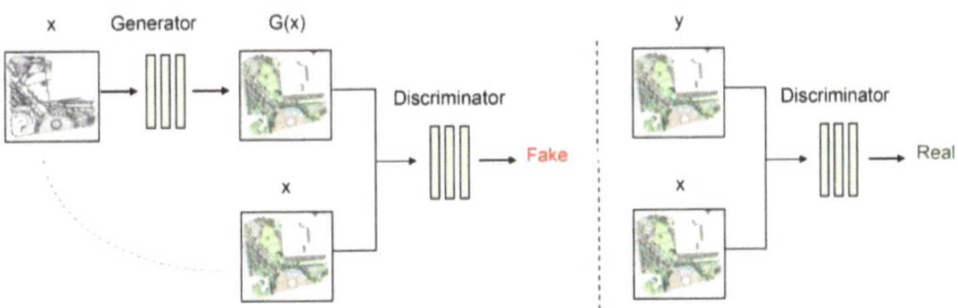

Figure 3. This is the workflow of Pix2pix.

3.3.2. Supervised Learning Working Principle and Process Based on CycleGAN

CycleGAN, a type of generative adversarial network (GAN), is utilized for unsupervised image generation. Unlike Pix2pix, CycleGAN does not necessitate paired data input to learn the mapping relationship from one image domain to another [35].

The framework of the CycleGAN algorithm, as illustrated in Figure 4, comprises two generators, Generator A and Generator B, and Discriminator B. The entire process encompasses two discrimination stages: (1) Ascertain the correlation between the result and input through Generator A and Generator B. (2) Compare the discriminator with the real reference scheme to judge whether the generated style is a design scheme or a hand-drawn line draft. In the first stage, the park line-draft plan is input into Generator A to obtain Result A (real image to fake image). Subsequently, Generator B generates the park line-draft scheme Result B (fake image to fake image) in reverse according to Result A. Through continuous iteration, the calculation error between the original line-draft plan and Result B is reduced, enabling the model to learn the implicit features in the image and generate the most realistic result. In the second stage, Result A is input into Discriminator B. After comparing it with the real design scheme, it is determined whether the output design scheme A is real or fake.

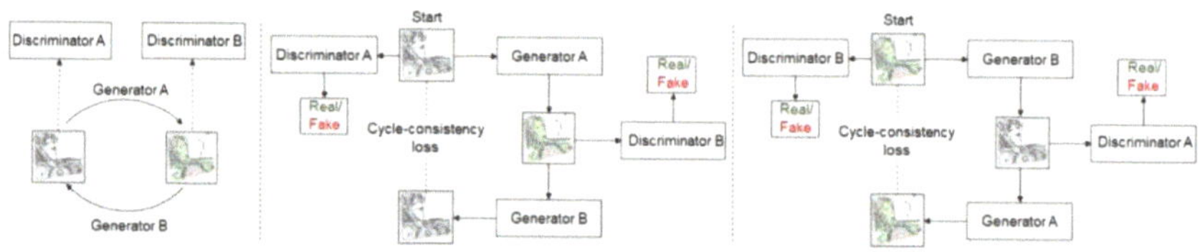

Figure 4. This is the workflow of CycleGAN.

3.4. Training

The entire training process, as depicted in Figure 5, comprises three stages: pretraining, data augmentation, and optimization training.

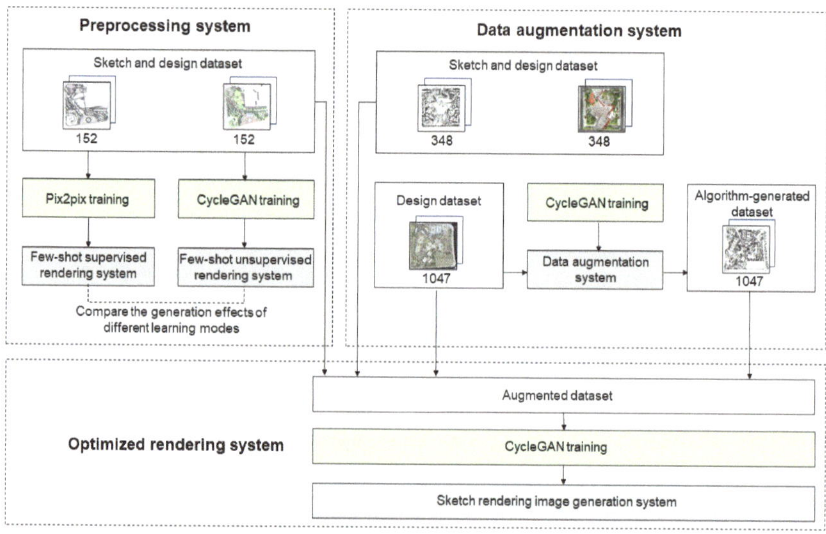

Figure 5. This is the framework for the optimized sketch rendering system, including the process from pretraining to data augmentation.

During the pretraining stage, the Pix2pix and CycleGAN algorithms are employed to train the hand-drawn sketch data, each with an equal sample size. The data are processed separately according to the format required by the different algorithms, leading to the initial construction of two small-sample line-draft plan rendering generation systems.

In the data augmentation stage, this study intends to use the CycleGAN algorithm as the foundation. It acquires 348 pairs of line-draft sketches and design drawings from various channels to construct a data augmentation system. Once the data augmentation system is established, we input 1047 design drawings, which lack corresponding line sketches, into the system and obtain 1047 black-and-white line sketches. These algorithmically generated black-and-white line sketches closely resemble human hand-drawn levels in terms of light and dark relationships, line strokes, and other aspects.

Finally, our data are augmented from the original 652 pairs of black-and-white line sketches and design scheme data to 1699 pairs of data samples. By retraining CycleGAN with the augmented dataset, we obtain the optimized design sketch rendering generation system.

3.5. Testing

To compare the efficiency of the generation systems built using different algorithms and varying sizes of data samples, this study selected five black-and-white line drafts as the test set. In the testing section, selected samples were inputted into the variously trained stages of CycleGAN and Pix2pix models, yielding post-test images.

From a landscape design perspective, the criteria for selecting test samples primarily included: (1) Park scale sufficient to clearly observe landscape elements such as paving and roads; (2) Rich and diverse vegetation planting methods in the samples; (3) The overall design encompassing a variety of spatial relationships. These were used to test the rendering systems and data enhancement models at each stage, as illustrated in Figure 6.

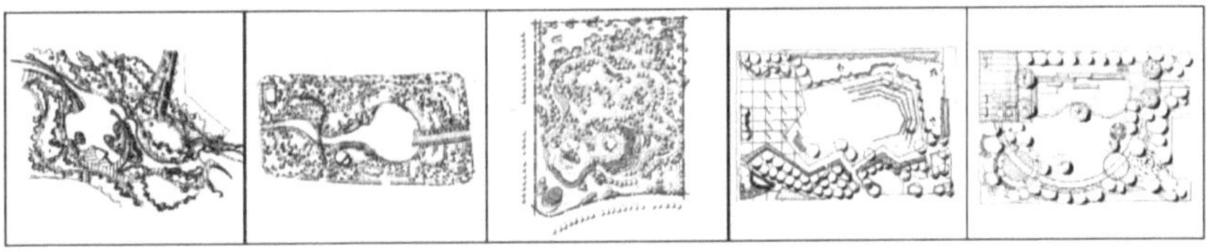

Figure 6. These are five representative hand-drawn sketches for testing.

The experimental samples selected are primarily categorized into medium-scale (samples a, b, c) and small-scale park line drafts (samples d, e), each exhibiting distinct characteristics. Planting: samples c and e primarily consist of point-shaped trees, while sample a involves a large cloud tree. The remaining samples comprise a mixed planting of cloud trees and point-shaped trees. Roads and Paving: samples a, b, and c all possess clear main ring roads and branch road structures, exhibiting pronounced spatial opening and closing characteristics. Conversely, samples d and e combine large-area paving and roads. Other Layout Elements: all five samples include lawns of varying areas. Apart from sample d, the rest of the samples contain water bodies.

4. Results and Analysis

To evaluate the performance of our developed model, we grouped and compared the test results of five samples. We assessed these results from a landscape design perspective using the following criteria: (1) Whether the color at the edges of the line drawings is clear; (2) Whether details like pavements, roads, and nodes are complete; (3) Whether the colors of trees, lawns, etc., are reasonable and diverse; (4) Whether the overall style is aesthetically pleasing and diverse.

This section primarily analyzes the results of two sets of comparative experiments. The first is a comparison of the small-sample rendering systems of the Pix2pix model and the CycleGAN model. The second is a comparison of the rendering systems of three different data volumes of the CycleGAN model.

4.1. Algorithm Comparison Evaluation

In the two small-sample line-draft plane rendering generation experiments, the Pix2pix algorithm and the CycleGAN algorithm exhibited a significant difference in the rendering results, as shown in Figure 7. The experimental results of line-draft rendering indicate that both algorithms have a certain degree of confusion in color expression, and the generated results are not accurate enough. However, the quality of the results generated by the CycleGAN model is significantly superior to that of the Pix2pix model.

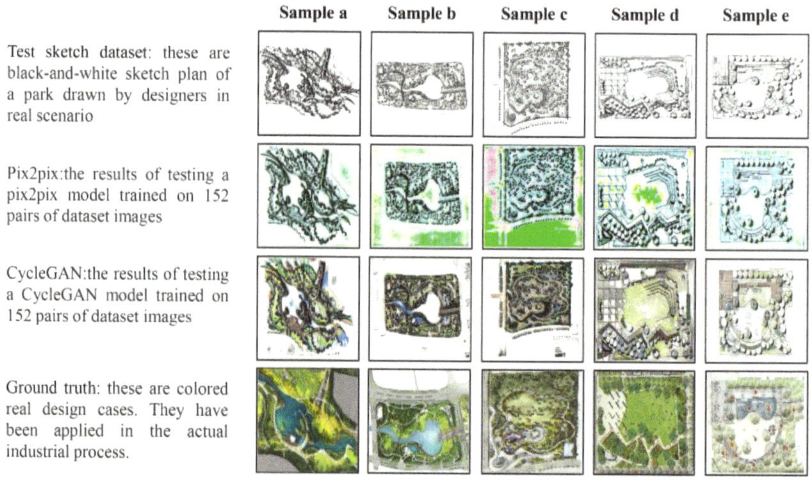

Figure 7. This is the comparison of the results generated by Pix2pix and CycleGAN.

In the small-sample test results of CycleGAN, it can be observed that this model can achieve different brightness and changes in the rendering processing of plants (such as samples a, b, c). Simultaneously, it has a certain distinction in the color of paving, water bodies, and lawns (such as samples b, d, e). However, it also has certain shortcomings: (1) It cannot clearly render the color of the water body (such as samples a, b, e). (2) Some trees appear inappropriately blue or purple (such as samples c and d). (3) The boundary color of some roads is blurred (such as samples a, b, c).

In contrast to CycleGAN, the rendering images generated by Pix2pix lack details and real textures, the colors are relatively uneven, and the possibility of blurring is much greater. These problems are because supervised learning needs to use a one-to-one dataset to let the computer learn the transformation logic in it. However, there is a certain gap between the hand-drawn line draft and the design drawing which cannot be completely aligned. Pix2pix makes it difficult to understand the connection between the two, and the results generated are of poor quality.

In the research of style transfer in the field of architectural design, related technologies, such as Pix2pix and CycleGAN, have become mainstream. However, few studies have compared and evaluated the results of the two algorithms in the field. Through the experiments and analysis of this study, the results of CycleGAN in line-draft rendering image generation are more stable and accurate than Pix2pix.

4.2. Data Volume Expansion Comparison Evaluation

Although our research indicates that CycleGAN's line-draft rendering is superior to Pix2pix, the generated results still exhibit issues, such as insufficient accuracy and a lack of diversity. Therefore, this study compared the rendering performance of the CycleGAN model under different sample volumes, as depicted in Figure 8. In these test results, models of three different sample sizes reflect differentiated style results.

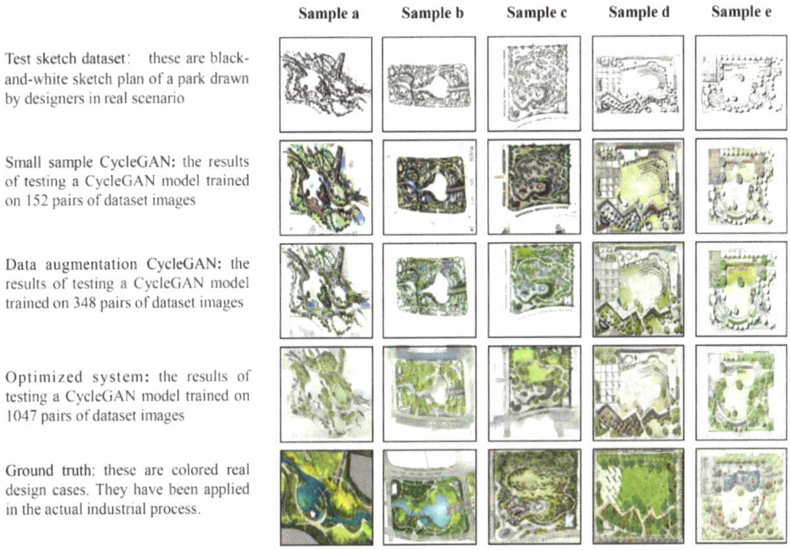

Figure 8. This is the comparison of results generated by the model before and after data augmentation.

In the flat rendering experiments of line drafts with two small samples, the Pix2pix algorithm and CycleGAN algorithm demonstrated significant differences in the rendering results. As seen in the comparison from Figure 8, in the landscape sketch rendering models trained with a dataset of 152 pairs, both algorithms exhibited some confusion in color representation and lacked precision in the generated results. However, CycleGAN showed markedly better performance than Pix2pix in terms of color accuracy and richness. Nonetheless, the CycleGAN model still made errors in judging certain detailed elements. The specific details are as follows:

Compared with the small-sample CycleGAN model mentioned earlier, the test results of the data expansion model have the following advantages: (1) It can better distinguish the color of road elements and their boundaries. (2) The expression of tree colors is more accurate, showing different brightness and saturation of green. However, it also has certain disadvantages: (1) Some cloud trees and point-shaped trees show mode collapse in rendering (such as samples a, d, e). (2) It is unable to render the color of water body elements accurately.

Moreover, our optimized system can more accurately and beautifully express rendering results for line drafts of different scales and has a certain style, mainly manifested as: (1) The rendering of the road structure is clear, primarily light gray. (2) The treatment of trees is shown as greens of different brightness, saturation, and transparency, and some lawns have a uniform gradient effect (such as samples a, c, e). (3) There is a certain distinction for the water body, and the rendering effect is significantly higher than the small sample, showing a blue–gray color (such as samples a and b). (4) The overall rendering effect shows a unique style of unified tone.

The comparison in Figure 8 shows that models trained with three different sizes of data samples exhibit varied stylistic results. The small-sample CycleGAN model trained with 152 data pairs showed diverse colors but some areas had colors inconsistent with reality. In contrast, the enhanced CycleGAN model trained with 348 data pairs had a fresher tone and a distinct style. Therefore, we trained an optimized model with 1047 data pairs for comparison. The results indicate that as the training data volume increases, the generated style leans towards a fresher color palette, with more accurate distinctions between roads and lawns, making this style more applicable in practical use. However, the model's ability to distinguish water bodies still needs further enhancement.

The main defects of this model are: (1) The clarity of the water body rendering color is insufficient. (2) The diversity of expressions for paving, nodes, etc., needs to be improved.

Following a comparative analysis, it has been observed that our generation results have significantly improved in several aspects after the data enhancement: (1) For identical plant elements, different brightness, saturation, and green textures can be generated in various positions rather than generating plants of different colors. (2) The interference of noise to the algorithm is relatively reduced, enabling it to accurately identify paving, water bodies, roads, etc., and distinguish color elements. (3) The rendering effect is more accurate and exhibits a certain style.

The experiment demonstrates that this study effectively expands the experimental samples using data enhancement methods, optimizes the model training effect, and generates more realistic and aesthetically pleasing park plan design drawings.

5. Discussion

This paper proposes a line-sketch rendering park design system based on Pix2pix and CycleGAN algorithms, which realizes the conversion from black-and-white flat line-sketch drawings to color texture rendering drawings. By utilizing a pretrained model to augment limited sample data, this approach overcomes the constraints of sparse and low-quality data in this field. It involves training an optimized model to enhance the accuracy and variety of generated images, thereby establishing an automated plan rendering system. Additionally, this research has advanced the evolution of human–computer collaborative design workflows within the landscape industry.

In related tasks, such as image coloring and style transfer, the GAN has been widely used and has in-depth research in the coloring of comic sketches and the transformation of different styles of art paintings [36,37]. In the design plan of the construction field, the related research was first implemented in the interior layout design [22]. At the same time, some researchers chose the object of multiple square enclosures composed of space in the Chinese Jiangnan Garden for generation rendering [38]. However, this series of plan rendering research focuses more on the functional area distribution of regular shapes and rarely uses natural shape plan sketches as objects. The field of park green space design, characterized by images rich in complex semantic content, presents a challenge for computers in interpreting the meaning of diverse lines. As a result, a comprehensive research paradigm for this domain has not yet been fully developed.

This study's proposed model enhances the rendering efficiency and can generate images with rich color features. It is capable, to a certain extent, of accurately differentiating various elements in park designs, leading to the production of effective design drawings. Diverging from the traditional approach of coloring based on the image block function and adding details and texture, our model discerns coloring rules from a multitude of hand-drawing design schemes. It extracts the interplay of the texture, shape, color tone, and placement of landscape elements. Consequently, the model can flexibly render plans based on input parameters, enabling the swift generation of results.

Most of the previous related research evaluated the effect of model-generated images from the perspective of computer science. In the research of using line drafts to generate Chinese paintings, Lin and others used the SR method to evaluate the noise of the generated images [39]. In the restoration of Thangka art, Li and others tested the FID distance between the generated image and the actual image and measured the effect of the generation through accuracy [40]. However, this study emphasizes the practicality of the results generated in the related design industry rather than simply conforming to the accuracy of the computer vision field. The results of our model are fresh in color and simple and atmospheric in the picture. This style is expressed as a low-saturation color in computer vision. However, in the actual design workflow, it can better help designers express their thinking and is more suitable for use as actual effect drawings. This study aims to facilitate effective communication between design teams and stakeholders by improving the rendering speed and quality of design drawings. The design process is a collaborative effort requiring feedback, and by accelerating rendering, we enable designers to present and share ideas more quickly. This speeds up feedback and iteration, enhancing the efficiency of the design process. Faster communication and iteration contribute to better quality and innovation in urban landscape design.

Our research has made progress in rendering design drawings, but it has limitations. The overall rendering effect depends on the quality of the drawings; rough or blurry drawings can lead to distorted results. Our method has yet to adapt to different design styles and aesthetics. Specific limitations include: (1) It is mainly used for small- to medium-scale parks, and may not be accurate for larger landscape designs. (2) It is unable to finely control data postenhancement. Future research could improve in several areas: First, vectorizing hand-drawn sketches to improve the rendering quality and adapt to different design stages and styles. Second, targeting landscape elements in design drawings for specific generation and optimization to enhance the accuracy and intelligence. Third, enhancing model interactivity, integrating with the designer's operational workflow for real-time rendering, quick design adjustments, and improved communication and experience.

While our results indicate that conditional generative adversarial networks are effective in rendering design sketches, this research has certain limitations: (1) The design schemes selected are primarily for small- and medium-sized parks, which may limit the model's ability to generate local nodes in larger-scale park designs accurately. (2) The enhancement of the algorithm does not allow for detailed constraints of the data. In future work, we aim to explore ways to refine the data enhancement methods and improve the algorithm. Our goal is to achieve multiscale plan rendering, thereby enhancing the rendering efficiency and diversifying styles. (3) Our algorithm exhibits limitations in generating vegetation, particularly in the crucial aspect of color diversity. This limitation is not just a technical issue, but also reflects the algorithm's inadequacy in understanding and reproducing the true richness and variety of colors in natural vegetation. Specifically, for vegetation colors other than green, our algorithm fails to capture the subtle nuances and diversity of plant colors found in nature. This issue touches on the profound challenge of bridging algorithmic design with the accurate representation of natural colors, necessitating a better balance between the development of algorithms and a deeper understanding of design theory.

6. Conclusions

This paper introduces a method leveraging generative adversarial networks (GANs) designed to rapidly produce rendered drawings from line-drawing plan designs. This method aids designers in swiftly conceptualizing design scenes. The research investigates the influence of the volume and quality of design data, as well as the impact of different algorithms on the generation results driven by algorithmic processes. Additionally, it involves the development of corresponding auxiliary design tools for evaluation.

This paper makes the following two breakthroughs:

Firstly, it addresses the critical challenge of data scarcity in this field. In response, the study introduces a data augmentation model based on generative adversarial networks (GANs) that effectively expands the range of small-sample hand-drawn line-sketch plan drawings.

Secondly, the prevalent color rendering techniques, which primarily depend on edge extraction, risk omitting crucial details in hand-drawn sketches. This loss is particularly evident in the design's structure and texture, significantly hindering designers' capacity to convey their fundamental concepts. To solve these problems, this study has developed a model for rendering directly against hand-drawn line drawings. The model aims to improve work efficiency and generate high-quality color-flat drawings that can be applied in industrial environments.

The experimental results demonstrate the main contributions of this study: (1) The development of a model based on Pix2pix and CycleGAN for rapidly generating diverse and rich design schemes from black-and-white hand-drawn sketches, significantly enhancing work efficiency. (2) The validation of the scheme's rationality and accuracy through objective evaluation, confirming its applicability in actual design processes and fostering the integration of artificial intelligence technology with landscape design. (3) Empirical evidence from data volume comparison underscores the importance of data augmentation in improving the model quality, leading to the creation of an optimized model with expanded data. (4) The use of hand-drawn sketches as experimental data aligns more closely with the practical application in landscape design processes. This study stands out for its emphasis on practical applications, in contrast to earlier research that mainly followed a computer science research paradigm with a sole focus on image accuracy. It primarily evaluates the generated results from a designer's viewpoint, highlighting its relevance to real-world scenarios. The results generated by the algorithm are expressed as low saturation in color saturation. However, in the actual design workflow, this artistic expression style is more conducive to the expression of designers' thinking and is more suitable for use as actual effect drawings. In the field of landscape design, this study enhances the rendering speed and quality of designs, improving collaboration and communication between designers and clients. While hand-drawn line drafts are indeed the primary embodiment of a designer's conceptualization in the tangible realm, the application of color plays a pivotal role in visual communication. For instance, in the realm of botanical landscaping, while line drafts delineate the hierarchical structure among plants, it is the application of color that critically conveys the harmony of the planting scheme. Similarly, in landscape design illustrations, line drafts establish the relational dynamics among design elements; however, the infusion of color, although not altering the design itself, significantly aids in the accentuation of design areas and focal points, thereby holding substantial value for both the designers and evaluators of multifaceted design proposals.

This research aims to enhance the rendering segment within the intelligent workflow of landscape design. The implementation of more efficient rendering technologies is expected to accelerate the iterative thinking process of designers and enable the vivid presentation of preliminary designs to stakeholders. By reducing the time spent on early design iterations, the overall speed and efficiency of the design process can be significantly improved.

This research initiates an inquiry into the intelligent rendering processes within landscape design. Addressing the challenges encountered in this study, future investigations will delve into specific areas, such as applying constrained color rendering to plant-specific projects and developing renderings in a variety of stylistic floor plans. These topics will direct our forthcoming research efforts, indicating a commitment to advancing the field of landscape design through the integration of innovative rendering techniques. To be more specific, in our future work, we plan to broaden the scope of our design schemes, incorporating the rendering of large-scale urban park line drawings. We will focus on controlling the details of pavements, buildings, and trees to ensure that the local colors in the output results are layered and the textures are richer. Additionally, the model will be fine-tuned to better adapt to various depths in plan sketch inputs, thereby enhancing its efficiency in providing inspiration for designers. We also intend to update related technical methods to achieve superior outcomes. Unlike previous research, this technological process is designed to integrate seamlessly with actual projects, assisting designers in rapidly developing ideas and providing timely feedback to users. This approach is likely to influence the operational modes and personnel composition of some companies. Many related factors arising from this new mode remain to be explored and discussed in future research.

Author Contributions: Conceptualization, Writing—Original Draft, R.C.; Funding acquisition, Supervision, J.Z.; Writing—Original Draft, Project administration, X.Y. (Xueqi Yao); Writing—review & editing, Visualization Y.L. and Y.H.; Data curation, Z.L. and H.L.; Methodology, Z.H. and X.Y. (Xingjian Yi). All authors have read and agreed to the published version of the manuscript.

Funding: This research was funded by the National Natural Science Foundation of China under Grant 52208041, the Key Laboratory of Ecology and Energy-saving Study of Dense Habitat (Tongji University), Ministry of Education, under Grant 20220110, and the Beijing High-Precision Discipline Project, Discipline of Ecological Environment of Urban and Rural Human Settlements.

Data Availability Statement: Dataset available on request from the authors. The raw data supporting the conclusions of this article will be made available by the authors on request.

Conflicts of Interest: Haoran Li was employed by the company China United Network Communications Group Co. The remaining authors declare that the research was conducted in the absence of any commercial or financial relationships that could be construed as a potential conflict of interest.

References

1. Tang, X. Research on Intelligent Landscape Design Based on Distributed Integrated Model. *Int. J. Semantic Web Inf. Syst.* **2023**, *19*, 1–19. [CrossRef]
2. Liu, M.; Nijhuis, S. Talking about landscape spaces. Towards a spatial-visual landscape design vocabulary. *Des. J.* **2022**, *25*, 263–281. [CrossRef]
3. Zhang, R.; Isola, P.; Efros, A.A. Colorful Image Colorization. In *Computer Vision—ECCV 2016*; Lecture Notes in Computer Science; Leibe, B., Matas, J., Sebe, N., Welling, M., Eds.; Springer International Publishing AG: Cham, Switzerland, 2016.
4. Zhang, R.; Zhu, J.-Y.; Isola, P.; Geng, X.; Lin, A.S.; Yu, T.; Efros, A.A. Real-time user-guided image colorization with learned deep priors. *ACM Trans. Graph.* **2017**, *36*, 1–11. [CrossRef]
5. Chu, N.S.-H.; Tai, C.-L. MoXi: Real-Time Ink Dispersion in Absorbent Paper. *ACM Trans. Graph.* **2005**, *24*, 504–511. [CrossRef]
6. Sýkora, D.; Dingliana, J.; Collins, S. LazyBrush: Flexible Painting Tool for Hand-drawn Cartoons. *Comput. Graph. Forum* **2009**, *28*, 599–608. [CrossRef]
7. Deshpande, A.; Lu, J.; Yeh, M.-C.; Chong, M.J.; Forsyth, D. Learning Diverse Image Colorization. In Proceedings of the 2017 IEEE Conference on Computer Vision and Pattern Recognition (CVPR), Honolulu, HI, USA, 21–26 July 2017; pp. 2877–2885.
8. Mouzon, T.; Pierre, F.; Berger, M.O. Joint CNN and Variational Model for Fullyautomatic Image Colorization. In *Scale Space and Variational Methods in Computer Vision. SSVM 2019*; Springer: Cham, Switzerland, 2019.
9. Fang, Y.; Deng, W.; Du, J.; Hu, J. Identity-aware CycleGAN for face photo-sketch synthesis and recognition. *Pattern Recognit.* **2020**, *102*, 107249. [CrossRef]
10. Teng, L.; Fu, Z.; Yao, Y. Interactive Translation in Echocardiography Training System With Enhanced Cycle-GAN. *IEEE Access* **2020**, *8*, 106147–106156. [CrossRef]
11. Peng, X.; Peng, S.; Hu, Q.; Peng, J.; Wang, J.; Liu, X.; Fan, J. Contour-enhanced CycleGAN framework for style transfer from scenery photos to Chinese landscape paintings. *Neural Comput. Appl.* **2022**, *34*, 18075–18096. [CrossRef]
12. Sun, L.; Chen, P.; Xiang, W.; Chen, P.; Gao, W.-Y.; Zhang, K.-J. SmartPaint: A co-creative drawing system based on generative adversarial networks. *Front. Inf. Technol. Electron. Eng.* **2019**, *20*, 1644–1656. [CrossRef]
13. Ren, H.; Li, J.; Gao, N. Two-Stage Sketch Colorization With Color Parsing. *IEEE Access* **2020**, *8*, 44599–44610. [CrossRef]
14. Wang, F.; Geng, S.; Zhang, D.; Zhou, M. Automatic colorization for Thangka sketch-based paintings. *Vis. Comput.* **2023**, *40*, 761–779. [CrossRef]
15. Chen, W.; Hays, J. SketchyGAN: Towards Diverse and Realistic Sketch to Image Synthesis. In Proceedings of the 2018 IEEE/CVF Conference on Computer Vision and Pattern Recognition (CVPR), Salt Lake City, UT, USA, 18–23 June 2018; pp. 9416–9425.
16. Wang, S.-Y.; Bau, D.; Zhu, J.-Y. Sketch Your Own GAN. In Proceedings of the IEEE/CVF International Conference on Computer Vision (ICCV), Montreal, QC, Canada, 10–17 October 2021.
17. Lee, J.; Kim, E.; Lee, Y.; Kim, D.; Chang, J.; Choo, J. Reference-Based Sketch Image Colorization using Augmented-Self Reference and Dense Semantic Correspondence. In Proceedings of the IEEE/CVF Conference on Computer Vision and Pattern Recognition (CVPR), Seattle, WA, USA, 13–19 June 2020.
18. Qian, W.; Xu, Y.; Li, H. A self-sparse generative adversarial network for autonomous early-stage design of architectural sketches. *Comput.-Aided Civ. Infrastruct. Eng.* **2022**, *37*, 612–628. [CrossRef]
19. Qian, W.; Yang, F.; Mei, H.; Li, H. Artificial intelligence-designer for high-rise building sketches with user preferences. *Eng. Struct.* **2023**, *275*, 115171. [CrossRef]
20. Zhao, L.; Song, D.; Chen, W.; Kang, Q. Coloring and fusing architectural sketches by combining a Y-shaped generative adversarial network and a denoising diffusion implicit model. *Comput.-Aided Civ. Infrastruct. Eng.* **2023**. [CrossRef]
21. Wu, W.; Fu, X.-M.; Tang, R.; Wang, Y.; Qi, Y.-H.; Liu, L. Data-driven interior plan generation for residential buildings. *ACM Trans. Graph.* **2019**, *38*, 1–12. [CrossRef]
22. Huang, W.; Yang, L.; Su, X. Human-Machine Combined Spatial Cognition. *New Archit.* **2021**, *1*, 4–10.
23. Yang, L. *Research on Automatic Generation of Youth Apartment Layout Based on Deep Learning*; South China University of Technology: Guangzhou, China, 2019.
24. Zhou, H.; Liu, H. Artificial Intelligence Assisted Design: Landscape Plan Recognition and Rendering Based on Deep Learning. *Chin. Landsc. Archit.* **2021**, *37*, 56–61.

25. Zou, Z.; Zhao, P.; Zhao, X. Virtual restoration of the colored paintings on weathered beams in the Forbidden City using multiple deep learning algorithms. *Adv. Eng. Inform.* **2021**, *50*, 101421. [CrossRef]
26. Sun, T.-H.; Lai, C.-H.; Wong, S.-K.; Wang, Y.-S. Adversarial Colorization of Icons Based on Structure and Color Conditions. In Proceedings of the 27th ACM International Conference on Multimedia (MM '19), Nice, France, 21–25 October 2019.
27. Aizawa, M.; Sei, Y.; Tahara, Y.; Orihara, R.; Ohsuga, A. Do You Like Sclera? Sclera-region Detection and Colorization for Anime Character Line Drawings. *Int. J. Netw. Distrib. Comput.* **2019**, *7*, 113–120. [CrossRef]
28. Golyadkin, M.; Makarov, I. Robust Manga Page Colorization via Coloring Latent Space. *IEEE Access* **2023**, *11*, 111581–111597. [CrossRef]
29. Li, X.; Li, H.; Wang, C.; Hu, X.; Zhang, W. Visual-attention GAN for interior sketch colourisation. *IET Image Process.* **2021**, *15*, 997–1007. [CrossRef]
30. Goodfellow, I.J.; Pouget-Abadie, J.; Mirza, M.; Xu, B.; Warde-Farley, D.; Ozair, S.; Courville, A.; Bengio, Y. Generative adversarial networks. *arXiv* **2014**, arXiv:1406.2661. Available online: https://arxiv.org/abs/1406.2661 (accessed on 13 November 2023). [CrossRef]
31. Cong, X. *Research on the Generation Design Method of Residential Area Strong Arrangement Scheme Based on CGAN*; Harbin Institute of Technology: Harbin, China, 2020.
32. Liu, Y.; Stoves, L.; Yang, Y. Urban Design Research Based on Conditional Generative Adversarial Networks. *Archit. J.* **2018**, *600*, 108–113.
33. Chen, R.; Zhao, J.; Yao, X.; Jiang, S.; He, Y.; Bao, B.; Luo, X.; Xu, S.; Wang, C. Generative Design of Outdoor Green Spaces Based on Generative Adversarial Networks. *Buildings* **2023**, *13*, 1083. [CrossRef]
34. Isola, P.; Zhu, J.-Y.; Zhou, T.; Efros, A.A. Image-to-Image Translation with Conditional Adverserial Networks. In Proceedings of the IEEE Conference on Computer Vision and Pattern Recognition (CVPR), Las Vegas, NV, USA, 27–30 June 2016.
35. Zhu, J.-Y.; Park, T.; Isola, P.; Efros, A.A. Unpaired Image-to-Image Translation using Cycle-Consistent Adversarial Networks. In Proceedings of the IEEE International Conference on Computer Vision (ICCV), Venice, Italy, 22–29 October 2017.
36. Gao, X.; Tian, Y.; Qi, Z. RPD-GAN: Learning to Draw Realistic Paintings with Generative Adversarial Network. *IEEE Trans. Image Process.* **2020**, *29*, 8706–8720. [CrossRef] [PubMed]
37. Lin, H.; Xu, C.; Liu, C. FAEC-GAN: An unsupervised face-to-anime translation based on edge enhancement and coordinate attention. *Comput. Animat. Virtual Worlds* **2023**, *34*, e2135. [CrossRef]
38. Liu, Y.; Fang, C.; Yang, Z.; Wang, X.; Zhou, Z.; Deng, Q.; Liang, L. Exploration on Machine Learning Layout Generation of Chinese Private Garden in Southern Yangtze. In Proceedings of the 2021 DigitalFUTURES; Yuan, P.F., Chai, H., Yan, C., Leach, N., Eds.; Springer: Singapore, 2022; pp. 35–44. [CrossRef]
39. Lin, D.; Wang, Y.; Xu, G.; Li, J.; Fu, K. Transform a Simple Sketch to a Chinese Painting by a Multiscale Deep Neural Network. *Algorithms* **2018**, *11*, 4. [CrossRef]
40. Li, H.; Fang, J.; Jia, Y.; Ji, L.; Chen, X.; Wang, N. Thangka Sketch Colorization Based on Multi-Level Adaptive-Instance-Normalized Color Fusion and Skip Connection Attention. *Electronics* **2023**, *12*, 1745. [CrossRef]

land

MDPI

Article

Influence of Urban Park Pathway Features on the Density and Intensity of Walking and Running Activities: A Case Study of Shanghai City

Junqi Chen, Zheng Tao, Wenrui Wu, Ling Wang and Dan Chen *

Department of Landscape Architecture, School of Design, Shanghai Jiao Tong University, Shanghai 200240, China; chen990619@sjtu.edu.cn (J.C.); zheng.tao@sjtu.edu.cn (Z.T.); wwlling@sjtu.edu.cn (L.W.)
* Correspondence: danchen.gator@sjtu.edu.cn

Abstract: Walking and running activities (W&RAs), encompassing strolling, slow walking, brisk walking, jogging, and running, hold significant importance as popular forms of exercise within urban parks. Recognized for their efficacy in promoting public health and preventing chronic diseases, understanding the nuanced impact of pathway features on W&RAs is crucial for advancing health-centric urban park planning. Based on extensive, high-frequency field observation data, we utilize multiple OLS regression models and univariate OLS regression models to investigate the relationship between urban park pathway features and variations in W&RAs, specifically examining activity density and intensity. Subsequently, we propose corresponding pathway optimization strategies. Our findings highlight the primary determinants, with vegetation coverage ratio, path type, and security facility density influencing activity density; and control value, time required to reach the nearest entrance, and pavement type influencing activity intensity. Significantly, increased vegetation coverage enhances density, while interconnected spaces and improved accessibility elevate intensity. In conclusion, our study delineates key features that merit prioritization, specifies their optimal ranges and proposes optimization design strategies for urban park pathways. By shedding light on these considerations, our research contributes valuable insights to the realm of health-oriented urban park planning and design.

Keywords: urban park; walking and running activities; field observation; activity density; activity intensity

Citation: Chen, J.; Tao, Z.; Wu, W.; Wang, L.; Chen, D. Influence of Urban Park Pathway Features on the Density and Intensity of Walking and Running Activities: A Case Study of Shanghai City. *Land* **2024**, *13*, 156. https://doi.org/10.3390/land13020156

Academic Editor: Eckart Lange

Received: 9 December 2023
Revised: 27 January 2024
Accepted: 28 January 2024
Published: 30 January 2024

1. Introduction

The World Health Organization (WHO) released the "Global Report on Urban Health" in 2016, indicating that chronic diseases constitute a primary health concern for contemporary urban areas. These conditions contribute to approximately 63% of the total mortality rate [1]. In contemporary society, numerous chronic diseases such as obesity and cardiovascular disorders pose a significant threat to human well-being and health. An increasing body of research indicates a close association between the occurrence of chronic diseases and the lack of physical activity(PA), as well as reduced exposure to natural environments [2,3].

Since the 1990s, rapid urbanization in China has led to a substantial increase in the urban population, posing increasingly severe challenges to urban environments and public health. The prevalence of chronic diseases, including obesity and cardiovascular disorders, has surged among urban residents [4]. In recent years, as living standards have improved, there has been a growing emphasis on both physical and mental health as a focal point of public concern. Initiatives such as "Healthy China", "Park Cities", and "National Fitness", coupled with a burgeoning body of public health research, underscore that engaging in physical activity (PA) is recognized as a crucial pathway to enhance public health [5–7]. Engaging in PA has become a pivotal response to national health initiatives, contributing significantly to the enhancement of public health. Among various activities, walking and running activities (W&RAs) have become popular and accessible forms of exercise due to their practicality, broad appeal, and ease of participation. Simultaneously, they stand out as common activities within urban parks and are closely linked to the improvement of health [8]. Research suggests that engaging in moderate-intensity brisk walking activities, accumulating 8–9 MET hours per week, can significantly improve cardiorespiratory health in healthy adults [9].

W&RAs typically take place within various pathways in urban parks. Pathways, proven to be significant park features in numerous studies, are closely associated with promoting park visitation and levels of PA [10–12]. Previous studies have indicated a correlation between various factors and W&RAs, including accessibility [13,14], spatial topology [15,16], spatial form [17–19], facilities [20,21], natural elements [22–26], aesthetic perception [27,28], and safety perception [29–31]. For instance, a study from Shanghai examined two large urban forest parks and explored the relationship between walking behavior and pathway topology by measuring variables such as integration, control value, and connectivity. It found that pathways with shorter average distances to park gates and more topologically accessible pathways were more preferred by visitors [32]. Another study from Chile explored the relationship between spatial place features of urban park pathways and walking behavior within the park. The study measured variables such as pathway width, pavement type, and vegetation coverage, and collected data on walking behavior. The findings revealed that factors such as increased vegetation, connectivity to activity areas, tranquility along pathways, and the presence of benches along the pathway significantly influenced the promotion of walking activities within the park. [33]. Additionally, a study from Harbin measured variables such as green view ratio and sky view ratio, exploring the association between aesthetic perception of park pathway spaces and W&RAs. It found that open skies and higher green view ratios significantly increased the intensity of these activities [27]. Furthermore, a literature review compared qualitative studies with previous quantitative research and emphasized the importance of park attributes reflecting spatial safety perception, such as the quantity of streetlights and safety facilities, in encouraging park use [34]. Based on the review of past research, we categorized these indicators into three dimensions: organizational relationships between spaces, place features within spaces, and subjective perception of spaces. The organizational dimension includes spatial accessibility and spatial topology, the place dimension includes spatial form, facilities, natural elements, and the perception dimension includes safety and aesthetics.

However, most previous studies have tended to focus on specific features influencing W&RAs, lacking a systematic discussion of multiple factors. For example, an Australian study explored the relative importance of micro-place features of urban park pathways in promoting walking among the elderly, finding that pathway slope, shaded trees, and pavement type played crucial roles in walking choices for the elderly [35]. However, this study only focused on place features within spaces and did not investigate the impact of organizational relationships between spaces and spatial perception on W&RAs. Therefore, we incorporate features from all three dimensions into regression models to comprehensively examine the influence of pathway features on W&RAs.

Activity density and activity intensity are two essential aspects used to measure different facets of W&RAs [27,33], reflecting the quantity and intensity of these activities, respectively. We hypothesize that various pathway features may influence different aspects of W&RAs, with certain features promoting more widespread and intense occurrences of these activities, while others may have the opposite effect. Identifying these key features is crucial for a comprehensive understanding of the impact of pathway features on W&RAs and providing better guidance for the planning and design of urban park pathways. Therefore, in this study, we concurrently examine both aspects, with activity density as the population density engaging in W&RAs within a sampling area in a day (daily W&RA density). Activity intensity is measured as the metabolic equivalent of tasks (METs) expended per capita in a day due to W&RAs (daily per capita W&RA METs), providing a holistic observation of the impact of pathway features on W&RAs.

Our study focuses on the pathway spaces within Xu Jiahui Park, Fu Xing Park, and Lujiazui Center Green in Shanghai as illustrative examples. Utilizing long-term, high-frequency field observation data, we constructed multiple OLS regression models and univariate OLS regression models for daily W&RA density and daily per capita W&RA METs. The models aimed to identify significant urban park pathway features influencing the density and intensity of W&RAs, as well as determining their optimal value ranges. The study addresses the following three research questions:

(1) Is there a significant correlation between urban park pathway features and the level of W&RAs?
(2) What are the significant urban park pathway features that influence the density and intensity of W&RAs?
(3) How do the effects and optimal value ranges of urban park pathway features differ concerning the density and intensity of W&RAs?

This study aims to systematically explore the relationship between urban park pathway features and the density and intensity of W&RAs from a micro-scale perspective. The study quantifies the impact of urban park pathway features on the density and intensity of W&RAs. Corresponding pathway optimization design strategies are proposed based on the quantitative results. The findings are anticipated to provide scientific support for the optimization design of urban park spaces with a focus on public health. This, in turn, aims to offer insights for urban park planning and design, contributing to the enhancement of public health.

2. Methodology

Figure 1 provides an illustrative overview of the research workflow. Based on the three research questions proposed, we selected 16 pathway plots from three parks as observation objects according to pre-survey and collected W&RA data and pathway feature data. We then tested the correlation between pathway features and W&RAs, then selected indicators with significant correlations for regression model construction. We first built a multiple linear regression model to observe the overall impact of pathway features on W&RAs. Subsequently, we constructed univariate regression models to determine the optimal range of each pathway feature in promoting W&RAs. This serves as the basis for proposing corresponding strategies for the optimization design of urban park pathway spaces to promote W&RAs.

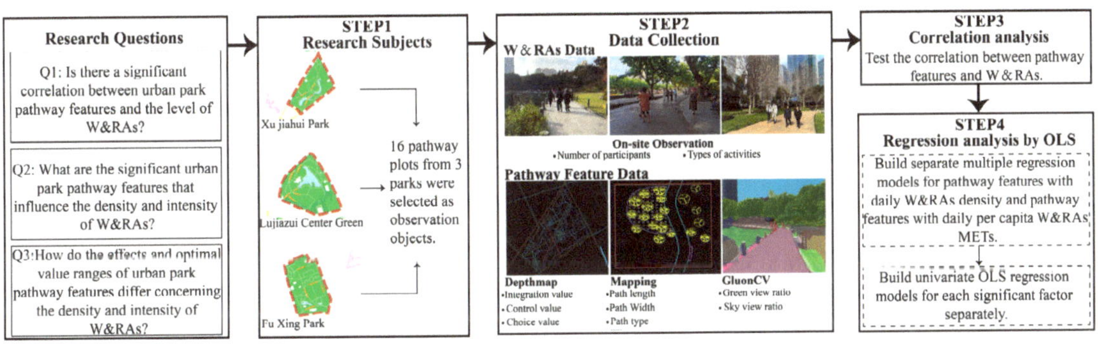

Figure 1. Workflow for the research.

2.1. Selection of Sample Parks and Plots

For this study, three urban parks located within the inner ring of Shanghai, characterized by substantial foot traffic and suitability for various physical activities, were chosen as sample parks: Xu Jiahui Park, Fu Xing Park, and Lujiazui Center Green. Following a preliminary investigation, a total of 5, 5, and 6 pathways conducive to W&RAs, respectively, were selected from these three parks for data collection and analysis (Figure 2). Selection criteria comprised the following:

(1) Plots should constitute a continuous part of the internal park pathway system.
(2) Plots should exhibit a smooth and continuous linear form.
(3) Plots should be devoid of outdoor elements like steps or stairs that could hinder W&RAs.
(4) The length of the plots should exceed 50 m.
(5) Plots should be uniformly distributed throughout the sample parks.

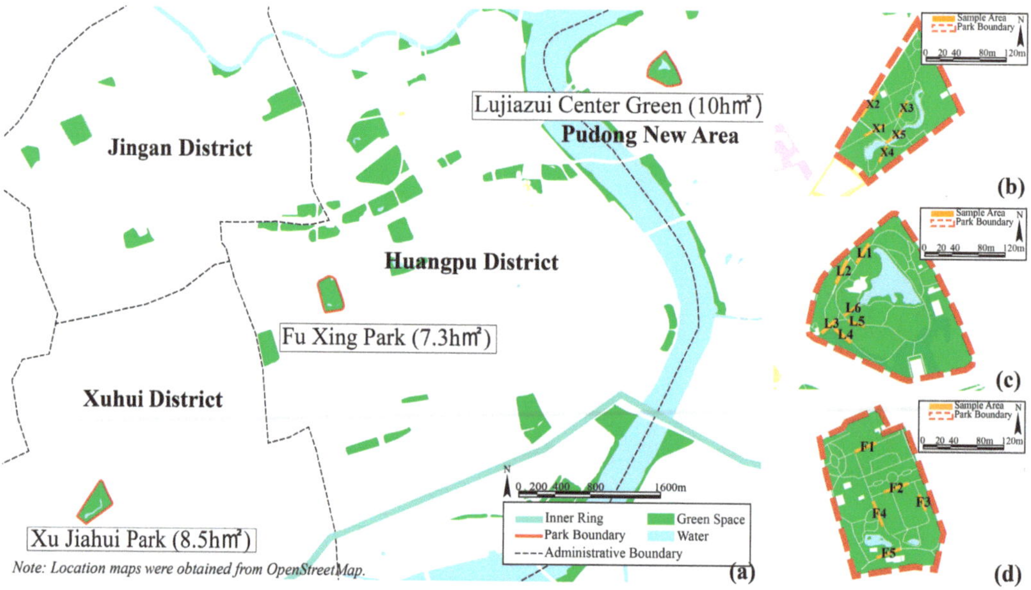

Figure 2. Park locations and spatial distribution of plots. (**a**) Park locations. (**b**) Distribution of plots in Xu Jiahui Park. (**c**) Distribution of plots in Fu Xing Park. (**d**) Distribution of plots in Lujiazui Center Green.

2.2. Selection of Pathway Features and Data Collection

PA is not only a physiological behavior but also a complex perceptual one. The willingness to engage in PA is influenced both directly by spatial environmental elements and indirectly by certain micro-space elements impacting subjective perception [36]. Integrating subjective perception factors with objective factors in green open spaces can overcome the limitations of describing PA from a single perspective. Therefore, building on relevant previous studies and incorporating on-site survey data, we selected indicators proven to be correlated with W&RAs in past research for regression model construction. These indicators were categorized into three aspects: spatial organization features, spatial place features, and spatial perception features. This categorization aimed to clearly demonstrate the ways of different indicators' influences and simultaneously examine the impact of features at different levels of urban park pathways on W&RAs (Table 1).

Table 1. Urban park pathway features' definitions and description.

Levels	Indicators	Specific Indicator Factors	Definition	Data Source	Calculation Method
Spatial Organization Features	Spatial topology	Integration value	The degree of spatial closeness to other spaces	OpenStreetMap	Average integration of each line segment in the plot
		Choice value	The number of times the space appears on the shortest topological path	OpenStreetMap	Average choice value of each line segment in the plot
		Control value	The reciprocal sum of the connection values to directly connected spaces	OpenStreetMap	Average control value of each line segment in the plot
	Accessibility	Time required to reach the nearest entrance	Time required to reach the nearest entrance at a speed of 1.5 m/s	Field Survey	Using a timer to record the time required for movement at a speed of 1.5 m/s
Spatial Place Features	Spatial form	Path width	/	OpenStreetMap	Extracted based on image maps from OpenStreetMap (m)
		Path length-to-width ratio	/	OpenStreetMap	Extracted based on image maps from OpenStreetMap
		Path type	/	Field Survey	Binary classification: Straight path = 0; Curved path = 1
	Natural elements	Vegetation structure	/	Field Survey	Ordered classification: Single-layer structure = 1; Double-layer structure = 2; Triple-layer structure = 3
		Vegetation coverage ratio	/	OpenStreetMap	Extracted based on image maps from OpenStreetMap: (Greening area of the plot/Plot area) (%)
		Water proximity	/	Field Survey	Binary classification: Waterside = 1; Non-waterside = 0
	Facilities	Pavement type	/	Field Survey	Ordered classification: Gravel paving = 1; Brick paving = 2; Block paving = 3; Concrete paving = 4; Asphalt paving = 5
		Density of seating	/	Field Survey	Number of seats in the plot/Plot area (units/m^2)
Spatial Perception Features	Aesthetic perception	Sky view ratio	Proportion of open sky in a person's horizontal field of view	Site photos	Image semantic segmentation: Open sky area/Photo image area (%)
		Green view ratio	Proportion of green plants in a person's horizontal field of view	Site photos	Image semantic segmentation: Plant greening area/Photo image area (%)
	Safety perception	Density of streetlights	/	Field Survey	Number of lamps in the plot/Plot area (units/m^2)
		Density of security facilities	/	Field Survey	Number of traffic signal lights, warning signs and cameras in the plot/Plot area (units/m^2)

Within the realm of spatial organization features, indicators like integration value, choice value, and control value draw upon the principles of spatial syntax theory. This theory quantifies and elucidates spatial structures, aiming to explore the correlation between spatial organization features and human society [37]. Established methods and indicators for urban-scale application have evolved within the framework of spatial syntax theory. Research indicates its adaptability to the exploration of the relationship between small-scale open spaces and physical activity [16], with specific applications in investigating the interplay between park pathway spatial composition and walking behavior [38]. Hence, we strategically employ this theory to delve into the spatial topology of urban park pathways. Utilizing DepthmapX v0.8.0 software to abstract the spatial composition of the park through the depiction of its axial map, we acquire essential indicators such as integration, choice value, and control value. Spatial perception features, including sky view ratio and green view ratio, are determined by invoking the GluonCV model, a deep learning toolkit in computer vision. This involves conducting image semantic segmentation on plot photos and calculating the proportion of pertinent features in the images. The remaining indicators are sourced through meticulous on-site surveys (Figure 3).

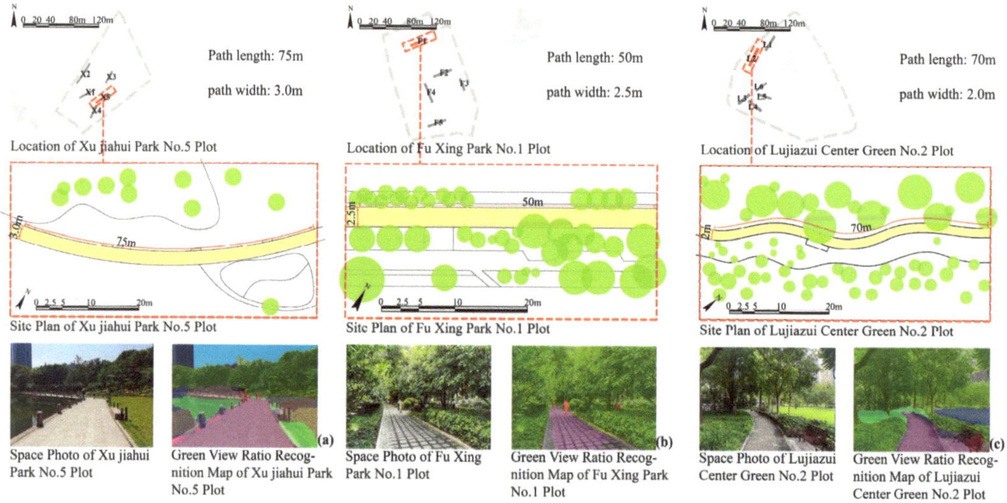

Note: Location maps were obtained from OpenStreetMap; Site plans were obtained through on-site surveys; Space photos were taken by the authors themselves.

Figure 3. Example of data collection for pathway features of plots: (**a**) Xu Jiahui Park No.5 Plot; (**b**) Fu Xing Park No.1 Plot; (**c**) Lujiazui Center Green No.2 Plot.

2.3. Data Collection of W&RAs and Calculation of Activity Levels

2.3.1. Data Collection of W&RAs

Our study conducted a total of 12 sets of on-site observations over 24 days from January 2021 to December 2021 to collect data on W&RAs within the three mentioned parks. Based on preliminary investigations, we selected 5, 5, and 6 plots in Xu Jiahui Park, Fu Xing Park, and Lujiazui Center Green, respectively, as the observation targets. Field observations were carried out on two clear days each month (including one weekday and one weekend day) for each park. The method employed for data collection primarily involved behavioral annotation, and the data collected included the types of W&RAs within each plot and the number of individuals engaging in each activity. The data collection periods were divided into five time slots: 6:00–8:00, 8:00–10:00, 11:00–13:00, 14:00–16:00, and 18:00–20:00. Single observations lasting 8 min were conducted during each time slot for each plot, aiming to comprehensively capture the real occurrences of W&RAs throughout the day. After a year of data collection, a total of 49,726 valid observation samples were collected. Participants were categorized into five activity types based on their different speeds: slow walking, strolling, brisk walking, jogging, and running. Specifically, slow walking refers to activities with an average speed of 4.0 km per hour, strolling at 4.5 km per hour, brisk walking at 5.6 km per hour, jogging at 7.0 km per hour, and running at 7.8 km per hour.

2.3.2. Calculation of W&RA Density and Intensity

Due to variations in plot size and notable differences in the number of participants within each plot, we employ the metrics of daily W&RA density and daily per capita W&RA METs to mitigate the impact of plot size and participant count on the assessment of W&RAs.

The formula for calculating daily W&RA density is as follows:

$$\text{Daily W\&RA density (persons/m}^2\text{)} = \text{Daily W\&RA participants (persons)/Plot area (m}^2\text{)} \quad (1)$$

The formula for calculating daily per capita W&RA METs is as follows:

$$\text{Daily per capita W\&RA METs (METs/person)} = \text{Cumulative daily W\&RA METs (METs)/Daily W\&RA participants (persons)} \quad (2)$$

Due to the challenge of real-time monitoring of the metabolic equivalents expended by all participants, we referenced the 2011 edition of the "Compendium of Physical Activities" (CPA) [39], which provides graded assessments of energy expenditure and intensity for physical activities. This classification was applied to the five types of activities involved in this study—strolling, slow walking, brisk walking, jogging, and running—to determine the corresponding metabolic equivalent values (Table 2). The cumulative daily W&RA METs are the product sum of the metabolic equivalents expended for different types of W&RAs and the corresponding total daily activity participants. The calculation formula is as follows:

$$\text{Cumulative daily W\&RA METs (METs)} = 3.0\ \text{(METs)} \times \text{Daily slow walking participants (persons)} + 3.5\ \text{(METs)} \times \text{Daily strolling participants (persons)} + 4.3\ \text{(METs)} \times \text{Daily brisk walking participants (persons)} + 7.0\ \text{(METs)} \times \text{Daily jogging participants (persons)} + 8.0\ \text{(METs)} \times \text{Daily running participants (persons)} \quad (3)$$

Table 2. Energy expenditure of walking and running activities in urban parks.

Activities	METs	Intensity
Slow walking	3	Moderate intensity ($3 \leq$ METs < 6)
Strolling	3.5	
Brisk walking	4.3	
Jogging	7	High intensity ($\geq$6METs)
Running	8	

2.4. *Statistical Analysis Methods*

We employed the ordinary least squares (OLS) model to investigate the overall impact of urban park pathway features on the density and intensity of W&RAs. The strength of this model lies in its simplicity, ease of interpretation, and ability to quantify the influence of each pathway feature on W&RAs. It helps establish a clear and interpretable relationship between pathway features and W&RAs, providing a concise analysis of how pathway features impact W&RAs. Prior to constructing regression models, a correlation analysis was conducted to examine the relationships between independent and dependent variables, with non-significant factors being excluded. The filtered results were then incorporated into the regression models, where pathway features served as independent variables, and daily W&RA density and daily per capita W&RA METs as dependent variables. Multiple linear regression models were established for each, and standardized coefficients were computed to assess the relative importance of each pathway feature on daily W&RA density and daily per capita W&RA METs (refer to Section 3.1 below). All independent variables passed the covariance test, with VIF values below 7.5.

To further understand the optimal ranges of pathway features promoting W&RA density and intensity, univariate linear regression models were constructed. Nine continuous variables from the pathway features were selected as independent variables, while daily W&RA density and daily per capita W&RA METs served as dependent variables. This facilitated the observation of the impact variations and optimal ranges of individual pathway features on W&RA density and intensity (refer to Section 3.2 below). Statistical significance for all models was set at $p < 0.05$. The entire analysis was conducted using SPSS 26.0.

3. Results

3.1. Analysis of the Impact of Pathway Features

We employed Pearson's correlation analysis, independent sample *t*-tests, and ANOVA analysis to identify pathway features significantly correlated with daily W&RA density as well as daily pèr capita W&RA METs, as shown in Tables 3 and 4. Features with significant correlations ($p < 0.05$) were selected to construct models for daily W&RA density and daily per capita W&RA METs.

Table 3. Pearson's correlation analysis of urban park pathway features with daily W&RA density and daily per capita W&RA METs.

	Daily W&RA Density		Daily per Capita W&RA METs	
	Pearson's Correlation	**Sig. (Two-Tailed)**	**Pearson's Correlation**	**Sig. (Two-Tailed)**
Integration value	0.154 **	0.002	0.227 **	0
Choice value	−0.371 **	0	0.08	0.116
Control value	−0.019	0.714	0.518 **	0
Time required to reach the nearest entrance	0.017	0.735	-0.410 **	0
Path width	−0.347 **	0	−0.09	0.077
Path length-to-width ratio	0.262 **	0	0.114 *	0.026
Vegetation coverage ratio	0.362 **	0	−0.04	0.437
Density of seating	−0.138 **	0.007	−0.151 **	0.003
Sky view ratio	0.254 **	0	−0.249 **	0
Green view ratio	−0.176 **	0.001	0.029	0.566
Density of streetlights	0.348 **	0	−0.140 **	0.006
Density of security facilities	0.334 **	0	0.191 **	0

Note: * Significant at the 0.05 level (two-tailed); ** Significant at the 0.01 level (two-tailed).

Table 4. Independent sample *t*-test and ANOVA of urban park pathway features with daily W&RA density and daily per capita W&RA METs.

Testing Method	Testing Variables	Pathway Features	t	F	Df	Sig.
Independent samples *t*-test	Daily W&RA density	Path type (straight)	−5.659	-	203.496	0.000
		Path type (curved)	-	-	-	-
		Water proximity (non-waterside)	−3.469	-	87.209	0.001
		Water proximity (waterside)	-	-	-	-
	Daily per capita W&RA METs	Path type (straight)	3.471	-	371.103	0.001
		Path type (curved)	-	-	-	-
		Water proximity (non-waterside)	5.049	-	248.369	0.000
		Water proximity (waterside)	-	-	-	-
ANOVA	Daily W&RA density	Vegetation structure	-	6.354	2	0.002
		Pavement type	-	0.127	4	0.972
	Daily per capita W&RA METs	Vegetation structure	-	27.849	2	0.000
		Pavement type	-	34.921	4	0.000

3.1.1. Results of the Daily W&RA Density Model

Thirteen pathway features with significant correlations ($p < 0.05$) were selected to construct the model for daily W&RA density. Due to the presence of multicollinearity among some features, we examined the correlation coefficients between each pair of independent variables and removed features with severe multicollinearity to avoid their impact on the model results. The final model incorporated 12 pathway features with a VIF < 7.5 and adjusted $R^2 = 0.430$. The specific results are presented in Table 5. The density of security facilities, the density of streetlights, integration value, path type, vegetation coverage ratio, and sky view ratio all exhibited a positive impact on daily W&RA density. Notably, the vegetation coverage ratio emerged as the most influential positive factor, while the positive impact of the sky view ratio was relatively weaker. Vegetation structure had a negative impact on daily W&RA density. Additionally, choice value, density of seating, green view ratio, waterfront proximity, and path width did not demonstrate statistical significance.

3.1.2. Results of the Daily per Capita W&RA METs Model

The final model for daily per capita W&RA METs incorporated 11 pathway features with a VIF < 7.5 and adjusted $R^2 = 0.589$. The detailed results are presented in Table 6. Control value, the density of security facilities, and pavement type exhibited positive effects on daily per capita W&RA METs. Control value emerged as the most influential positive factor, while the density of security facilities showed a relatively weaker positive impact. Five features, namely time required to reach the nearest entrance, path length-to-width ratio, sky view ratio, density of seating, and path type, demonstrated negative effects on daily per capita W&RA METs. The negative impact of time required to reach the nearest entrance was the strongest, while the negative impact of path type was the weakest. Additionally, the density of streetlights, waterfront proximity, and vegetation structure did not exhibit statistical significance.

3.2. Analysis of the Optimal Range of Pathway Features

The above regression models explored the pathway features influencing the density and intensity of W&RAs, along with their variations in impact strength. To further identify the optimal range of pathway features significantly affecting W&RA density and intensity, we established univariate OLS regression models for the nine continuous variable features that demonstrated significance (Table 7). The optimal range of each feature was observed through scatter plots (Figure 4).

3.2.1. Results of the Univariate OLS Regression Models for Spatial Organization Features

The model results (Figure 4a) indicate a weak positive correlation between integration value and daily W&RA density. Daily W&RA density shows a slow increase with the improvement of integration value.

Control value demonstrates an overall positive correlation with daily per capita W&RA METs. When the control value ranges from 0 to 1.3, daily per capita W&RA METs increase with the rise of the control value. As the control value ranges from 1.3 to 2.5, daily per capita W&RA METs exhibit an extremely gradual decline with the increase of the control value. When the control value exceeds 2.5, daily per capita W&RA METs rapidly increase with the ascent of the control value.

Time required to reach the nearest entrance is negatively correlated with overall daily per capita W&RA METs. When the time required to reach the nearest entrance remains between 0 and 100 s, the daily per capita W&RA METs decrease with the increase in the time required. When the time required to reach the nearest entrance exceeds 100 s, daily per capita W&RA METs remain relatively constant.

Table 5. Results of the regression model for daily W&RA density.

Levels	Indicators	Specific Indicator Factors	Unstandardized Coefficient		Standardized Coefficient	t	Significance	95.0% Confidence Interval for B		Covariance Statistics
			B	Standard Error	Beta			Lower Limit	Upper Limit	VIF
Spatial Organization Features	Spatial topology	(constant)	−0.594	0.210		−2.826	0.005	−1.007	−0.181	
		Integration value	0.227	0.075	0.205	3.039	0.003	0.080	0.375	2.965
		Control value	0.000	0.000	0.095	1.153	0.250	0.000	0.000	4.452
Spatial Place Features	Spatial form	Path width	−0.049	0.030	−0.148	−1.628	0.104	−0.109	0.010	5.371
		Path type	0.153	0.047	0.232	3.278	0.001	0.061	0.244	3.263
	Natural elements	Vegetation structure	−0.059	0.025	−0.136	−2.411	0.016	−0.108	−0.011	2.075
		Vegetation coverage ratio	0.997	0.178	0.413	5.600	0.000	0.647	1.348	3.533
		Water proximity	−0.048	0.073	−0.058	−0.662	0.508	−0.191	0.095	4.941
	Facilities	Density of seating	3.161	1.913	0.136	1.652	0.099	−0.602	6.923	4.388
Spatial Perception Features	Aesthetic perception	Sky view ratio	2.283	0.738	0.154	3.093	0.002	0.832	3.735	1.618
		Green view ratio	−0.072	0.109	−0.036	−0.662	0.508	−0.287	0.142	1.926
	Safety perception	Density of streetlights	5.858	2.194	0.195	2.670	0.008	1.544	10.173	3.467
		Density of security facilities	14.745	3.511	0.210	4.199	0.000	7.841	21.650	1.631

Table 6. Results of the regression model for daily per capita W&RA METs.

Levels	Indicators	Specific Indicator Factors	Unstandardized Coefficient		Standardized Coefficient	t	Significance	95.0% Confidence Interval for B		Covariance Statistics
			B	Standard Error	Beta			Lower Limit	Upper Limit	VIF
Spatial Organization Features	Spatial topology	(constant)	3.491	0.093		37.724	0.000	3.309	3.673	
		Control value	0.286	0.026	0.533	11.035	0.000	0.235	0.336	2.112
	Accessibility	Time required to reach the nearest entrance	−0.002	0.000	−0.290	−6.035	0.000	−0.003	−0.002	2.092
Spatial Place Features	Spatial form	Path length-to-width ratio	−0.007	0.002	−0.196	−4.600	0.000	−0.010	−0.004	1.646
		Path type	−0.081	0.040	−0.107	−2.022	0.044	−0.159	−0.002	2.547
	Natural elements	Vegetation structure	−0.028	0.023	−0.055	−1.226	0.221	−0.072	0.017	1.847
		Water proximity	0.101	0.057	0.105	1.764	0.079	−0.012	0.213	3.222
	Facilities	Density of seating	−4.649	1.217	−0.174	−3.819	0.000	−7.042	−2.255	1.886
		Pavement type	0.079	0.016	0.247	4.800	0.000	0.047	0.111	2.396
Spatial Perception Features	Aesthetic perception	Sky view ratio	−2.844	0.739	−0.168	−3.851	0.000	−4.296	−1.392	1.721
	Safety perception	Density of security facilities	19.207	3.656	0.239	5.254	0.000	12.018	26.397	1.879
		Density of streetlights	−1.299	2.041	−0.038	−0.637	0.525	−5.311	2.714	3.186

Table 7. Results of the univariate OLS regression models for daily W&RA density and daily per capita W&RA METs.

Implicit Variable	Independent Variable	Models	Adjusted R^2	Constant	B			Sig.
					B1	B2	B3	
Daily W&RA density	Integration value	$y = 0.206e^{0.491x}$	0.046	0.206	0.491	/	/	0.000
	Vegetation coverage ratio	$y = 0.073e^{2.059x}$	0.175	0.073	2.059	/	/	0.000
	Density of streetlights	$y = 0.546 - 47.841x + 3428.438x^2 - 52750.227x^3$	0.192	0.546	−47.841	3428.438	−52,750.227	0.000
	Density of security facilities	$y = 0.465 - 41.773x + 4324.538x^2$	0.198	0.465	−41.773	4324.538	/	0.000
	Sky view ratio	$y = 0.485 - 13.522x + 369.067x^2 - 1703.033x^3$	0.141	0.485	−13.522	369.067	−1703.033	0.000
Daily per capita W&RA METs	Control value	$y = 2.942 + 1.581x - 1.016x^2 + 0.195x^3$	0.417	2.942	1.581	−1.016	0.195	0.000
	Time required to reach the nearest entrance	$y = 4.071 - 0.009x + 0.0000366x^2$	0.199	4.071	−0.009	0.0000366	/	0.000
	Path length-to-width ratio	$y = 3.847 - 0.046x + 0.003x^2 - 0.0000405x^3$	0.057	3.847	−0.046	0.003	−0.0000405	0.000
	Density of seating	$y = 3.685 + 35.959x - 2147.82x^2 + 26971.594x^3$	0.067	3.685	35.959	−2147.82	26,971.594	0.000
	Density of security facilities	$y = 3.516 + 92.475x - 1990.5x^2 - 207502.067x^3$	0.201	3.516	92.475	−1990.5	−207,502.07	0.000
	Sky view ratio	$y = 3.827 - 13.472x + 173.261x^2 - 662.562x^3$	0.080	3.827	−13.472	173.261	−662.562	0.000

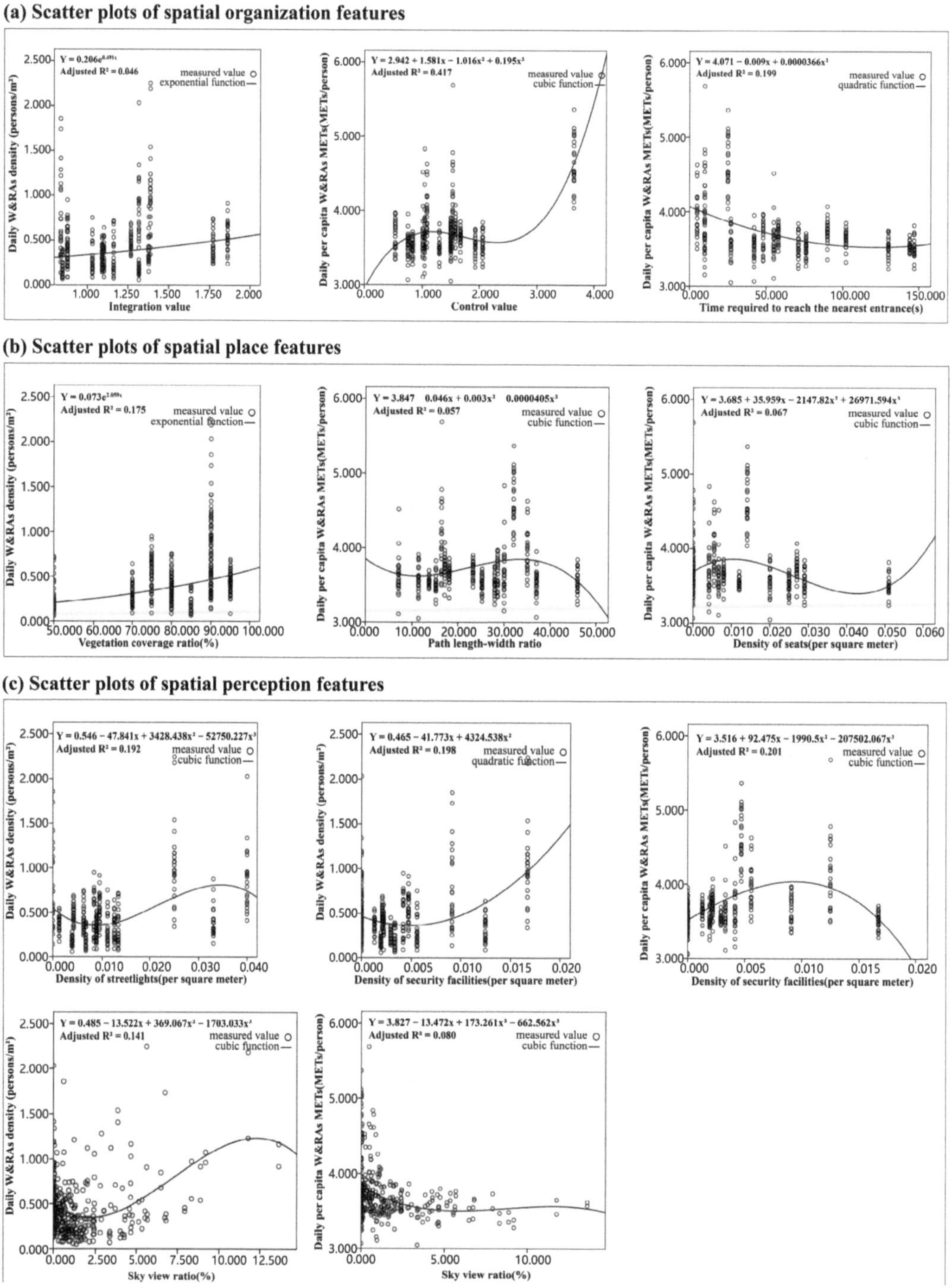

Figure 4. Scatter plots for each pathway feature. (**a**) Scatter plots of spatial organization features. (**b**) Scatter plots of spatial place features. (**c**) Scatter plots of spatial perception features.

3.2.2. Results of the Univariate OLS Regression Models for Spatial Place Features

Vegetation coverage ratio demonstrates a positive correlation with daily W&RA density. Within different ranges of values, the impact of vegetation coverage ratio on daily W&RA density varies. When the vegetation coverage ratio is <0.75, the increase in daily W&RA density is relatively slow. However, when the vegetation coverage ratio exceeds 0.75, the daily W&RA density shows a relatively rapid increase.

Path length-to-width ratio is negatively correlated with overall daily per capita W&RA METs. When the path length-to-width ratio is less than 10, the daily per capita W&RA METs decrease with the increase in the path length-to-width ratio. When the path length-to-width ratio is between 10 and 35, the daily per capita W&RA METs increase with the rise in path length-to-width ratio, reaching its peak at a ratio of 35. For ratios greater than 35, the daily per capita W&RA METs rapidly decrease with an increase in path length-to-width ratio.

Density of seating is negatively correlated with overall daily per capita W&RA METs. When the density of seating is less than 0.01 per square meter, daily per capita W&RA METs increase with the rise in the density of seating, reaching its peak at a density of 0.01 per square meter. When the density of seating ranges from 0.01 per square meter to 0.045 per square meter, daily per capita W&RA METs decrease with the increase in seating density, reaching a minimum at a density of 0.045 per square meter (Figure 4b).

3.2.3. Results of the Univariate OLS Regression Models for Spatial Perception Features

Density of streetlights is positively correlated with overall daily W&RA density. When the density of streetlights is less than 0.01 per square meter, the daily W&RA density exhibits a gradual decline with increasing streetlight density. With streetlight density ranging from 0.01 to 0.035 per square meter, daily W&RA density gradually increases, reaching its peak at a density of 0.035 per square meter. However, when the streetlight density exceeds 0.035 per square meter, daily W&RA density declines again with an increase in streetlight density.

Density of security facilities significantly influences both daily W&RA density and daily per capita W&RA METs. The impact of security facility density on daily W&RA density is generally positive. When the density of security facilities is less than 0.005 per square meter, daily W&RA density experiences an extremely slow decline with an increase in the density of security facilities. However, when the density of security facilities exceeds 0.005 per square meter, the daily W&RA density rapidly increases. On the other hand, the density of security facilities has an overall negative correlation with the daily per capita W&RA METs. When the density of security facilities is less than 0.01 per square meter, the daily per capita W&RA METs increase with rising safety facility density, reaching its peak at a density of 0.01 per square meter. Conversely, when the density of security facilities exceeds 0.01 per square meter, daily per capita W&RA METs decrease with an increase in safety facility density.

Sky view ratio also significantly influences both daily W&RA density and daily per capita W&RA METs. There is a positive overall correlation trend between the sky view ratio and daily W&RA density. When the sky view ratio is less than 0.025, daily W&RA density decreases with an increase in the sky view ratio. However, when the sky view ratio is between 0.025 and 0.125, the daily W&RA density rapidly increases with an increase in the sky view ratio. The sky view ratio has a weak negative correlation with the daily per capita W&RA METs. When the sky view ratio is less than 0.05, daily per capita W&RA METs show a gradual decline with increasing sky view ratio. When the sky view ratio exceeds 0.05, the daily per capita W&RA METs remain relatively stable with an increase in the sky view ratio (Figure 4c).

4. Discussion

Figure 5 provides an illustrative overview of the discussion structure. We first discussed the differential impact of factors influencing daily W&RA density and daily per capita W&RA METs, comparing and contrasting their effects. Through this comparative discussion, we identified the most crucial influencing factors. Utilizing scatter plot results, we further explored the optimal ranges for these key factors in promoting W&RAs. Based on these findings, we ultimately proposed pathway optimization strategies to enhance W&RAs, aiming to provide insights for health-oriented urban park planning and design.

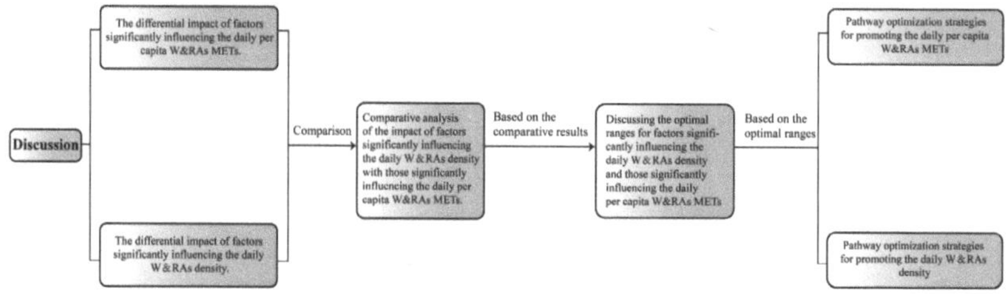

Figure 5. Discussion Summary Flowchart.

4.1. Factors Influencing Daily W&RA Density

Regarding the factors influencing daily W&RA density, pathway features significantly affecting the density are arranged in descending order based on their impact. These features include vegetation coverage ratio (Beta = 0.413), path type (Beta = 0.232), density of security facilities (Beta = 0.210), integration value (Beta = 0.205), density of streetlights (Beta = 0.195), sky view ratio (Beta = 0.154), and vegetation structure (Beta = −0.136). Among these, vegetation coverage ratio, path type, density of security facilities, integration value, density of streetlights, and sky view ratio demonstrate a positive impact on daily W&RA density. Higher vegetation coverage indicates a greener space, and existing research suggests that green spaces can alleviate stress, soothe emotions, and motivate people to engage in W&RAs [40]. Furthermore, the occurrence of W&RAs is associated with the quality of vegetation. Spaces with higher-quality vegetation promote the occurrence of W&RAs more effectively [22], indicating a close relationship between the quantity and quality of vegetation and W&RAs. To encourage the occurrence of W&RAs, attention should be paid not only to increasing vegetation coverage but also to vegetation selection, combination, and daily maintenance to enhance the quality of green spaces. Daily W&RA density is positively influenced by path type, suggesting that curved paths are more likely to attract W&RAs. This is because the observed W&RAs in our study mainly involve moderate-intensity activities such as strolling and slow walking, with the primary purpose of relaxation. Compared to high-intensity activities such as jogging and running, which focus more on exercise and fitness, these moderate-intensity activities prioritize the landscape and touring experience during the activity. Curved paths offer more varied scenic changes on both sides than straight paths, creating a richer visual experience for participants [41]. Both the density of security facilities and the density of streetlights have a positive impact on daily W&RA density, indicating that spaces with sufficient security facilities and good lighting are more likely to attract W&RA participants. Adequate lighting and well-equipped security facilities reduce potential safety risks during activities, boost the psychological sense of security for participants, and stimulate the willingness to engage in W&RAs [20]. Further studies indicate that W&RAs are more likely to occur in spaces with multiple lighting sources and diverse colors [42]. Integration value also has a positive impact on daily W&RA density. Higher integration value indicates better accessibility [15], a prerequisite for attracting participants to enter the space [43]. Additionally, the sky view ratio positively influences daily W&RA density. A larger visible sky range increases the quantity of W&RAs, and as the distance of W&RAs increases, the influence of a large visible sky range on daily W&RA density becomes stronger. This is because an open sky can help individuals alleviate psychological and physical stress, thereby improving quality of life [44].

Vegetation structure exhibits a negative impact on daily W&RA density, indicating that a complex vegetation structure is unfavorable for the occurrence of W&RAs. A possible explanation is that a complex vegetation structure may create a spatial environment with high canopy closure, obstructing lines of sight during activities [45]. Such spaces are associated with higher crime rates [46], compromising the safety of visitors. Conversely, a simple vegetation structure is more likely to attract W&RAs, such as open lawns and tree-lined squares. This is because open spaces provide clear lines of sight, accommodating a greater number of activities, and thereby enhancing the willingness of individuals to engage in W&RAs [33].

In addition, choice value, density of seating, green view ratio, water proximity, and path width did not exhibit significance in influencing daily W&RA density. The regression results for green view ratio, water proximity, and path width differed from expectations. Vegetation, as a critical natural element in parks, is generally believed to enhance people's willingness to engage in PA [23–26]. In our study, green view ratio did not significantly affect daily W&RA density. However, the correlation analysis revealed a negative correlation between green view ratio and daily W&RA density, possibly due to the high levels of green view ratio in the plots (≥45%). An excessively high green view ratio may lead to a sense of insecurity among visitors [47], hindering the occurrence of W&RAs. Moreover, the green view ratio shows significant seasonal variations, and its impact on W&RAs may fluctuate. Studies in different regions [27,47] suggest that the impact of the green view ratio can be either positive or negative, influenced primarily by climate and seasonal changes. Water proximity is generally believed to attract more activities [26], but it did not significantly affect daily W&RA density in our study. The reason for this result may be that the water bodies selected in our study are relatively small and often appear in the form of artificial water features such as fountains and cascades, lacking significant appeal for W&RAs. Another explanation could be the abundant water resources in the Shanghai region, where the differences in water features among parks are not pronounced. In regions with scarce water resources, water features tend to have a stronger attraction for visitors [48]. Path width did not significantly affect daily W&RA density in our study. Previous research has indicated a significant positive influence of path width on W&RAs, with studies demonstrating a noticeable relationship between path width and the quantity of W&RAs typically involving differences of 5 m or more [49]. The selected width range for path segments in our study was 2.0–5.0 m, suggesting that subtle differences in path width are unlikely to impact the occurrence of W&RAs. Further research is needed to explore the path width intervals that result in significant differences in W&RA density.

4.2. Factors Influencing Daily per Capita W&RA METs

In terms of factors influencing daily per capita W&RA METs, the significant pathway features are arranged in descending order based on their impact. They are control value (Beta = 0.533), time required to reach the nearest entrance (Beta = −0.290), pavement type (Beta = 0.247), density of security facilities (Beta = 0.239), path length–width ratio (Beta = −0.196), density of seating (Beta = −0.174), sky view ratio (Beta = −0.168), and path type (Beta = −0.107). Among these, three features—control value, pavement type, and density of security facilities—demonstrate a positive impact on daily per capita W&RA METs. A higher control value indicates greater spatial importance, a central position within the spatial system, and closer connections with the surrounding space. Consequently, it is more likely to enhance the intensity of W&RAs [15]. The regression results for pavement type indicate that asphalt-paved paths contribute to higher daily per capita W&RA METs, suggesting that asphalt paving is more conducive to elevating the level of W&RAs compared to other paving materials. This finding aligns with previous research; Deleen et al. and Ettema emphasized the role of the "comfort" of path surfaces in influencing the frequency of running and the attractiveness of routes [17,18]. Similarly, the density of security facilities exhibits a significant positive impact, indicating that an ample provision of security facilities promotes higher levels of W&RAs. Security facilities create a secure, clean, and well-maintained environment, providing individuals with a sense of psychological security and are considered essential for an ideal running space [17]. In a Danish study, 33.4% of respondents believed that security facilities such as signs and traffic lights play a role in promoting W&RAs levels [21], supporting the conclusions drawn in our study.

Five pathway features, including time required to reach the nearest entrance, path length–width ratio, density of seating, sky view ratio, and path type, exhibit a negative impact. This suggests that the mentioned park features are not conducive to high-intensity W&RAs. The negative impact of the time required to reach the nearest entrance implies that high-intensity W&RAs tend to occur near the park entrance. High-intensity W&RAs, characterized by continuity, require spaces with clear pathways and minimal disturbances [50]. Our observations reveal that dedicated running tracks are typically situated along the periphery of parks. This layout is designed to minimize interference from other activities within the park on running exercises. The significant negative impact of path length–width ratio on daily per capita W&RA METs concurs with the findings of Hou [51]. Long and narrow paths contribute to a monotonous exercise experience, reducing the appeal and diversity of activities. In contrast, spacious paths provide more activity space, lower safety risks, and are more conducive to high-intensity activities [19]. The negative impact of the density of seating on daily per capita W&RA METs suggests that an abundance of seating decreases the intensity of W&RAs in the space. This is because seating, benches, and similar resting facilities offer visitors more opportunities to pause and engage in diverse activities, such as sitting, chatting, or playing instruments, maintaining the intensity of W&RAs at a lower level [25]. The unexpected negative impact of the sky view ratio contradicts prior research that demonstrated a positive association between the sky view ratio and PA levels as well as mental well-being [44]. Paths with higher intensity activities often have a greater sky view ratio, providing both a spacious exercise experience and increased sunlight for improved microclimates [27]. The differences between our study and previous conclusions may be influenced by vegetation and climatic conditions. For instance, in colder climates, open environments are believed to promote W&RAs [52], while in hotter regions, unshaded environments are less conducive to such activities [53]. The negative impact of path type implies that straight paths enhance the intensity of W&RAs. This is because when running along a curved path, individuals tend to lean towards the center of the curve, requiring additional centripetal force. At the same metabolic energy expenditure level, running speed on curved paths is significantly lower than on straight paths [54]. Straight paths not only minimize energy consumption but also provide clear directional guidance, making them more attractive for high-intensity W&RAs.

Additionally, density of streetlights, water proximity, and vegetation structure did not exhibit significance regarding daily per capita W&RA METs. The impact of streetlight density on daily per capita W&RA METs did not reach statistical significance. Through on-site observations, we noted that high-intensity W&RAs such as jogging and running often occurred between 14:00 and 16:00. Perception of lighting among park visitors may be influenced by the time of day, potentially contributing to the non-significant regression results. The effects of vegetation structure and water proximity were also found to be non-significant. This could be attributed to the regulation of vegetation's impact on W&RAs by factors such as seasonal changes and climatic conditions [27]. Despite the belief that abundant vegetation and water features may enhance people's willingness to engage in activities [55], they do not appear to significantly influence the intensity of W&RAs.

4.3. Comparison of Factors Influencing Daily W&RA Density and Daily per Capita W&RA METs

Upon examination of the standardized coefficients of the influencing factors in the regression model, it is evident that, overall, daily W&RA density is primarily influenced by natural elements and safety perception features. Spatial safety is a prerequisite for visitors engaging in activities within a space [29]. Lower spatial safety, such as insufficient security facilities, can heighten people's alertness and potentially pose risks to users, thereby reducing the willingness of individuals to participate in W&RAs [56]. Additionally, aesthetically pleasing natural elements have been proven to have a positive correlation with W&RAs. The larger the proportion of natural elements, the higher users rate their aesthetic appeal, facilitating the occurrence of W&RAs [28]. In contrast, daily per capita W&RA METs are more significantly influenced by spatial topology, spatial form, and facilities. In other words, spatial organizational relationships and spatial place features dominate the intensity of W&RAs within the space. It is evident that spatial safety and beautiful natural scenery are crucial for enhancing visitors' willingness to engage in W&RAs. To increase the intensity of such activities, there is a need for greater attention to spatial organization, spatial form optimization, and the configuration of functional facilities.

Among the significant influencing factors, we observed that three pathway features, namely path type, sky view ratio, and density of security facilities, simultaneously exert significant effects on both daily W&RA density and daily per capita W&RA METs. Specifically, path type and sky view ratio exhibit a significant positive influence on daily W&RA density. However, they manifest a significantly negative impact on daily per capita W&RA METs, indicating that a winding road and expansive skies promote the quantity of W&RAs while maintaining the intensity of W&RAs at a lower level. This is conducive to accommodating activities with a larger number of participants and lower intensity, such as strolling and slow walking. Therefore, in the spatial design of urban park paths, emphasis should be placed on creating winding paths and open scenic views. The density of security facilities shows a significant positive impact on both daily W&RA density and daily per capita W&RA METs. This implies that an ample provision of security facilities not only attracts more W&RA participants but also enhances the intensity of these activities. Once again, this reaffirms that spatial safety is a prerequisite for visitors engaging in activities within a space [29]. Regardless of the design of paths, safety should be a primary consideration. Specifically, crime-related safety can be enhanced by increasing the number of cameras and reducing visual blind spots created by tall trees [30]. Environmental safety can be ensured through the configuration of security facilities such as traffic lights and signs, as well as amenities like paved surfaces, handrails, and fences [31].

4.4. Optimal Range of Factors Influencing Daily W&RA Density and Daily per Capita W&RA METs

Our study discusses the differences in the impact of various pathway features on the density and intensity of W&RAs at a global level. Building upon this, the study further constructs univariate OLS regression models to explore the optimal range of each pathway feature. The results reveal that the trends and intensities of the effects of these features on the density and intensity of W&RAs in univariate OLS regression models are consistent with the global model results. However, factors such as control value, path length-to-width ratio, density of seating, density of streetlights, density of security facilities, and sky view ratio do not exhibit a linear impact on the density and intensity of W&RAs. Instead, they show specific positive and negative impact intervals.

The control value exhibits a basic positive correlation with daily per capita W&RA METs. Since there is only one survey plot where the control value exceeds 2.5, the model results can only reflect the trend of daily per capita W&RA METs when the control value is within a range from 0 to 3.0. The positive impact interval is from 0 to 1.3, and the negative impact interval is from 1.3 to 2.5. A higher spatial control value implies a greater influence of that space on adjacent areas, designating it as a primary space within the spatial system [15]. The model results indicate that both excessively high and low spatial control values are unfavorable for enhancing the intensity of W&RAs. Therefore, we infer that spaces with high levels of W&RA intensity are more likely to occur in sub-central spaces within the spatial system, where spaces with a moderate number of adjacent spaces. Taking the example of an urban park in a high-density urban area of 7–10 hectares, conducive to high-intensity W&RAs, its suitable control value is around 1.3.

Path length-to-width ratio exhibits a predominantly negative correlation with daily per capita W&RA METs. Since there is only one survey plot where the path length-to-width ratio is less than 10, the model results can only reflect the trend of daily per capita W&RA METs when the ratio ranges from 10 to 50. The positive impact interval is from 10 to 35, and the negative impact interval is from 35 to 50. The model results indicate that when path length-to-width ratio is maintained within the 10–35 range, the intensity of W&RAs within the space can be sustained at a higher level. Additionally, when the path length-to-width ratio reaches around 35, the intensity of W&RAs within the space can reach its highest level, further promoting activities such as jogging and running. However, when path length-to-width ratio exceeds 35, daily per capita W&RA METs within the space rapidly decline, hindering the enhancement of activity intensity. This result provides an appropriate path length-to-width ratio conducive to increasing W&RA intensity. For example, for a 5 m-wide path, its length should ideally not exceed 175 m to avoid causing a fatiguing exercise experience.

Density of seating demonstrates a primarily negative correlation with daily per capita W&RA METs. Since there is only one survey plot where the seat density exceeds 0.03 per square meter, the model results can only reflect the trend of daily per capita W&RA METs when seat density is less than or equal to 0.03 per square meter. The positive impact interval is from 0.00 to 0.01 per square meter, and the negative impact interval is from 0.01 to 0.03 per square meter. The model results indicate that a lower density of seating in the space is more conducive to enhancing W&RA intensity. For example, for an urban park in a high-density urban area covering 7–10 hectares, the suitable seating density for spaces promoting high-intensity W&RAs is approximately 0.01 per square meter.

Density of streetlights exhibits a primarily positive correlation with daily W&RA density. The model results reflect the trend of daily W&RA density when streetlight density is in the range of 0.00–0.04 per square meter. The positive impact interval is from 0.010 to 0.035 per square meter, while the negative impact interval is 0.00–0.01 per square meter and 0.035–0.040 per square meter. The model results indicate that insufficient or excessive streetlight density reduces willingness to engage in W&RAs. The presence of a moderate number of streetlights in a space is more conducive to the occurrence of W&RAs. Taking an urban park in a high-density urban area covering 7–10 hectares as an example, the suitable streetlight density for spaces promoting the quantity of W&RAs is approximately 0.035 per square meter.

Both density of security facilities and sky view ratio have simultaneous effects on daily W&RA density as well as daily per capita W&RA METs. The density of security facilities is generally positively correlated with daily W&RA density and negatively correlated with daily per capita W&RA METs. The model results reflect the trends of daily W&RA density and daily per capita W&RA METs when the density of security facilities is in the range of 0.00–0.02 per square meter. The results indicate that when the density of security facilities is in the range of 0.010–0.013 per square meter, both W&RA density and intensity can reach higher levels. Within this range, it is possible to increase the quantity of W&RAs on the basis of a relatively small impact on W&RA intensity. Taking an urban park in a high-density urban area covering 7–10 hectares as an example, maintaining the density of security facilities within the range of 0.010–0.013 per square meter can promote the quantity of W&RAs while maintaining W&RA intensity at a higher level.

Sky view ratio is generally positively correlated with daily W&RA density and negatively correlated with daily per capita W&RA METs. As the overall level of sky view ratio in the surveyed plots is relatively low, the model results can only reflect the trends of daily W&RA density and daily per capita W&RA METs when the sky view ratio is low (0.000–0.125). The results indicate that when the sky view ratio is in the range of 0.035–0.125, both the quantity and intensity of W&RAs can reach higher levels, allowing for a rapid increase in W&RAs quantity with a relatively small impact on W&RA intensity. Using an urban park in a high-density urban area covering 7–10 hectares as an example, maintaining the sky view ratio within the range of 0.035–0.125 can simultaneously promote the number and intensity of W&RAs. However, further exploration is needed to understand the impact of high levels of sky view ratio on W&RA density and intensity.

Our study, through the construction of univariate OLS regression models and the analysis of scatter plots, explores the optimal range of various pathway features for the density and intensity of W&RAs. The optimal ranges for the mentioned features can serve as references for the spatial design of urban parks in high-density urban areas in Asia. However, due to the limitations of the surveyed parks and plots, these optimal ranges may not be applicable to other countries or regions. In the future, it is essential to strengthen cross-regional comparative studies on urban park pathway features under different climatic conditions.

4.5. Pathway Optimization Strategies

Under the premise of ensuring the multi-functional use of urban parks, optimization and enhancement strategies are proposed for pathway spaces oriented towards W&RAs. Based on quantitative analysis clues, selected pathway plots within the park are used for design optimization demonstrations. By setting specific pathway plots the plot to be optimized and reference pathway plots, optimization strategies are proposed through the comparison of key pathway features.

4.5.1. Pathway Optimization Strategies to Promote W&RA Density

Based on the discussion of pathway features that significantly impact daily W&RA density and their optimal ranges, selected plots within the same park with significant differences in W&RA density are chosen for optimization and transformation demonstrations. Using plot 4 in Xujiahui Park as the plot to be optimized and plot 5 as the reference plot, optimization strategies are proposed to promote W&RA density without changing their location conditions and morphological features. These strategies include increasing spatial openness, providing appropriate lighting and safety facilities and simplifying vegetation structure to create visually clear and aesthetically appealing pathway spaces, attracting more W&RA participants (Table 8).

Table 8. Comparison of key indicators between Plot 4 and Plot 5 in Xujiahui Park.

Plot to be optimized		 Plot 4 in Xujiahui Park (Length: 75 m. Width: 2 m. Area: 120 m^2)	
Reference plot		 Plot 5 in Xujiahui Park (Length: 75 m. Width: 3 m. Area: 120 m^2)	
		Plot to Be Optimized	**Reference Plot**
		Plot 4	Plot 5
Daily W&RA density (people/square meter)		0.42	1.12
Comparison of Key Indicators	Vegetation coverage ratio	0.9	0.9
	Path type	straight	curved
	Density of security facilities *	0	0.017
	Integration value	1.38	1.39
	Density of streetlights(units/square meter) *	0.025	0.033
	Sky view ratio *	0.006	0.129
	Vegetation structure *	triple-layer structure	double-layer structure

Note: * Indicates the main adjusted indicators.

(1) Increase spatial openness to create an open activity field of vision

This is mainly achieved through adjustments in roadside trees, selecting combinations of deciduous and evergreen trees with higher branching points to create as open a top boundary surface as possible. Based on quantitative analysis results, it is recommended that the sky view ratio be maintained within the range of 3.5–12.5% in Xujiahui Park sample 4 to ensure clear visibility for people engaged in W&RA activities. Simultaneously, enhance pruning and management of plants to avoid excessive growth of roadside trees leading to excessively low sky view ratio.

(2) Equip lighting and safety facilities to ensure activity safety

Combine appropriate lighting and safety facilities along the pathway border. Based on quantitative analysis results, we recommend adding 1 group of streetlights and 2 groups of safety facilities in Xujiahui Park sample 4, maintaining the streetlight density and safety facility density at $0.035/m^2$ and $0.013/m^2$, respectively, to ensure sufficient lighting and safety for the pathway. Additionally, attention should be paid to the choice of light source color and brightness, using a mix of warm and cool light colors and increasing lighting brightness to promote W&RAs [20].

(3) Simplify vegetation structure to enhance visual connections

Based on quantitative analysis results, we suggest appropriately reducing the proportion of shrub greenery and increasing the area of lawns and ground cover in Xujiahui Park sample 4. This creates a double-layered vegetation structure dominated by trees and ground cover, forming a more open facade space to strengthen visual connections with other spaces and improve pathway accessibility.

4.5.2. Pathway Optimization Strategies to Promote W&RA Intensity

Based on the discussion of pathway features and optimal ranges that significantly impact daily per capita W&RA METs, we selected plots from the same park with significant differences in W&RA intensity for optimization and transformation demonstrations. Using Lujiazui Central Green Plot 5 as the plot to be optimized and Lujiazui Central Green Plot 2 as the reference plot, we proposed strategies to enhance the intensity of W&RAs on the pathway without changing their location conditions and morphological characteristics. The strategies include enhancing the ground pavement for W&RAs, adjusting the ratio of safety facilities and seats, increasing visual length of pathways, and providing ample tree shade, creating a pathway space with some shelter and conducive to high-intensity W&RAs (Table 9).

(1) Enhancing the Ground Pavement for W&RAs

Based on quantitative analysis results, we recommend replacing the brick pavement with asphalt pavement in Lujiazui Central Green Plot 5. Simultaneously, with the use of spray-painted mile markers and a sports guidance system on the ground pavement, distances between different points in the park and information on energy consumption during activities are visualized. This encourages users to engage in W&RAs, providing a better exercise experience.

(2) Adjusting the Ratio of Safety Facilities and Seats to Minimize Disturbance

Appropriately reduce the number of existing seats and add a suitable number of safety facilities along the pathway, such as monitoring devices and railings, to enhance the safety of W&RAs. Based on quantitative analysis results, we recommend reducing the number of seats in Lujiazui Central Green Plot 5 to three groups and adding two groups of safety facilities, maintaining seat density and safety facility density at 0.01 per square meter. This reduces interference with high-intensity W&RAs while enhancing safety.

(3) Increasing Visual Length and Tree Shade for Pathways, Creating a Positive Activity Experience

Achieve this by adding plants along the pathway to partially obstruct the line of sight, reducing the visible range of the pathway. This creates a visual illusion of a long and narrow pathway, reducing sky view ratio and causing a visual illusion of continuous movement space for high-intensity walkers and runners. Based on quantitative analysis results, we recommend maintaining a length-to-width ratio of around 35:1 for the pathway in Lujiazui Central Green Plot 5. This involves planting a row of tall deciduous trees along the pathway, 2 m away from the pathway centerline in the direction of pathway travel. This enhances the length-to-width ratio without affecting the multi-purpose use of the pathway. It is important to choose tree species with a small trunk diameter, high branching points, dense foliage, and a large crown, providing more tree shade while minimizing occupation of the pathway ground space.

Table 9. Comparison of key indicators between Plot 2 and Plot 5 in Lujiazui Center Green.

Plot to be optimized		 Lujiazui Central Green Plot 5 (Length: 60 m. Width: 4.5 m. Area: 300 m^2)	
Reference plot		 Lujiazui Central Green Plot 2 (Length: 70 m. Width: 2 m. Area: 180 m^2)	
		Plot to Be Optimized	**Reference Plot**
		Plot 5	Plot 2
	Daily per capita W&RA METs (METs/people)	3.53	4.52
	Control value	1.01	1.08
	Time required to reach the nearest entrance (s)	55	5
	Pavement type *	Brick paving	Asphalt paving
Comparison of Key Indicators	Density of security facilities(units/square meter) *	0.003	0.012
	Path length-to-width ratio *	13.3	35
	Density of seating (units/square meter) *	0.014	0.008
	Sky view ratio *	0.124	0.051
	Path type	curved	curved

Note: * Indicates the main adjusted indicators.

4.6. Research Contributions and Limitations

Our study makes significant contributions on three fronts. Firstly, it conducted year-long field observations for data collection, thereby mitigating errors stemming from variables such as climate, seasons, day–night cycles, weekdays, and weekends. The extensive dataset facilitated a thorough exploration of the influence of urban park pathway features on W&RAs, incorporating both objective and subjective aspects associated with spatial organization, place, and perception. Secondly, the study delves into the impact of urban park pathway features on W&RAs from a micro-perspective, identifying optimal ranges of pathway features with a focus on promoting health. This addresses a gap in prior research that often lacked a nuanced examination of specific physical activity types and internal features of green open spaces. Lastly, the study concurrently investigates the effects of urban park pathway features on both the density and intensity of W&RAs, striving for a more comprehensive understanding of how these features shape such activities and avoiding potential discrepancies arising from a singular focus.

Nevertheless, the study is subject to certain limitations. Firstly, its cross-sectional design constrains the establishment of causal relationships. Secondly, the study exclusively examines the overall impact of urban park pathway features on the density and intensity of W&RAs at a global level. In reality, variations in the temporal and spatial utilization of pathways within urban parks exist. Subsequent research is imperative to categorize different types of pathways within urban parks, discerning differences in influencing factors across various spatial types. Thirdly, the study is confined to three urban parks in Shanghai, and variations in park construction, living habits, and other factors across different regions and countries may yield disparate results. Therefore, the findings of our study may not be universally applicable, and future research should amplify cross-regional comparative studies on park spaces.

5. Conclusions

This study focused on pathways within three urban parks in Shanghai, utilizing mathematical models to investigate the influence of urban park pathway features on the density and intensity of W&RAs and proposing specific pathway optimization strategies. The research aimed to establish optimal ranges for various pathway features and to elucidate the underlying mechanisms through which these features impact W&RAs. The findings contribute quantifiable insights to inform the construction of high-quality urban park pathways. The study discerned that daily W&RA density is predominantly shaped by natural elements and safety perception features, while daily per capita W&RA METs are chiefly influenced by spatial topology, spatial form, and facilities. Noteworthy factors attracting higher W&RAs participation encompass abundant vegetation cover, curved paths, sufficient security facilities and lighting, good accessibility, open skies, and simple vegetation structures. Conversely, elements enhancing the intensity of W&RAs include connectivity to the surroundings, asphalt pavement, ample security facilities, proximity to entrances, spacious paths, a limited number of seats, less open skies, and straight paths. Therefore, prioritizing these influential factors in the design of urban park pathways is crucial for fostering W&RAs, promoting public health, and creating versatile exercise environments that cater to diverse citizen needs. The study's conclusions provide valuable insights and foundational guidance for the design and development of urban park spaces in densely populated Asian cities, ultimately contributing to the enhancement of public health.

Author Contributions: Conceptualization, J.C., D.C. and L.W.; methodology, J.C.; software, J.C.; validation, Z.T. and W.W.; formal analysis, J.C.; investigation, J.C. and Z.T.; data curation, J.C.; writing—original draft preparation, J.C.; writing—review and editing, Z.T.; visualization, J.C.; supervision, D.C. and L.W.; funding acquisition, D.C. All authors have read and agreed to the published version of the manuscript.

Funding: This research was funded by National Natural Science Foundation of China, grant number 32001361.

Data Availability Statement: The raw data supporting the conclusions of this article will be made available by the authors on request.

Conflicts of Interest: The authors declare no conflicts of interest.

References

1. World Health Organization. Global Report on Urban Health. Available online: https://www.who.int/publications/i/item/9789241565271 (accessed on 3 September 2023).
2. Jackson, L.E. The relationship of urban design to human health and condition. *Landsc. Urban Plan.* **2003**, *64*, 191–200. [CrossRef]
3. Warburton, D.E.; Charlesworth, S.; Ivey, A.; Nettlefold, L.; Bredin, S.S. A systematic review of the evidence for Canada's Physical Activity Guidelines for Adults. *Int. J. Behav. Nutr. Phys. Act.* **2010**, *7*, 39. [CrossRef]
4. World Health Organization. Global Health Risks: Mortality and Burden of Disease Attributable to Selected Major Risks. Available online: https://www.who.int/publications/i/item/9789241563871 (accessed on 3 September 2023).
5. Boarnet, M.G. About this issue: Planning's role in building healthy cities: An introduction to the special issue. *J. Am. Plan. Assoc.* **2006**, *72*, 5–9. [CrossRef]
6. Piercy, K.L.; Troiano, R.P.; Ballard, R.M.; Carlson, S.A.; Fulton, J.E.; Galuska, D.A.; George, S.M.; Olson, R.D. The physical activity guidelines for Americans. *JAMA* **2018**, *320*, 2020–2028. [CrossRef]
7. Xiao, M. Strengthening Physical Activity and Preventing Chronic Diseases Viewing from the Perspective of "Healthy China". *Sport Sci. Technol.* **2020**, *3*, 18–20. [CrossRef]
8. Althoff, T.; Sosič, R.; Hicks, J.L.; King, A.C.; Delp, S.L.; Leskovec, J. Large-scale physical activity data reveal worldwide activity inequality. *Nature* **2017**, *547*, 336–339. [CrossRef]
9. Garber, C.E.; Blissmer, B.; Deschenes, M.R.; Franklin, B.A.; Lamonte, M.J.; Lee, I.M.; Nieman, D.C.; Swain, D.P. Quantity and quality of exercise for developing and maintaining cardiorespiratory, musculoskeletal, and neuromotor fitness in apparently healthy adults: Guidance for prescribing exercise. *Med. Sci. Sports Exerc.* **2017**, *43*, 1334–1359. [CrossRef]
10. Schmidt, T.; Kerr, J.; Schipperijn, J. Associations between neighborhood open space features and walking and social interaction in older adults—A mixed methods study. *Geriatrics* **2019**, *4*, 41. [CrossRef]
11. Veitch, J.; Ball, K.; Rivera, E.; Loh, V.; Deforche, B.; Best, K.; Timperio, A. What entices older adults to parks? Identification of park features that encourage park visitation, physical activity, and social interaction. *Landsc. Urban Plan.* **2022**, *217*, 104254. [CrossRef]
12. Van Puyvelde, A.; Deforche, B.; Mertens, L.; Rivera, E.; Van Dyck, D.; Veitch, J.; Poppe, L. Park features that encourage park visitation among older adults: A qualitative study. *Urban For. Urban Green.* **2023**, *86*, 128026. [CrossRef]
13. Kowaleski-Jones, L.; Fan, J.X.; Wen, M.; Hanson, H. Neighborhood context and youth physical activity: Differential associations by gender and age. *Am. J. Health Promot.* **2017**, *31*, 426–434. [CrossRef] [PubMed]
14. Kaczynski, A.T.; Besenyi, G.M.; Stanis, S.A.; Koohsari, M.J.; Oestman, K.B.; Bergstrom, R.; Potwarka, L.R.; Reis, R.S. Are park proximity and park features related to park use and park-based physical activity among adults? Variations by multiple socio-demographic characteristics. *Int. J. Behav. Nutr. Phys. Act.* **2014**, *11*, 146. [CrossRef] [PubMed]
15. Hou, Y.; Zhao, X.; Zhang, B. Significance Analysis between Morning Exercise and Spatial Organization Characteristics of Urban Park—Taking 4 Urban Parks in Harbin for Example. *Landsc. Archit.* **2017**, *2*, 109–116. [CrossRef]
16. Koohsari, M.J.; Kaczynski, A.T.; Mcormack, G.R.; Sugiyama, T. Using space syntax to assess the built environment for physical activity: Applications to research on parks and public open spaces. *Leis. Sci.* **2014**, *36*, 206–216. [CrossRef]
17. Deelen, I.; Janssen, M.; Vos, S.; Kamphuis, C.B.; Ettema, D. Attractive running environments for all? A cross-sectional study on physical environmental characteristics and runners' motives and attitudes, in relation to the experience of the running environment. *BMC Public Health* **2019**, *19*, 366. [CrossRef] [PubMed]
18. Ettema, D. Runnable cities: How does the running environment influence perceived attractiveness, restorativeness, and running frequency? *Environ. Behav.* **2016**, *48*, 1127–1147. [CrossRef]
19. Zhai, Y.; Li, D.; Wu, C.; Wu, H. Urban park facility use and intensity of seniors' physical activity—An examination combining accelerometer and GPS tracking. *Landsc. Urban Plan.* **2021**, *205*, 103950. [CrossRef]
20. Hu, B.; Zhao, J. Factors promoting nature-based outdoor recreation during the daytime and evening. *J. Outdoor Recreat. Tour.* **2022**, *40*, 100572. [CrossRef]
21. Schipperijn, J.; Bentsen, P.; Troelsen, J.; Toftager, M.; Stigsdotter, U.K. Associations between physical activity and characteristics of urban green space. *Urban For. Urban Green.* **2013**, *12*, 109–116. [CrossRef]
22. Chiang, Y.C.; Weng, P.Y.; Li, D.; Ho, L.C. Quantity and quality: Role of vegetation in park visitation and physical activity. *Landsc. Ecol. Eng.* **2023**, *19*, 337–350. [CrossRef]
23. Rivera, E.; Timperio, A.; Loh, V.H.; Deforche, B.; Veitch, J. Critical factors influencing adolescents' active and social park use: A qualitative study using walk-along interviews. *Urban For. Urban Green.* **2021**, *58*, 126948. [CrossRef]
24. Christian, H.; Lester, L.; Trost, S.G.; Trapp, G.; Schipperijn, J.; Boruff, B.; Maitland, C.; Jeemi, Z.; Rosenberg, M.; Barber, P.; et al. Shade coverage, ultraviolet radiation and children's physical activity in early childhood education and care. *Int. J. Public Health* **2019**, *64*, 1325–1333. [CrossRef]
25. Chen, Y.; Liu, X.; Gao, W.; Wang, R.Y.; Li, Y.; Tu, W. Emerging social media data on measuring urban park use. *Urban For. Urban Green.* **2018**, *31*, 130–141. [CrossRef]
26. Zhang, S.; Zhou, W. Recreational visits to urban parks and factors affecting park visits: Evidence from geotagged social media data. *Landsc. Urban Plan.* **2018**, *180*, 27–35. [CrossRef]

27. Zhao, X.; Hou, Y.; Qiu, X.; Lv, F. The Study on Morphological Characteristics of Pathway in Urban Parks Based on Walking and Running Capacity—A Case of Harbin City. *Chin. Landsc. Archit.* **2019**, *6*, 12–17. [CrossRef]
28. Brownson, R.C.; Baker, E.A.; Housemann, R.A.; Brennan, L.K.; Bacak, S.J. Environmental and policy determinants of physical activity in the United States. *Am. J. Public Health* **2001**, *91*, 1995–2003. [CrossRef]
29. Weimann, H.; Rylander, L.; van den Bosch, M.A.; Albin, M.; Skärbäck, E.; Grahn, P.; Björk, J. Perception of safety is a prerequisite for the association between neighborhood green qualities and physical activity: Results from a cross-sectional study in Sweden. *Health Place* **2017**, *45*, 124–130. [CrossRef] [PubMed]
30. Foster, S.; Giles-Corti, B. The built environment, neighborhood crime and constrained physical activity: An exploration of inconsistent findings. *Prev. Med.* **2008**, *47*, 241–251. [CrossRef] [PubMed]
31. Molnar, B.E.; Gortmaker, S.L.; Bull, F.C.; Buka, S.L. Unsafe to play? Neighborhood disorder and lack of safety predict reduced physical activity among urban children and adolescents. *Am. J. Health Promot.* **2004**, *18*, 378–386. [CrossRef] [PubMed]
32. Zhai, Y.; Baran, P.K.; Wu, C. Can trail spatial attributes predict trail use level in urban forest park? An examination integrating GPS data and space syntax theory. *Urban For. Urban Green.* **2018**, *29*, 171–182. [CrossRef]
33. Paydar, M.; Kamani Fard, A.; Gárate Navarrete, V. Design Characteristics, Visual Qualities, and Walking Behavior in an Urban Park Setting. *Land* **2023**, *12*, 1838. [CrossRef]
34. McCormack, G.R.; Rock, M.; Toohey, A.M.; Hignell, D. Characteristics of urban parks associated with park use and physical activity: A review of qualitative research. *Health Place* **2010**, *16*, 712–726. [CrossRef] [PubMed]
35. Paudel, C.; Timperio, A.; Loh, V.; Deforche, B.; Salmon, J.; Veitch, J. Understanding the relative importance of micro-level design characteristics of walking paths in parks to promote walking among older adults. *Urban For. Urban Green.* **2023**, *89*, 128129. [CrossRef]
36. Ma, M.; Mugerauer, B.; Cai, Z. Research on the Determinants of Urban Open Green Space Design Affecting the Physical Activity from the Perspective of Health. *Landsc. Archit.* **2018**, *4*, 92–97. [CrossRef]
37. Penn, A. Space syntax and spatial cognition: Or why the axial line? *Environ. Behav.* **2003**, *35*, 30–65. [CrossRef]
38. Zhai, Y.; Baran, P. Application of space syntax theory in study of urban parks and walking. In Proceedings of the Ninth International Space Syntax Symposium 2013, Seoul, Republic of Korea, 31 October–3 November 2013; Sejong University Press: Seoul, Republic of Korea, 2013; Volume 32, pp. 1–13. Available online: http://sss9sejong.or.kr/paperpdf/gusd/SSS9_2013_REF032_P.pdf (accessed on 17 January 2024.).
39. Ainsworth, B.E.; Haskell, W.L.; Herrmann, S.D.; Meckes, N.; Bassett, D.R., Jr.; Tudor-Locke, C.; Greer, J.L.; Vezina, J.; Whitt-Glover, M.C.; Leon, A.S. 2011 Compendium of Physical Activities: A second update of codes and MET values. *Med. Sci. Sports Exerc.* **2011**, *43*, 1575–1581. [CrossRef]
40. Liu, H.; Li, F.; Li, J.; Zhang, Y. The relationships between urban parks, residents' physical activity, and mental health benefits: A case study from Beijing, China. *J. Environ. Manag.* **2017**, *190*, 223–230. [CrossRef] [PubMed]
41. Shayestefar, M.; Pazhouhanfar, M.; van Oel, C.; Grahn, P. Exploring the Influence of the Visual Attributes of Kaplan's Preference Matrix in the Assessment of Urban Parks: A Discrete Choice Analysis. *Sustainability* **2022**, *14*, 7357. [CrossRef]
42. Hagerhall, C.M. Clustering predictors of landscape preference in the traditional Swedish cultural landscape: Prospect-refuge, mystery, age and management. *J. Environ. Psychol.* **2000**, *20*, 83–90. [CrossRef]
43. Wang, B.; Zhen, F.; Zhang, H. The Dynamic Changes of Urban Space-time Activity and Activity Zoning Based on Check-in Data in Sina Web. *Sci. Geogr. Sin.* **2015**, *2*, 151–160. [CrossRef]
44. He, H.; Lin, X.; Yang, Y.; Lu, Y. Association of street greenery and physical activity in older adults: A novel study using pedestrian-centered photographs. *Urban For. Urban Green.* **2020**, *55*, 126789. [CrossRef]
45. Nasiri, M.; Pourmajidian, M.R. Effects of vegetation type and horizontal curve radius on the rate of tree pruning to provide line of sight on main access roads. *J. For. Sci.* **2014**, *60*, 208–211. [CrossRef]
46. Badiora, A.I.; Dada, O.T.; Adebara, T.M. Correlates of crime and environmental design in a Nigerian international tourist attraction site. *J. Outdoor Recreat. Tour.* **2021**, *35*, 100392. [CrossRef]
47. Mak, B.K.; Jim, C.Y. Examining fear-evoking factors in urban parks in Hong Kong. *Landsc. Urban Plan.* **2018**, *171*, 42–56. [CrossRef]
48. Lyu, F.; Zhang, L. Using multi-source big data to understand the factors affecting urban park use in Wuhan. *Urban For. Urban Green.* **2019**, *43*, 126367. [CrossRef]
49. Kim, S.; Choi, J.; Kim, Y. Determining the sidewalk pavement width by using pedestrian discomfort levels and movement characteristics. *KSCE J. Civ. Eng.* **2011**, *15*, 883–889. [CrossRef]
50. Schuurman, N.; Rosenkrantz, L.; Lear, S.A. Environmental preferences and concerns of recreational road runners. *Int. J. Environ. Res. Public Health* **2021**, *18*, 6268. [CrossRef] [PubMed]
51. Hou, Y. Research on Park Characteristics Identification and Optimization Pattern Based on Leisure Physical Activity. Ph.D. Thesis, Harbin Institute of Technology, Harbin, China, 2019. [CrossRef]
52. Zhao, X.; Bian, Q.; Hou, Y.; Zhang, B. A Research on the Correlation between Physical Activity Performance and Thermal Comfortable of Urban Park in Cold Region. *Chin. Landsc. Archit.* **2019**, *4*, 80–85. [CrossRef]
53. Huang, H.; Wei, L. Environment Factors to Park Thermal Comfort of the Elderly in Hot and Humid Areas. *J. Chin. Urban For.* **2023**, *1*, 13–19. [CrossRef]
54. Taboga, P.; Kram, R. Modelling the effect of curves on distance running performance. *PeerJ* **2019**, *7*, e8222. [CrossRef]

55. Petrunoff, N.A.; Edney, S.; Yi, N.X.; Dickens, B.L.; Joel, K.R.; Xin, W.N.; Sia, A.; Leong, D.; van Dam, R.M.; Cook, A.R.; et al. Associations of park features with park use and park-based physical activity in an urban environment in Asia: A cross-sectional study. *Health Place* **2022**, *75*, 102790. [CrossRef] [PubMed]
56. Soma, Y.; Tsunoda, K.; Kitano, N.; Jindo, T.; Tsuji, T.; Saghazadeh, M.; Okura, T. Relationship between built environment attributes and physical function in Japanese community-dwelling older adults. *Geriatr. Gerontol. Int.* **2017**, *17*, 382–390. [CrossRef] [PubMed]

land

MDPI

Article

Exploring the Impact of Multimodal Access on Property and Land Economies in Shanghai's Inner Ring Districts: Leveraging Advanced Spatial Analysis Techniques

Wei He [1], Ruqing Zhao [2,*] and Shu Gao [3]

1 School of Architecture and Allied Art, Guangzhou Academy of Fine Arts, Guangzhou 510006, China; gafahew@gzarts.edu.cn
2 Department of Geoscience and Natural Resource Management, Faculty of Science, University of Copenhagen, 1870 Copenhagen, Denmark
3 Texas Department of Transportation, Austin, TX 78744, USA; sgao7@utexas.edu
* Correspondence: kzt283@alumni.ku.dk

Abstract: This study explores the impact of accessibility on property pricing and land economies by advanced spatial analysis techniques, focusing on Shanghai as a representative metropolis. Despite the impact of metro systems on residential property values, which has been frequently assessed, a research gap exists in understanding this phenomenon in Asian, particularly Chinese, urban contexts. Addressing this gap is crucial for shaping effective urban land use policy and improving the land economy rationally in China and similar settings facing urban challenges. To assess the impact of metro station accessibility on property prices in Shanghai, with extensive rail transit, and to deeply explore the overall impact of land value varieties driven by metro on urban development, we conducted a comprehensive analysis, with discussion about future aspirations for land planning and management along with landscape and facility design, and measures to improve land economy. The procedures involved creating neighborhood centroids to represent accessibility and using the Euclidean distance analysis to determine the shortest paths to metro stations. Our evaluation incorporated a hedonic pricing model, considering variables like neighborhood characteristics, housing attributes, and socio-economic factors. Advanced spatial analysis encompassing Ordinary Least Squares (OLS) regression and XGBoost analysis were employed to explore spatial effects, and Geographically Weighted Regression (GWR) helped examine spatial patterns and address autocorrelation challenges. Results revealed a negative association between distance to metro station and property prices, indicating a non-linear and spatially clustered relationship and heterogeneous spatial pattern. We dissected the non-linear results in detail, which complemented the conclusion in existing research. This study provides valuable insights into the dynamic interplay between metro accessibility and housing market behaviors in a significant Asian urban context, offering targeted suggestions for urban planners and governors to decide on more reasonable land use planning and management strategies, along with landscape and infrastructure design, to promote not only the healthy growth of the real estate market but also the sustainable urban development in China and similar regions.

Keywords: land economic; land management; property price; accessibility; hedonic model; spatial analysis techniques; Shanghai

Citation: He, W.; Zhao, R.; Gao, S. Exploring the Impact of Multimodal Access on Property and Land Economies in Shanghai's Inner Ring Districts: Leveraging Advanced Spatial Analysis Techniques. *Land* **2024**, *13*, 311. https://doi.org/10.3390/land13030311

Academic Editors: Xiwei Shen, Yang Song, Niall Kirkwood, Mary Padua, Bo Zhang and Rosalea Monacella

Received: 1 February 2024
Revised: 25 February 2024
Accepted: 27 February 2024
Published: 29 February 2024

1. Introduction

Urban rail transit, encompassing metro and train, stands out as a vital mode of public transportation due to its high speed, efficiency, and safety advantages, garnering greater scholarly attention compared to other investigated transport modes [1,2]. Metro commuting efficiently accommodates more passengers than motor vehicles, mitigating urban traffic congestion caused by sprawl and rapid development. This fosters an efficient lifestyle, enabling increased social engagement and easier employment access [3]. However, whether the rail transit system has a favorable influence on residential property prices has not yet been drawn to an exact conclusion by prior studies. Comprehending this influence is crucial for the development of land and social economies, as it can inform decisions related to urban land planning and management.

To date, the research on the relationship between rail transit accessibility and real estate value has shown mixed results in the literature, ranging from positive to negligible or negative effects [4]. Gatzlaff and Smith (1993) claimed that the disparity in the results of the empirical studies can be attributed to local factors in each city [5]. Among the studies, North America and Europe have received the most attention. In contrast, limited studies focused on developing countries because of transport investment constraints and the lack of available and reliable data [6]. Many researchers found that metro accessibility had a profitable impact on property value. Bajic (1983) surveyed Toronto and found that the establishment of the subway boosted the average home market value by $2237 [7]. Benjamin and Sirmans (1996) evaluated the impact of the proximity to Buffalo, NY, on residential property values. Houses within a quarter-mile radius received a premium of 2–5% of the median for urban homes [8]. Dorantes, Paez and Vassallo (2011) assessed the impact of proximity to metro stations in Madrid, Spain, showing that metro accessibility had a positive impact on property value and was more pronounced at sale [9]. Mayor et al. (2012) found that in Dublin, a prominent increase showed in prices when the property was near rapid transit DART or light rail Luas [10]. However, some studies found that the effect was not significant. Gatzlaff and Smith (1993) examined the impact of the development of the Miami metro system on the property value near its station location, indicating that the rail system had little effect on residential value [5]. Moreover, Vessali (1996) suggested that while accessibility brought an increment to average residential property value, several complementary factors, such as supportive local land use policies and existing high-density demand development, needed to be considered [11]. Others came to the opposite conclusion. They claimed that proximity to metro stations had a negative impact [12,13].

Discussions based on the Western context may not completely fit Asian cities with high development density and high metro usage rates, where more diverse results may arise. In recent years, some relevant studies have covered prime metropolitan areas in Asia, aiming at providing practical insights for decision-making process of urban land planning and management in the specific context of Asia, to better meet the demands of urban development and contribute to land and socio-economic growth. Bae et al. (2003) investigated the effects of the opening of the new metro line in Seoul on nearby property values and claimed that metro station accessibility had a statistically significant impact on residential prices, consistent with the expected effects observed in previous studies [14]. Chin et al. (2020) studied the same project in Seoul, saying that in some blocks, the setting of the metro line was related to higher increases in apartments, but in other neighborhoods, a reducing trend in price was shown [15]. However, Diao et al. (2016) indicated negative external effects on property values due to railway noise based on research in Singapore [16], while Anantsuksomsri and Tontisirin (2015) studied the Bangkok metropolitan area and found that the closer distance to large traffic stops meant a lower property value [17]. Some studies have been carried out in China. Zheng and Kahn (2008) surveyed Beijing and claimed that public transport infrastructure, including the metro, has been taken into account in property prices [18]. Li (2017) also examined Beijing and concluded that consumers were willing to pay more to use rail transit in congested areas [19]. Yang et al. (2020) documented that metro accessibility affected property prices in a non-linear manner in Shenzhen [20]. Although Shanghai possesses a very well-developed rail transit system, and its dependence on rail transit can be seen to a high extent, research in Shanghai is still quite limited. One of the studies that analyzed the construction of Line 6 in Pudong District showed that the transportation improvement to the CBD employment center brought about an appreciation of property prices in real estate along Line 6 [21]. The average appreciation was 3.75%, comparable to the data in a previous study in Toronto [7]. It also dissected that the most increases happened in remote suburbs and lower-income communities. However, the study lacks the scale and validity to fully demonstrate how the rail transit system performs in the Shanghai context. There is also research that is available to control the specific characteristics related to real estate to better identify the value of improved accessibility incorporated into property prices.

To fill the knowledge gaps indicated above, we applied a hedonic price model to investigate the impact of metro accessibility on property prices in Urban Shanghai to obtain a broad view and precise conclusion. We aim to offer a profound insight into metro system planning and a reference for the exploitation of real estate and other types of land use to facilitate balanced land and socio-economic development. By applying methods of advanced spatial analysis consisting of Ordinary Least Squares (OLS) regression, XGBoost analysis, and Geographically Weighted Regression (GWR), we approached to evaluate: (1) the relationships between metro accessibility and property prices in Urban Shanghai, and (2) the spatial variation of the influence.

2. Materials and Methods

2.1. Study Area

Shanghai is widely considered as a city with a single-center structure [22], and is divided into three concentric rings according to the inner and outer rings, including the inner city, expanded inner city and suburbs [23,24]. Our study area is the inner city, popularly known as Urban Shanghai, including districts of Yangpu, Hongkou, Putuo, Changning, Xuhui, Huangpu and Jing'an (see Figure 1). This is the core hinterland of Shanghai, covering an area of 289.59 square kilometers, and is home to 6.88 million people. The broad area and large population establish a solid foundation for investigation because it can provide various samples for the real estate market. As a metropolitan area, Shanghai has perfect facilities, well-developed industry and commerce, and a diverse cultural life, all of which are factors that influence the trend of property prices. The rich construction background and large database enable us to point to more accurate results and make more reasonable suggestions for the future planning of facility construction and property distribution in the city.

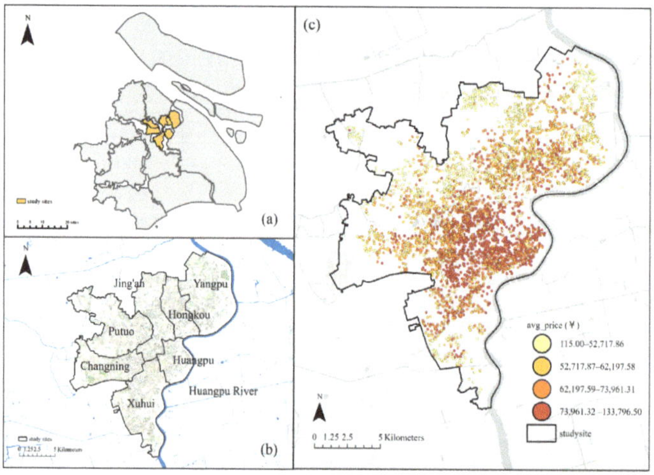

Figure 1. Study area. (**a**) Shanghai inner ring area. (**b**) The names and locations of districts within Shanghai Inner Ring and Huangpu River. (**c**) The average property prices.

2.2. Data Sources

Table 1 presents a comprehensive overview of the data utilized in this project within the Shanghai Inner Ring region, encompassing administrative boundary data, building data, property prices, metro station travel flow data, and Shanghai GDP data. All data underwent processing within the WGS 1984 UTM Zone 51N coordinate system. The table provides details on each data source, including the data name, year, resolution, usage, and the corresponding source link.

Table 1. Data Sources.

Data Name	Year	Usage	Data Source
Shanghai GDP	2019–2020	Calculating neighborhoods' social-economic attributes	http://www.dsac.cn/DataProduct (accessed on 12 November 2023)
Shanghai buildings	2020	Extracting building attributes	https://www.amap.com/ (accessed on 12 November 2023)
Shanghai AOI—buildings, roads, water systems, subways, administrative divisions	2020	Dividing neighborhoods, facilities and boundaries	https://www.amap.com/ (accessed on 12 November 2023)
Shanghai property price transaction data	2019–2020	Calculating neighborhoods' property prices	https://m.anjuke.com/bj/ (accessed on 12 November 2023)

2.3. Framework

Figure 2 depicts the research flow of this project, in which the chart on the left is the ensemble of the database we needed, and the one on the right is a flow of processing and analyzing data. First, to have an integrative simulation of property prices, the hedonic model proposed by Rosen (1974) was used [25]. Facility data were extracted, and the accessibility data were calculated based on neighborhood units within the boundary of Shanghai's inner ring to build up the hedonic model. Second, a series of correlation analyses including typical linear model Ordinary Least Squares (OLS) and non-linear XGBoost were performed, before which the Pearson correlation test and VIF were conducted to check collinearity. To explore the spatial autocorrelation between variables and the influences of hedonic attributes, we applied Geographically Weighed Regression with an advanced step of checking Moran's I. With these three models, we evaluated comprehensively the relationship between accessibility to metro stations and property values.

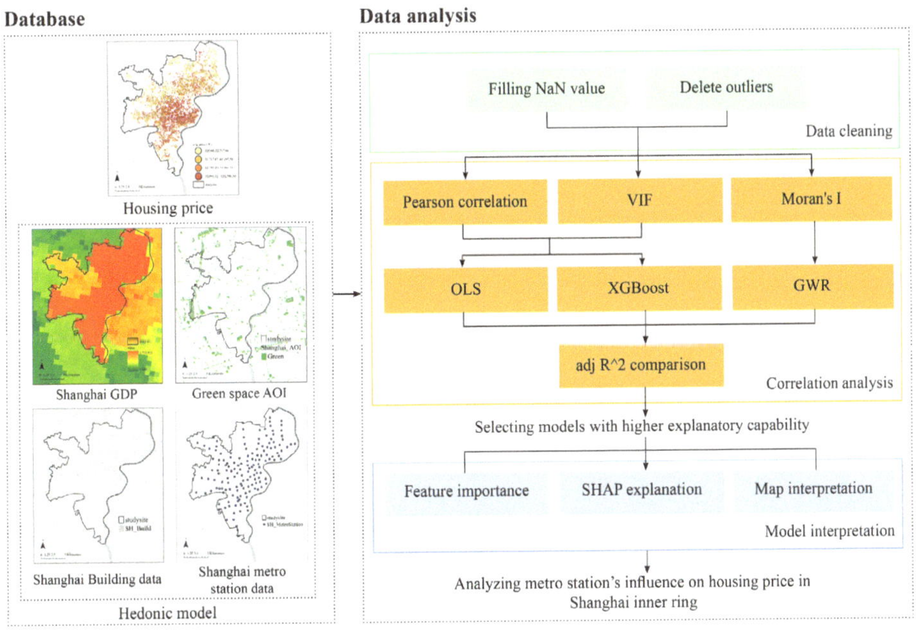

Figure 2. Proposed Flowchart of This Research.

2.4. Hedonic Model

The hedonic model has been applied in most previous studies to test the acquisition of property values in housing studies. The earliest study, conducted by Dewees (1976), analyzed the relationship between rail travel costs and residential property values [26]. The hedonic model claims that the cost of a specific property can be unveiled through a set of implicit features. A fundamental assumption of the model is that the residential market is determined by a series of choices made by consumers and producers under market conditions [27], and the variations in real estate prices can be explained by the buyer's inclination to invest in various attributes that impact the property's value. The series of characteristics that affect the property price is composed of the housing structure and the external environment. The property price hedonic model applied in this project is expressed as the following Equation (1), which is the most commonly used specification in hedonic property price models:

$$P = \beta_0 + \beta_H H + \beta_N N + \beta_S S + \varepsilon \tag{1}$$

where P is a vector of property prices, H is a matrix of housing structural attributes, N is a matrix of neighborhood attributes, and S is a matrix of socio-economic attributes. β_0 is the constant term vector, β_H, β_N as well as β_S are matrices of the corresponding parameters, and ε is a vector of error terms. In our Hedonic model, V (property price) was taken as the dependent variable, and 12 implicit variables in total were employed, categorized as I (housing structural attributes including number of floors and number of rooms), E1 (neighborhood attributes composed of accessibility from residential areas to metro stations, schools, grocery stores, etc.), E2 (socio-economic attributes mainly with GDP data), were taken as explanatory variables.

We used data for the Shanghai road network and points of interest (POI). POI includes metro stations, schools, grocery stores, green spaces, restaurants, hospitals, and shopping malls. To aggregate accessibility from residential areas to facilities, centroids for each neighborhood were generated. To calculate the shortest paths, we used the Euclidean distance which can provide a visual representation of accessibility and be easily computed using geographic coordinates fitted in with many geographic information system (GIS) tools and technologies:

$$d_{ij} = \sqrt{(x_i - x_j)^2 + (y_i - y_j)^2} \tag{2}$$

where: x_i and y_i = X and Y coordinates of point i with a plane projection. To gain a comprehensive understanding of metro stations, apart from accessibility data, the study delved into flow levels and the number of employees within a 10 min walking distance of each station. These metrics were further categorized into low, medium, and high levels based on their distribution, providing a nuanced perspective on the dynamics of metro station utilization and surrounding employment concentrations.

2.5. Correlation Analysis

A set of three models was employed to assess the relation between neighborhoods' shortest distance to metro stations and average property prices: (1) typical linear model Ordinary Least Squares (OLS); (2) non-linear XGBoost which applies a tree-based algorithm for predicting and interpreting data; (3) Geographical Weighed Regression where the spatial autocorrelation between variables is considered.

2.5.1. Ordinary Least Squares (OLS)

The OLS method is the simplest and most applied method among common regression models for analyzing the correlations between two or more variables. The theoretical basis of OLS is that the model coefficients assume that the sample regression model is closest to the observed value, and the coefficients are constant relative to the location. OLS and the coefficient estimation matrix are represented by Equation (3):

$$y = X\beta + \varepsilon \tag{3}$$

where y is the dependent variable, X is the independent variable, ε is the deviation while estimating the coefficients.

However, the drawback of this method in spatial modeling is that the value of the dependent variable estimated by it has to serve the whole study area; and, in different parts of the region, the value is estimated the same [28,29].

2.5.2. XGBoost

XGBoost is a tree-based boosting machine learning method which has been recently adopted to explain the non-linear relationship between built environment features and travel. Compared with other traditional models, XGBoost has the advantage of high speed and precision, and it is not influenced by collinearity, which means we can contain all the variables though some of them share high correlation [30,31]. The mathematics of XGBoost can be simplified as follows [32]:

For a given dataset $D = \{(x_i,\ y_i)\}$, n instances and m features fit a function that can best estimate the response variable $\hat{y}_i$ based on explanatory variables x_i. The prediction waterlogging depth ($\hat{y}_i$) after K times iteration of sample i is calculated by (4):

$$\hat{y}_i = \sum_{k=1}^{K} f_k(x_i) = {y_i}^{(K-1)} + f_k(x_i) \tag{4}$$

where K is the number of iterations and $f_k(x_i)$ is the tree model after k-th iteration.

The objective function of XGBoost is written as (5) and (6):

$$L = \sum_{i=1}^{n} 1(y_i, \hat{y}_i) + \sum_{k=1}^{K} \Omega(f_k) \tag{5}$$

$$\Omega(f_k) = \gamma T + \frac{1}{2}\lambda|\omega|^2 \tag{6}$$

where n is the amount of data to improve the k-th tree. $1(y_i, \hat{y}_i)$ is the training loss with target y_i and prediction $\hat{y}_i$. Training loss should be minimized by RMSE. T is the number of leaves and $|\omega|$ is the weight of the leaf i. γ and λ are all hyperparameters.

The K-th base learner function is calculated by summing all weights of leaves (7) and (8):

$$f_k(x_i) = \eta \sum_{I=1}^{T} \omega_{ik} I \tag{7}$$

$$I = \{i | q(x_i) = j\} \tag{8}$$

ω ranges from 0 to 1, to control the learning rate of iteration to avoid over-fitting. $q(x_i)$ reflects the specific leaf node. I is the sample set of j nodes. To interpret XGBoost, we applied Shapely Additive Explanations (SHAP). SHAP is a Python package for model interpretation. For each explanatory variable, this package returns a SHAP value, which indicates each explanatory variable's influence on the response variable, either intensity or magnitude. The mathematics of SHAP value are as follows (9):

$$g(z') = \theta_0 + \sum_{j=1}^{M} \theta_j z\prime_j \tag{9}$$

where g is each explanatory variable, M is the total amount of explanatory variables and θ_j is feature j's attribution.

2.5.3. Geographically Weighted Regression (GWR)

Considering the difficulty of simultaneously analyzing all kinds of influence levels and properly dealing with the autocorrelation between different factors, Geographically Weighted Regression (GWR) was adopted to analyze its spatial pattern. This method was first developed by Fotheringham, Charlton, and Brunsdon, professors at British Universities, in 1970 [33]. The first contribution of GWR can be traced back to research in the field of transportation by Du and Mulley (2006, 2007) to compare the impact of hedonic pricing and local modeling methods on land value enhancement [34,35].

GWR is a method for modeling spatial heterogeneity to achieve higher accuracy in analyzing location-affected correlations, which indicates that in each geographical coordinate, the correlation between the dependent and independent variables is different according to the variety of coefficients of the model on the point [33,36]. The coefficients of the model can be estimated at any point in the place. Within this method, observations around points in the place can be used to estimate the model coefficients at each point, while coefficients with closer observations are given greater weight. It is worth noting that each coefficient of GWR has a value with a sign. The GWR linear multivariate regression model is represented by Equation (10):

$$y_i(u) = \beta_{0i}(u) + \beta_{1i}(u)x_{1i}\beta_{2i}(u)x_{2i} + \beta_{mi}(u)x_{mi} \quad (10)$$

where y is the vector of the observed value, x is the matrix of the independent variables and β is the estimated coefficient vector.

Before delving into the correlation analysis, a pre-processing process was initiated. This involved subjecting the dataset to scrutiny through both Pearson correlation and variance inflation factor (VIF) assessments, aimed at forestalling collinearity among explanatory variables. The meticulous identification and mitigation of collinearity are essential steps in preserving the integrity and interpretability of the subsequent analyses.

Furthermore, to enhance the robustness of the dataset, a stringent approach was taken to address outliers. Employing z-scores with a threshold set at 3, outliers were systematically identified and subsequently removed. This meticulous outlier removal process contributes to the reliability of the subsequent statistical analyses by mitigating the undue influence of extreme values. Additionally, addressing missing data is another crucial facet of data preparation. The strategy employed involved filling in missing values with the mean of each respective column, ensuring a balanced and representative dataset.

3. Result

3.1. Hedonic Model

At the core of this investigation lies the dependent variable, the average property price around which an intricate web of explanatory variables revolves. The selection of these variables was guided by the hedonic model, categorizing them into three overarching themes: structural attributes, neighborhood attributes, and socio-economic attributes (see Table 2).

To ensure a comprehensive analysis, data pertaining to the centroids of all neighborhoods, totaling 4627, was meticulously gathered for examination, with structural housing attributes, including the number of floors and number of rooms, as explanatory variables. From Figure 3, both variables were relatively randomly distributed.

Table 2. Statistics of explanatory variables.

	Variable Name	Description	Mean	Std. Dev.	Data Type	Unit
Structural attributes	Floor	Average number of floors per neighborhood	4	2	Continuous	
	Number of rooms	Average number of rooms per neighborhood	2	1	Continuous	
Neighborhood attributes	Distance to restaurants	Shortest Euclidean distance between each neighborhood and restaurants	719.89	411.94	Continuous	meter
	Distance to groceries	Shortest Euclidean distance between each neighborhood and groceries	707.32	408.88	Continuous	meter
	Distance to schools	Shortest Euclidean distance between each neighborhood and schools	258.97	164.36	Continuous	meter
	Distance to hospitals	Shortest Euclidean distance between each neighborhood and hospitals	514.03	290.34	Continuous	meter
	Distance to greenery	Shortest Euclidean distance between each neighborhood and green space	452.62	254.16	Continuous	meter
	Distance to shopping centers	Shortest Euclidean distance between each neighborhood and shopping centers	1998.49	1020.38	Continuous	meter
	Flow level	Average flow level on the day of metro stations	3	1	Ordinal	
	Distance to metro stations	Shortest Euclidean distance between each neighborhood and metro stations	575.91	318.43	Continuous	meter
	Number of employments	Number of employment points within walking distance of each metro station	48.16	32.49	Continuous	
Social-economic attributes	GDP	Average GDP per neighborhood	598,218.69	497,457.74	Continuous	¥

For neighborhood attributes, we calculated the Euclidean distances between facilities (restaurants, grocery stores, schools, hospitals, greenery space, shopping centers, and metro stations) and neighborhoods as accessibility input. Besides the accessibility data to obtain details of metro stations, passenger flow levels and the number of employment centers were calculated, as shown in Figure 4. From the maps, metro stations located in the center of Shanghai's Inner Ring region have higher passenger flow levels and a larger number of available employment centers. Neighborhoods near these metro stations have relatively shorter Euclidean distances to the stations.

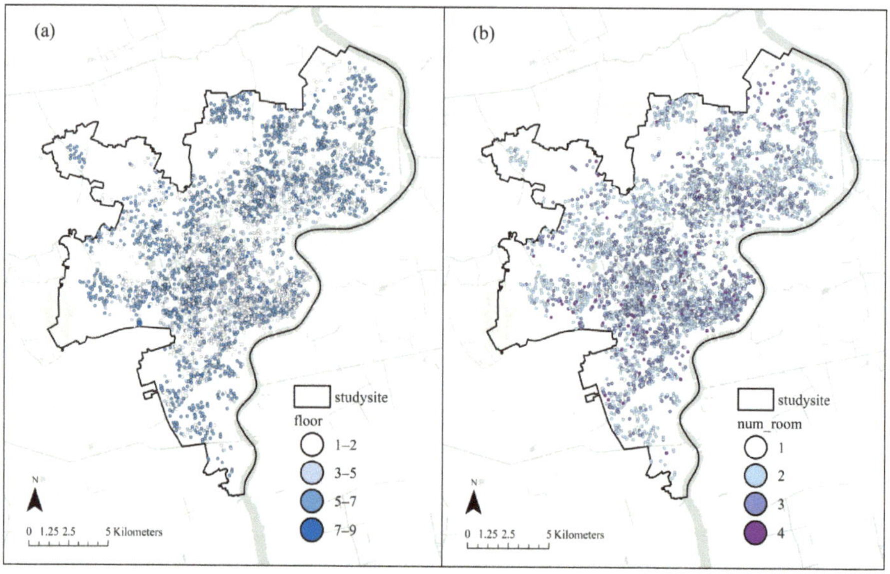

Figure 3. (**a**) Illustration of number of floors per neighborhood. (**b**) Illustration of number of rooms per neighborhood.

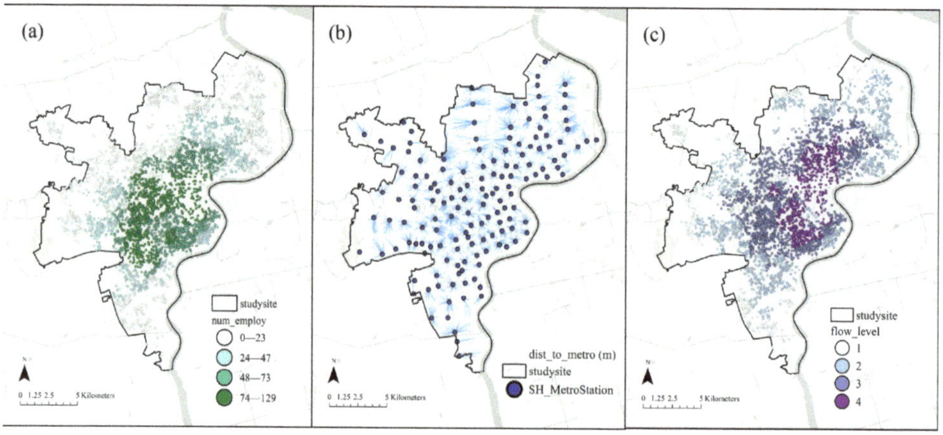

Figure 4. (**a**) Number of available employment centers per metro station. (**b**) Example of shortest distance to metro stations. (**c**) Passenger flow level per metro station.

In addition to the data related to accessibility and housing construction, to enhance the model accuracy, we collected GDP data of each neighborhood as economic factors that have the potential to influence the result. As shown in Figure 5, central areas along the Huangpu River have the highest average GDP, and neighborhoods in the west are in relatively low GDP areas within Shanghai's Inner Ring region.

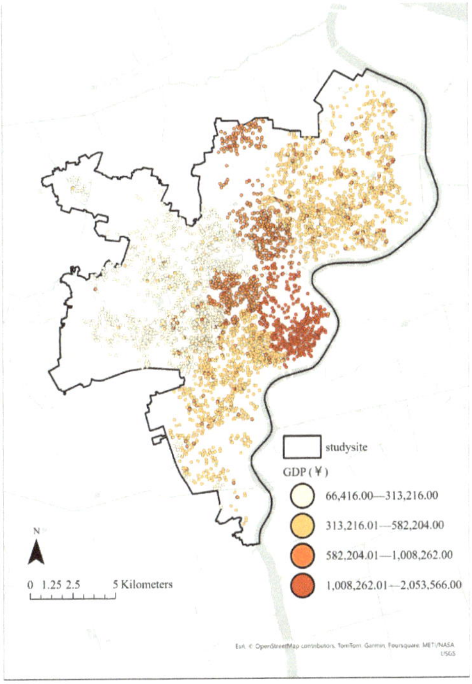

Figure 5. Illustration of GDP per neighborhood.

3.2. Correlation Analysis

Through data-cleaning, a total of 4170 neighborhoods emerged as the refined dataset, ready for a more nuanced and accurate exploration of the relationships between explanatory variables and average property prices. Before doing correlation analysis, the Pearson correlation test and VIF were conducted to check collinearity. Pearson results and VIF were all below 0.7, indicating that no collinearity existed in the input dataset.

Moran's I was checked before implementing the Geographically Weighted Regression model (GWR). The result of 0.50 with a p-value less than 0.05 indicated that the residuals were significantly clustered. The involvement of GWR could effectively solve this problem and better explain the relation between our target variables.

3.2.1. OLS & XGBoost Model

Table 3 serves as a comprehensive repository delineating the myriad explanatory variables employed in this research endeavor, accompanied by their corresponding metrics. This tabular exposition offers a profound insight into the dynamics of average property prices within the analytical frameworks of both Ordinary Least Squares (OLS) and XGBoost models.

Table 3. Correlation Analysis Results.

Variables/Models	OLS		XGBoost
	Coefficient	$p > \|t\|$	Feature Importance
Intercept	0.425	0.0 ***	/
	Structural attributes		
floor	−0.032	0.0 ***	0.090
Num_room	0.085	0.0 ***	0.060
	Neighborhood attributes		
dist_rest	−0.067	0.0 ***	0.085
dist_grocery	0.045	0.0 ***	0.045
dist_school	−0.027	0.009 **	0.035
dist_hospital	−0.017	0.08 **	0.030
dist_greenery	−0.014	0.138	0.022
dist_shopping	−0.037	0.0 ***	0.112 *
Flow_level	−0.0001	0.989	0.037
dist_metro	−0.053	0.0 ***	0.053
num_employ	0.183	0.0 ***	0.307 ***
	Social-economic attributes		
GDP	0.099	0.0 ***	0.124 **
R^2 (adj.)	0.242		0.520

Note: For OLS, significance level = 0.01: ***, 0.05: **, 0.1: *; For XGBoost, significance levels = highest feature importance: ***, 2nd highest feature importance: **, 3rd highest feature importance: *.

In dissecting the results derived from the OLS analysis, a discernible and statistically significant relationship was unearthed for most variables associated with average property prices. An intriguing pattern emerged as variables representing the shortest distances, apart from proximity to groceries, predominantly manifested negative impacts on property prices. Noteworthy among the metro station features was the revelation that a number of employees were within convenient walking distance.

Transitioning to the XGBoost model (see Figure 6), a distinct hierarchy of influential features emerged. Foremost among them was the number of employees in the vicinity of metro stations, followed closely by local GDP levels, and, subsequently, the distance to shopping centers. Corresponding with the OLS findings, most variables regarding the shortest distances showcased a positive influence on average property prices in the XGBoost model. However, a salient departure from this trend was observed in the case of the shortest distance to grocery stores.

A pivotal divergence between the two models lies in the higher R-squared value associated with the XGBoost results. When employing the XGBoost model, the elevated R-squared of 0.520 signifies a more nuanced and non-linear relationship between explanatory variables and average property prices, compared to the R-squared of 0.242 derived from OLS model. This underscores the superior capability of XGBoost in capturing intricate and complex relationships within the dataset, which is in stark contrast to the inherently linear nature of the OLS model. The XGBoost model's capacity to discern subtle nuances and non-linear patterns signifies a substantial advancement in predictive modeling, which can better capture the intricate association between factors and prices.

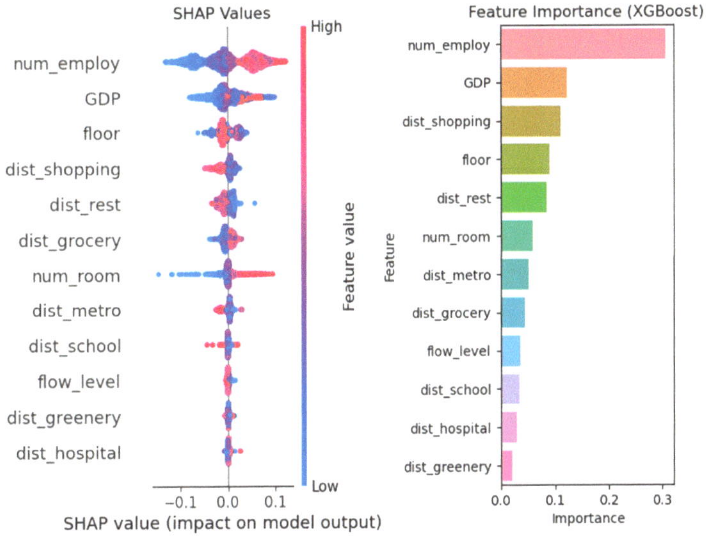

Figure 6. Correlation Matrix of explanatory variables.

3.2.2. GWR Analysis

After detecting the non-linear statistics, a Geographically Weighted Regression (GWR) model was utilized to better examine the correlation between the proximity of neighborhoods to metro stations and their average property prices. The model yielded an adjusted R-squared of 0.623, indicating a superior explanatory power compared to other models, particularly when accounting for spatial disparities, which evidently revealed the non-linear and spatially clustered relationship and heterogeneous spatial pattern.

In Figure 7, the coefficients for the shortest distance to metro stations in each neighborhood are depicted, revealing evident spatial heterogeneity. Notably, neighborhoods in the central areas along the Huangpu River, such as the Jing'an district and the Huangpu district, exhibited positive coefficients, suggesting that, as the distance to metro stations increased, average property prices also rose. Spatial clusters were demonstrated. Surprisingly, these areas were usually surrounded by neighborhoods where closer proximity to metro stations was associated with higher property prices. This trend represented a reduction in distance to metro stations, corresponding to a significant increase in property prices. As one moves farther away from the core area, the impact of the distance to metro stations on property prices gradually stabilizes.

As shown in Figure 8, according to the coefficients of the distance to the metro in the GWR analysis outcome, we divided the data into five categories. Then, we analyzed the average values of the main factors that influence property prices the most on the basis of the XGBoost model. Among the variables with negative interpretation, the average distance to the metro is quite stable, which is about 505 to 550 m, but varies in the positive interpretation from around 440 to 550 m. This means that to affect real estate value, various factors are inseparably interconnected, such as the number of employees, the distance to a shopping center and the GDP, each of which has a great effect on property prices.

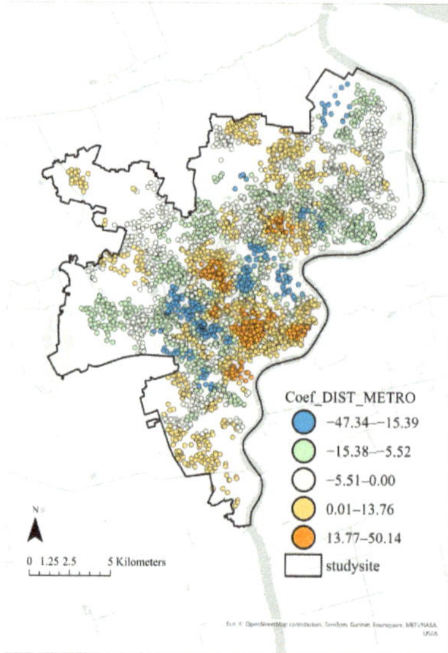

Figure 7. GWR regression coefficients of shortest distance to metro stations.

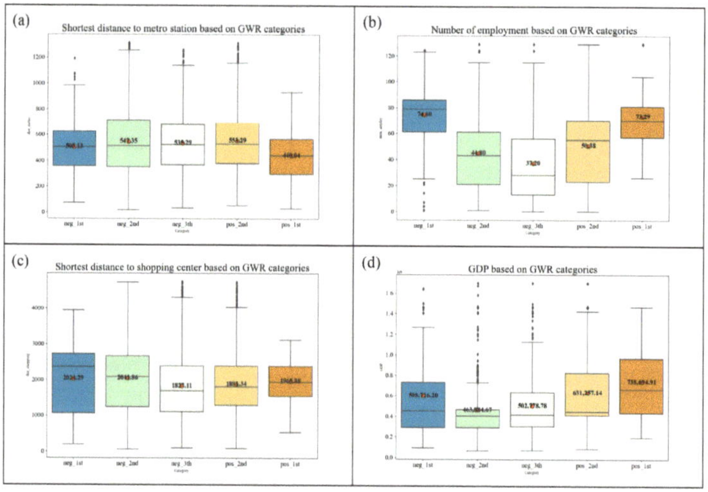

Figure 8. Distribution of coefficients average values of the main factors. (**a**) Shortest distance to metro station based on GWR categories. (**b**) Number of employment based on GWR categories. (**c**) Shortest distance to shopping center based on GWR categories. (**d**) GDP based on GWR categories.

4. Discussion

4.1. Overall Impact of Metro Accessibility on Property Prices: Non-Linear and Spatially Heterogeneous

Our result showed that the accessibility to metro stations was positively related to property prices, which was consistent with many previous studies [7,9,10,26]. Meanwhile, the relationship between the variables displayed a non-linear pattern and showed a spatial aggregation. The spatial autocorrelation had a certain impact on the property price. In the central areas along the Huangpu River, such as the Jing'an district and the Hongkou district, average property prices rose as the distance to metro stations increased. Conversely, in the neighborhoods surrounding the central area, closer proximity to metro stations was associated with higher property prices. On the other hand, in the districts farther away from the core area, the impact of distance to the metro station on property prices gradually stabilized. Similarly, Bowes and Ihlanfeldt (2001) indicated in their study that the impact of railway stations depended on their distance from the central business district [1]. We also suggested the significance of service facilities in residential areas, as our findings highlighted that accessibility to amenities like restaurants, shopping malls, and schools had a positive impact on property values. A study conducted in Seoul by Bae, Jun and Park (2003) found consistent results, indicating that individuals tended to reside in sub-center areas characterized by a higher concentration of recreational and commercial services [14].

There may be several reasons to explain the results of our study. Proximity to metro services has the potential to improve nearby property values, but metro accessibility alone may be insufficient to bring about significant changes. Other factors, such as economic stimulus, land use policy, and development subsidies should be taken into account when evaluating the influences of metro services on property prices [5]. Accessibility to metro stations also has some adverse effects on nearby communities. Therefore, the comprehensive impacts may be mixed, and the trend in property prices ultimately depends on which factor dominates.

In the central area along the Huangpu River, such as the Jing'an district and the Huangpu district, the city density is relatively higher than that of the surrounding neighborhoods. The bustling business and complex personnel, leading to noise, increased the crime rate and visual intrusion [37], directly a reducing living comfort. When purchasing properties, the downsides considered outweighed the advantages of convenience. As mentioned before (see Figure 5), central areas along the Huangpu River have the highest average GDP and there is an office building cluster in this district, where commuters living nearby may choose to walk and ride bikes (shared or private) to avoid the metro crowds during rush hours. However, in the areas surrounding the core, which is a little farther, traffic is often very congested and blocked, and parking facilities are in short supply. To avoid these two annoying troubles, people may take the metro as a more relaxing option. In a study conducted by Li (2017), it was found that the intersection of multiple metro lines, as well as the 30 min metro to work, results in an additional premium for property prices [19]. Additionally, the superior location of the central area along the Huangpu River, with a view of the river and the convenience of refined facilities, leads to almost the highest property price in the whole of Shanghai. People living in the central area are more likely to belong to the high-income group, who are more inclined to choose more convenient and comfortable private trip modes. Conversely, low-income groups are not affluent enough to afford cars and taxis, and hence prefer public transport like the metro, and always tend to walk or cycle to metro stations. Prior studies have also found that increased metro accessibility did not help property price growth in affluent communities [1,38], but Nelson (1992) claimed this improvement will accelerate the capitalization of low-income communities [39]. He tested the impact of the Atlanta Metropolitan Regional Rapid Transit Authority (MARTA) in an area of DeKalb County and found that the proximity to the metro is positively correlated with the property value in lower-income neighborhoods, with the opposite occurring in high-income areas. People in high-income areas (central areas) were less dependent on the metro, and even public transportation, than people in lower-income areas. Similarly, He (2020) proved that the property value premium due to rail accessibility was more significant in the suburbs than in urban areas in Hong Kong [40].

Besides, according to previous research, metro proximity and other related attributes have a different impact on property prices and rental prices [41]. Due to the expensive living costs in the central core, more people who reside here simply choose to rent rather than buy. A considerable amount of university students or new graduates may dwell in the central area and choose off-campus housing for rent [42]. The monetary effect of improved metro accessibility may somehow be reflected in rent rather than the selling price [43]. Another possible reason is that the strengths of improving accessibility may lie in accelerating the sale of homes rather than increasing their value [35].

4.2. Limitations of Our Current Study

We employed three models to gain convincing conclusions and explained the non-stationarity between metro accessibility and property value, i.e., the spatially existing autocorrelation. Although many previous studies have discussed the correlation between property prices and metro (rail transit) accessibility, most were concentrated in mature Western economies, few in the context of developing countries, Asia, and China. Since urban densities and population structures vary from country to country, we cannot fully apply conclusions and suggestions in urban planning from Western country research. Meanwhile, current research on the impact of the metro on the surrounding property prices seems to focus on a certain rail line or discuss it in general (including intercity high-speed rail, etc.). Therefore, we sought to fill these vacancies. The results of this study can provide some ideas for the comprehensive evaluation of metro system construction, including the manifestation of construction and operation costs in external real estate benefits. During metro line planning, the station location can be reasonably arranged to balance the discrepancy of property prices between different living areas of the city and make the urban development of Shanghai more all-round. Exploring lessons from Shanghai as a representative metropolis can offer meaningful insights not only for Chinese cities but also for other fast-growing urban areas across the globe.

Nevertheless, there are several limitations. Since the property transaction data we collected were based on records from different years, certain spatial and temporal variation factors may not be considered in the model. Moreover, the diversification of property prices in the economic environment, as well as market adjustments, might potentially bias our research results. To obtain more precise results, spatio-temporal consistency and longitudinal studies should be incorporated. We have applied a series of variables about the external environment and housing structure in our study to help analyze the impact of metro accessibility on property prices. Whereas in the housing structure, we simply encompassed the number of floors and the number of rooms, attributes like building typology, construction standards, structural technical condition and facility equipment are also significant to real estate values. In addition, the area and age of the properties that would contribute to the fluctuations in prices were neglected, especially in the context of multiple administrative regions, where development initiated and peaked in different periods in each area, greatly affecting the values of the real estate [44]. Properties that were built in different eras were equipped with varying floor area ratios and green space ratios in the residential region, which can also affect real estate prices. Moreover, there may be some other potential covariates influencing the values such as the level of rail service, network connectivity, service coverage, and station facilities [4]. In addition, as mentioned above, many commuters choose to ride from home to the metro station. With the development of the sharing economy, shared bikes may provide them with more convenience, which may also increase the influence of metro accessibility on property prices [45]. Moreover, when defining accessibility variables, we applied the Euclidean distance which simply calculated the straight distance between starting and ending points. The simple assessment of accessibility did not consider the practical state of the street and walking path networks, so it is insufficient to reflect the fact of travel costs or barriers. For further studies, accessibility measures with higher robustness and diversity are suggested.

4.3. Prospects for Future Study

Since we have discovered that spatial autocorrelation and heterogeneity can be elements that affected property values, in the future, the related factors can be fixed while applying a statistic model. For instance, as the outcomes showed, the number of employees in the vicinity of metro stations and local GDP levels are two important factors affecting the non-linear. The research in the next phase will be expected to involve or fix more socio-economic factors comprising a matrix of the proportion of different race or ethnic groups of people, the median household income and unemployment rate of the census tract [27]. Moreover, apart from the existing characteristics popularly used, natural amenities, historical amenities and modern amenities can be supplemented. Among them, natural amenities are always linked with forest coverage (NDVI), Slope, Index of NOX as well as CO.

We hope that the outcomes of our research can provide a reference in the field of property and rail transit, even in a broader scope of the whole urban land use planning and management, as well as land and socio-economic development. In metropolia in developing countries such as Shanghai, the optimization of urban planning and management, including the reasonable land use, transport system and amenity setup, along with emphasis on improving landscape design, is for the sake of promoting local economic growth and improving the quality of residential life. To achieve these goals, attracting real estate development investment is an effective approach [46–48]. In Shanghai, the current urban center is over-concentrated [49]. The government needs to ensure more fairly accessible facility services, address the issue of affordable housing, disperse the traffic flow, and balance the regional development by establishing multiple employment centers and more considerate land use planning and land management. Though we have reached the conclusion of a positive relationship between metro accessibility and real estate values in general, complying with the non-linear pattern, there is potential to explore more deeply, such as the distance thresholds where metro accessibility has an impact on property values. This can inform the land use decision-making of the transition zone between properties and metro stations. In the future, when investigating and evaluating the construction distribution and investment effect of rail transit systems or other infrastructure, we should consider various aspects, not sticking to the results reflected in the real estate price. The improvement of metro accessibility not only has direct influence on property prices, but also a profound impact on urban land planning, design and management. Optimizing the layout of the metro network needs to take into account future urban development orientation to apply the optimal solution of land use, while the balance between the supply and the demand of land should also be considered to avoid the unstable land economic situation. Objectively, the high-efficient allocation of land, along with perpetual land management, the popularization of high-quality public transport systems and the improvement of landscape and facility services, are of great significance to urban development, leading to the agglomeration of high-level employment opportunities, growth of land economy, and promotion of the overall development of cities [50]. Apart from emphasizing accessibility to metro stations, future research can also incorporate additional accessibility to amenities such as parks, grocery stores, and schools [51], while also accounting for negative forces brought on by the proximity, such as vacant lands, crime rates, congestion, air contamination and noise pollution [1,52]. Not only the site selection of the metro stations but also the surrounding land planning and its management, as well as maintenance, are important factors affecting the property price. The measured planning of the land use around both metro stations and real estate, such as commercial, education and public facilities, can have a valid impact on land value and the property price level of the whole area.

5. Conclusions

The metro has great potential to enhance the value of residential property by improving convenience and saving transportation budgets. However, whether the metro system has a favorable effect on residential property prices has not reached a unified conclusion through prior studies. Furthermore, Shanghai, as a typical example of a flourishing metropolis in Asia, with a highly dependent and well-developed metro system, has great research value, but few studies based on it have been conducted. To bridge gaps in knowledge, our objective is to gain a comprehensive understanding of the relationship between property prices and metro accessibility in Urban Shanghai and to further explore the spatial pattern of this correlation, which is expected to offer insight into urban land use planning along with land management, and suitable landscape and facility design.

We adopted the Euclidean distance between the metro station and the property to measure metro accessibility. Then, we applied the hedonic model proposed by Rosen

(1974) [25], using the Ordinary Least Squares (OLS) model and the XGBoost model, which is a relatively new analytical method, to identify the statistical relationship between metro accessibility and property prices. Considering the difficulty in simultaneously analyzing the various levels of influence and correctly handling the autocorrelation between different factors, the spatial pattern was analyzed by Geographically Weighted Regression (GWR).

Overall, our OLS and XGBoost results indicated that the distance to the metro station was negatively associated with property prices. In other words, the value of land around metro stations can be raised with the improvement of metro accessibility. This provides great economic potential for land development and attracts investors and developers to participate in the active development of urban land economy. Specifically, XGBoost showed a higher r-square, indicating a non-linear association between the explanatory variables and property prices. The GWR model examined the heterogeneous spatial pattern between the proximity of properties and metro stations and their average property prices. The outcomes of the model indicated that communities in the central area, such as the Jing'an and Hongkou districts, exhibited positive coefficients, suggesting that average property prices rose as the distance to the metro station increased. Conversely, in the surrounding neighborhoods, closer proximity to metro stations meant higher property prices. In addition, our results also indicated that the accessibility variable performance had mainly positive effects, except for groceries, emphasizing the importance of service facilities around residential areas in the formation of property prices. We recognized the complex relationship between property prices and metro accessibility, and provided a certain reference for the urban planning of Shanghai to promote the land and social economies. In the future, more in-depth research is needed to fully understand this correlation and to provide more specific guidance for future urban development regarding land use planning and management with the consideration of land economic growth.

Author Contributions: Conceptualization, W.H. and R.Z.; methodology, R.Z. and S.G.; data collection, W.H.; data curation, W.H.; formal analysis, R.Z. and W.H.; visualization, R.Z. and S.G.; project administration, W.H.; supervision, W.H.; software, R.Z. and S.G.; validation, W.H.; writing—original draft, W.H. and R.Z.; writing—review and editing, W.H., R.Z. and S.G. All authors have read and agreed to the published version of the manuscript.

Funding: This research received no external funding.

Data Availability Statement: The raw data supporting the conclusions of this article will be made available by the authors on request.

Conflicts of Interest: The authors have no relevant financial or non-financial competing interests.

References

1. Bowes, D.R.; Ihlanfeldt, K.R. Identifying the Impacts of Rail Transit Stations on Residential Property Values. *J. Urban Econ.* **2001**, *50*, 1–25. [CrossRef]
2. Lin, B.; Du, Z. Can Urban Rail Transit Curb Automobile Energy Consumption? *Energy Policy* **2017**, *105*, 120–127. [CrossRef]
3. Song, Y.; Zhong, S.; Lee, C.; Xu, M. Causal Effect of New Bus Rapid Transit on Non-Work-Related Activities. *J. Plan. Educ. Res.* **2023**. [CrossRef]
4. Debrezion, G.; Pels, E.; Rietveld, P. The Impact of Railway Stations on Residential and Commercial Property Value: A Meta-Analysis. *J. Real Estate Finan. Econ.* **2007**, *35*, 161–180. [CrossRef]
5. Gatzlaff, D.H.; Smith, M.T. The Impact of the Miami Metrorail on the Value of Residences near Station Locations. *Land Econ.* **1993**, *69*, 54–66. [CrossRef]
6. Chalermpong, S. Rail Transit and Residential Land Use in Developing Countries: Hedonic Study of Residential Property Prices in Bangkok, Thailand. *Transp. Res. Rec.* **2007**, *2038*, 111–119. [CrossRef]
7. Bajic, V. The Effects of a New Subway Line on Housing Prices in Metropolitan Toronto. *Urban Stud.* **1983**, *20*, 147–158. [CrossRef]
8. Benjamin, J.; Sirmans, G.S. Mass Transportation, Apartment Rent and Property Values. *J. Real Estate Res.* **1996**, *12*, 1–8. [CrossRef]
9. Dorantes, L.M.; Paez, A.; Vassallo, J.M. Analysis of House Prices to Assess Economic Impacts of New Public Transport Infrastructure: Madrid Metro Line 12. *Transp. Res. Rec.* **2011**, *2245*, 131–139. [CrossRef]
10. Mayor, K.; Lyons, S.; Duffy, D.; Tol, R.S.J. A Hedonic Analysis of the Value of Rail Transport in the Greater Dublin Area. *J. Transp. Econ. Policy JTEP* **2012**, *46*, 239–261.
11. Vessali, K.V. Land Use Impacts of Rapid Transit: A Review of the Empirical Literature. *Berkeley Plan. J.* **1996**, *11*, 71–105. [CrossRef]

12. Portnov, B.A.; Genkin, B.; Barzilay, B. Investigating the Effect of Train Proximity on Apartment Prices: Haifa, Israel as a Case Study. *J. Real Estate Res.* **2009**, *31*, 371–396.
13. Ibeas, Á.; Cordera, R.; dell'Olio, L.; Coppola, P.; Dominguez, A. Modelling Transport and Real-Estate Values Interactions in Urban Systems. *J. Transp. Geogr.* **2012**, *24*, 370–382. [CrossRef]
14. Bae, C.-H.C.; Jun, M.-J.; Park, H. The Impact of Seoul's Subway Line 5 on Residential Property Values. *Transp. Policy* **2003**, *10*, 85–94. [CrossRef]
15. Chin, S.; Kahn, M.E.; Moon, H.R. Estimating the Gains from New Rail Transit Investment: A Machine Learning Tree Approach. *Real Estate Econ.* **2020**, *48*, 886–914. [CrossRef]
16. Diao, M.; Qin, Y.; Sing, T.F. Negative Externalities of Rail Noise and Housing Values: Evidence from the Cessation of Railway Operations in Singapore. *Real Estate Econ.* **2016**, *44*, 878–917. [CrossRef]
17. Anantsuksomsri, S.; Tontisirin, N. The Impacts of Mass Transit Improvements on Residential Land Development Values: Evidence from the Bangkok Metropolitan Region. *Urban Policy Res.* **2015**, *33*, 195–216. [CrossRef]
18. Zheng, S.; Kahn, M.E. Land and Residential Property Markets in a Booming Economy: New Evidence from Beijing. *J. Urban Econ.* **2008**, *63*, 743–757. [CrossRef]
19. Li, T. The Value of Access to Rail Transit in a Congested City: Evidence from Housing Prices in Beijing 2017. *Real Estate Econ.* **2020**, *48*, 556–598. [CrossRef]
20. Yang, L.; Chen, Y.; Xu, N.; Zhao, R.; Chau, K.W.; Hong, S. Place-Varying Impacts of Urban Rail Transit on Property Prices in Shenzhen, China: Insights for Value Capture. *Sustain. Cities Soc.* **2020**, *58*, 102140. [CrossRef]
21. Zhou, Z.; Chen, H.; Han, L.; Zhang, A. The Effect of a Subway on House Prices: Evidence from Shanghai. *Real Estate Econ.* **2021**, *49*, 199–234. [CrossRef]
22. Li, H.; Wei, Y.H.D.; Huang, Z. Urban Land Expansion and Spatial Dynamics in Globalizing Shanghai. *Sustainability* **2014**, *6*, 8856–8875. [CrossRef]
23. Jim, C.Y.; Chen, W.Y. Consumption Preferences and Environmental Externalities: A Hedonic Analysis of the Housing Market in Guangzhou. *Geoforum* **2007**, *38*, 414–431. [CrossRef]
24. Chen, J.; Hao, Q. The Impacts of Distance to CBD on Housing Prices in Shanghai: A Hedonic Analysis. *J. Chin. Econ. Bus. Stud.* **2008**, *6*, 291–302. [CrossRef]
25. Rosen, S. Hedonic Prices and Implicit Markets: Product Differentiation in Pure Competition. *J. Political Econ.* **1974**, *82*, 34–55. [CrossRef]
26. Dewees, D.N. The Effect of a Subway on Residential Property Values in Toronto. *J. Urban Econ.* **1976**, *3*, 357–369. [CrossRef]
27. Li, H.; Wei, Y.D.; Yu, Z.; Tian, G. Amenity, Accessibility and Housing Values in Metropolitan USA: A Study of Salt Lake County, Utah. *Cities* **2016**, *59*, 113–125. [CrossRef]
28. Deilami, K.; Kamruzzaman, M.; Hayes, J.F. Correlation or Causality between Land Cover Patterns and the Urban Heat Island Effect? Evidence from Brisbane, Australia. *Remote Sens.* **2016**, *8*, 716. [CrossRef]
29. Li, X.; Zhou, Y.; Asrar, G.R.; Imhoff, M.; Li, X. The Surface Urban Heat Island Response to Urban Expansion: A Panel Analysis for the Conterminous United States. *Sci. Total Environ.* **2017**, *605–606*, 426–435. [CrossRef]
30. Chen, T.; Guestrin, C. XGBoost: A Scalable Tree Boosting System. In Proceedings of the 22nd ACM SIGKDD International Conference on Knowledge Discovery and Data Mining; Association for Computing Machinery, New York, NY, USA, 13 August 2016; pp. 785–794.
31. Parsa, A.B.; Movahedi, A.; Taghipour, H.; Derrible, S.; Mohammadian, A.K. Toward Safer Highways, Application of XGBoost and SHAP for Real-Time Accident Detection and Feature Analysis. *Accid. Anal. Prev.* **2020**, *136*, 105405. [CrossRef]
32. Wang, M.; Li, Y.; Yuan, H.; Zhou, S.; Wang, Y.; Adnan Ikram, R.M.; Li, J. An XGBoost-SHAP Approach to Quantifying Morphological Impact on Urban Flooding Susceptibility. *Ecol. Indic.* **2023**, *156*, 111137. [CrossRef]
33. Fotheringham, S.; Charlton, M.; Brunsdon, C. Spatial Variations in School Performance: A Local Analysis Using Geographically Weighted Regression. *Geogr. Environ. Model.* **2001**, *5*, 43–66. [CrossRef]
34. Du, H.; Mulley, C. Relationship between Transport Accessibility and Land Value: Local Model Approach with Geographically Weighted Regression. *Transp. Res. Rec.* **2006**, *1977*, 197–205. [CrossRef]
35. Du, H.; Mulley, C. The Short-Term Land Value Impacts of Urban Rail Transit: Quantitative Evidence from Sunderland, UK. *Land Use Policy* **2007**, *24*, 223–233. [CrossRef]
36. Szymanowski, M.; Kryza, M. Local Regression Models for Spatial Interpolation of Urban Heat Island—An Example from Wrocław, SW Poland. *Theor. Appl. Clim.* **2012**, *108*, 53–71. [CrossRef]
37. Diaz, R.B. *Impacts of Rail Transit on Property Values*; American Public Transportation Association: Toronto, ON, Canada, 1999.
38. Hess, D.B.; Almeida, T.M. Impact of Proximity to Light Rail Rapid Transit on Station-Area Property Values in Buffalo, New York. *Urban Stud.* **2007**, *44*, 1041–1068. [CrossRef]
39. Nelson, A.C. Effects of elevated heavy-rail transit stations on house prices with respect to neighborhood income. *Transp. Res. Rec.* **1992**, *1359*, 127–132.
40. He, S.Y. Regional Impact of Rail Network Accessibility on Residential Property Price: Modelling Spatial Heterogeneous Capitalisation Effects in Hong Kong. *Transp. Res. Part A Policy Pract.* **2020**, *135*, 244–263. [CrossRef]
41. Efthymiou, D.; Antoniou, C. How Do Transport Infrastructure and Policies Affect House Prices and Rents? Evidence from Athens, Greece. *Transp. Res. Part A Policy Pract.* **2013**, *52*, 1–22. [CrossRef]

42. Song, Y.; Lee, C.; Tao, Z.; Lee, R.J.; Newman, G.; Ding, Y.; Jessica, F.; Sohn, W. COVID-19 and Campus Users: A Longitudinal and Place-Based Study of University Mobilities in Texas. *Sustain. Cities Soc.* **2023**, *96*, 104656. [CrossRef]
43. Gadziński, J.; Radzimski, A. The First Rapid Tram Line in Poland: How Has It Affected Travel Behaviours, Housing Choices and Satisfaction, and Apartment Prices? *J. Transp. Geogr.* **2016**, *54*, 451–463. [CrossRef]
44. Pan, Q. The Impacts of an Urban Light Rail System on Residential Property Values: A Case Study of the Houston METRORail Transit Line. *Transp. Plan. Technol.* **2013**, *36*, 145–169. [CrossRef]
45. El-Geneidy, A.; van Lierop, D.; Wasfi, R. Do People Value Bicycle Sharing? A Multilevel Longitudinal Analysis Capturing the Impact of Bicycle Sharing on Residential Sales in Montreal, Canada. *Transp. Policy* **2016**, *51*, 174–181. [CrossRef]
46. He, S.; Wu, F. China's Emerging Neoliberal Urbanism: Perspectives from Urban Redevelopment. *Antipode* **2009**, *41*, 282–304. [CrossRef]
47. Mihaescu, O.; Hofe, R. The Impact of Brownfields on Residential Property Values in Cincinnati, Ohio: A Spatial Hedonic Approach. *J. Reg. Anal. Policy* **2012**, *42*, 223–236.
48. Wu, Q.; Cheng, J.; Chen, G.; Hammel, D.J.; Wu, X. Socio-Spatial Differentiation and Residential Segregation in the Chinese City Based on the 2000 Community-Level Census Data: A Case Study of the Inner City of Nanjing. *Cities* **2014**, *39*, 109–119. [CrossRef]
49. Yang, Y.-R.; Chang, C. An Urban Regeneration Regime in China: A Case Study of Urban Redevelopment in Shanghai's Taipingqiao Area. *Urban Stud.* **2007**, *44*, 1809–1826. [CrossRef]
50. Glaeser, E.L.; Kominers, S.D.; Luca, M.; Naik, N. Big Data and Big Cities: The Promises and Limitations of Improved Measures of Urban Life. *Econ. Inq.* **2018**, *56*, 114–137. [CrossRef]
51. Song, Y.; Lee, S.; Park, A.; Lee, C. COVID-19 Impacts on Non-Work Travel Patterns: A Place-Based Investigation Using Smartphone Mobility Data. *Environ. Plan. B Urban Anal. City Sci.* **2022**, *50*, 239980832211249. [CrossRef]
52. Zhu, R.; Newman, G.; Li, D.; Song, Y.; Zou, L. Associations between Vacant Urban Lands and Public Health Outcomes in Growing and Shrinking Cities. *Urban For. Urban Green.* **2023**, *89*, 128127. [CrossRef] [PubMed]

land

Article

A Semantic Analysis Method of Public Public Built Environment and Its Landscape Based on Big Data Technology: Kimbell Art Museum as Example

Zhongzhong Zeng [1,*], Meizhu Wang [1], Dingyi Liu [1], Xuan Yu [1] and Bo Zhang [2]

[1] School of Architecture and Design, Beijing Jiaotong University, Beijing 100080, China; 22126450@bjtu.edu.cn (M.W.); 18126377@bjtu.edu.cn (D.L.); yx@mail.deeparchi.com (X.Y.)
[2] Horticulture and Landscape Architecture Department, Oklahoma State University, Stillwater, OK 74078-1015, USA; b.zhang@okstate.edu
* Correspondence: zzzeng@bjtu.edu.cn

Abstract: Based on big data, a new public space evaluation method is proposed. Using programming technology to collect visitor reviews from the travel website TripAdvisor to build a database, based on the data of 99,240 words in 1573 visitor reviews in 10 years, the connection between data and reality is established through systematic data classification and visualization. Following an assessment of the Kimbell Art Museum's functionality, architectural design, and landscape design, along with visitor feedback, a new evaluation methodology was formulated for application to public buildings with landscapes. By utilizing the unique advantages of big data, it provides convenient and efficient analysis methods for public spaces with similar data foundations and opens the way for the optimization of the built environment in the information age.

Keywords: space evaluation method; big data mining; public architecture evaluation; Kimbell Art Museum

Citation: Zeng, Z.; Wang, M.; Liu, D.; Yu, X.; Zhang, B. A Semantic Analysis Method of Public Public Built Environment and Its Landscape Based on Big Data Technology: Kimbell Art Museum as Example. *Land* **2024**, *13*, 655. https://doi.org/10.3390/land13050655

Academic Editor: Thomas Panagopoulos

Received: 12 April 2024
Revised: 5 May 2024
Accepted: 7 May 2024
Published: 10 May 2024

1. Introduction

A knowledge base representing semantic relations between concepts in a network is termed a semantic network or frame network [1]. The concept of "semantic networks" in propositional calculus had its embryonic origins as early as 1956 (Richard H. Richens). Rooted in Computational Linguistics, research emerged utilizing phrase-structure grammar algorithms for sentence generation (Victor Yngve, 1960) [2]. Sheldon Klein and Robert F. Simmons encapsulated it as "a method for controlling the sense of what was generated by respecting the semantic dependencies of words as they occurred in the text". Subsequently, through practical research efforts led by numerous scholars, including M. Ross Quillian, semantic network analysis gradually formalized its framework [3]. According to visual analysis, semantic networks have the capacity to be automatically extracted from unstructured textual data, serving as a platform for visual text analysis. It also involved modeling semantic relationships (Sowa, 1991) and visualizing patterns of labeled nodes and edges (Di Battista, 1999) [1]. In recent years, this research has evolved towards social semantic networks [4].

Currently, a single public building post-occupancy evaluation based on user reviews is a scientific method [5–8]. Since the early 21st century, the development of big data technology has promoted the gradual application of semantic network analysis in post-occupancy evaluation, which has the advantages of sufficient data, rich diversity, and spontaneity [9].

In the past 20 years, there have been more than one thousand articles on semantic network analysis [10], but the focused fields are limited: in addition to technology research, this method is mainly used in the landscape planning of scenic spots (accounting for

24.6%) [11,12], whereas in the field of architectural science and engineering, related articles only account for 1.18% [13,14], and the research usually stops at the data sorting level, such as semantic classification and word frequency extraction, with no mature evaluation system that can establish connections between the complex network semantic data and actual usage based on a continuous logic [15,16]. This article summarizes a method of identifying problems from data and converting them into design guidance, making the structural evaluation open; therefore, the traditional top–bottom presupposition is replaced by bottom–top spontaneity, making up for the shortcomings of traditional methods such as lack of soft experience and visitors' spontaneity. The method proposed in this article truly based public building post-occupancy evaluation on the visitors' experiences.

Python was adopted to acquire, filter, classify, and visualize the data [17]. By analyzing the internal relation of the data, this article summarizes public building design guidance based on semantic network analysis [18]. The classification, visualization, and analysis approaches adopted are all original ones proposed by the study team, and they addressed the issue of a lack of visitors' soft experience analysis that haunts existing semantic analysis methods and a scientific method of establishing analysis logic with big data semantic text as the data source was proposed [19]. A pure visitor perspective was adopted to explore architectural characteristics. The research subject can be built public buildings that have been put into use for a certain period, available for public review and with good landscapes, such as libraries, art galleries, etc. [20]. Urban scenic spots in major cities are predominantly occupied by museum-type public buildings, making them prime candidates for researching assessment methods, as indicated by evaluation rankings (Figure 1) [21].

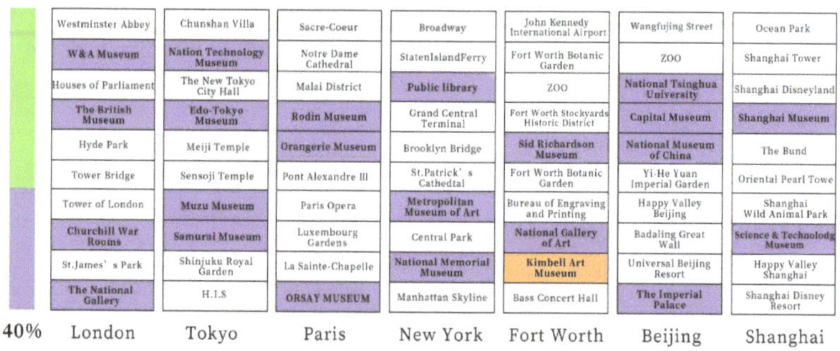

Figure 1. Urban Scenic Spots Evaluation Ranking.

Evaluation systems based on computer technology have all developed from independent case analysis to a standard system [9]. Under this background, this article takes the Kimbell Art Museum in Fort Worth, Texas, USA as the subject, based on the reviews gained through big data python, builds a network data and programming technology-based evaluation system that uses semantic classification and word frequency extraction as the basis for analysis to dig deep into the systematic spatial–temporal connection of behavior, objects, and emotions inside the building. The Kimbell Art Museum was chosen for study because of its distinctive blend of public functionality, curated indoor spaces, and meticulously maintained outdoor landscapes [22]. Analyzing these elements enables the assessment of how functional activities, indoor spaces, and outdoor environments influence visitors in public buildings, leading to the development of a comprehensive evaluation framework [23]. The richness of big data enables the study team to deeply explore the correlation mechanism behind the use of buildings, therefore forming a semantic network evaluation framework for public buildings and their landscapes, which can guide and optimize the operation and design of the Kimbell Art Museum and similar buildings and open up new development directions for such research.

2. Materials and Methods

The method proposed in this study classified the effective data into three categories: human emotions (emotion), human behavior (behavior) and objects in the building that interact with people (object) and established a one-to-one corresponding relationship between visitors' reviews and the building. Following the Gephi network visualization logic mechanism, basic data analysis (overall word frequency analysis), temporal correlation analysis (analysis by month), spatial correlation analysis (spatial correlation analysis), and correlation analysis with others (Gephi correlation mechanism analysis) were performed on the three categories of reviews [24]. Data analysis covers the basic experience and attention of visitors, the influence of time factors such as seasons on the evaluation and landscape, and the overall relationship between the focus of the reviews in specific spaces and emotion, behavior, and object [25]. The data analysis provided a thorough evaluation of the building's advantages, disadvantages, and future development trends, which is of theoretical support for the selection of the building's development focus, seasonal activity planning, space quality improvement, and interaction detail improvement (Figure 2).

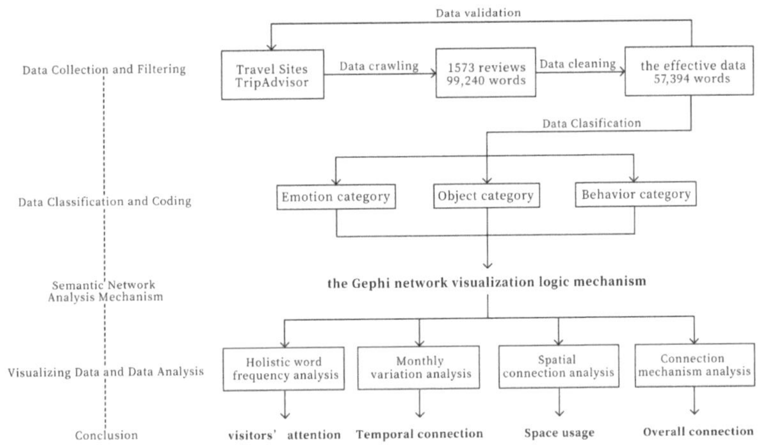

Figure 2. Research Framework.

2.1. Data Collection and Filtering

The data source of this method was the reviews published by visitors on a tourism website after visiting the building. Four advantages that distinguish this method from traditional data research have been given full play in this study (Table 1) [26]:

1. Large amount of data: 1573 reviews were collected with a total of 99,240 words;
2. Data accumulation: the data covers a period of 9 years from January 2011 to December 2019;
3. Diversified data: In addition to the overall visiting experience reviews, single-data content also includes tour time, ratings, and user information;
4. Data spontaneity: the content is not guided by questionnaire questions or interviews, so it can provide a subjective evaluation from many aspects.

Since there is a lot of secondary information in the data, meaningless conjunctions, pronouns such as such as "And", non-associated references such as "TripAdvisor (TripAdvisor)", basic information such as "Kimbell (Kimbell), and words that appear too few times were screened out, leaving 45,732 words as valid data (53,508 words were screened out) (Figure 3).

Table 1. Data Collection via Big Data vs. Traditional Approaches.

	Survey	Questionnaire	Big Data
Amount of Data	Usually above thousands of words, based on times of actual survey	Usually above thousands of words, based on valid samples	Usually above tens of thousands words, based on data source content
Time Span	Days, based on survey time and frequency	Weeks, based on number of samples and where they are collected	Years, based on when there was the data source
Closeness to Research	Subjective records point to objective problems and phenomena, facilitating the identification of research directions	The questionnaire design points to the research purpose, and the structure is unified, facilitating analysis	The network data points to the user experience. The data is complex and needs to be processed before analysis
Spontaneity	Based on objective phenomena	Interviewees passively answers items on the questionnaire	Users publish reviews spontaneously
Scope of Research	Subjective screening, there might be unexpected research results	The questionnaire design directly points to research expectation with few other possibilities	Analyzing complex data, there might be unexpected research results
Cost Effectiveness	Human labor, time, and facilities	Human labor, time, and effort on questionnaire design	Network and programming technology, no other resources are required

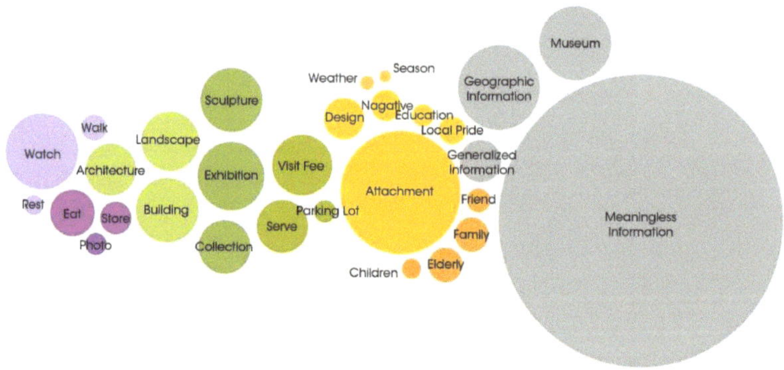

Figure 3. Data Classification and Coding.

2.2. Data Collection and Filtering

The valid data was divided into three categories: behavior, object, and emotion and 20 subcategories, which realized comprehensive and specific evaluation from various perspectives (Figure 4):

1. Behavior (reflecting the behavior of visitors in the art museum): photo-shooting, arrival, visit, dining, and other behaviors;
2. Object (reflecting what visitors observe in the museum): landscape, architecture, activities, exhibits, glossary, dining area, and services;
3. Emotion (reflecting the emotional feedback of visitors in the art museum): positive emotion, negative emotion, sense of design, season, time, architectural perception, and crowd type.

Tags according to the characteristics of the building were attached to the 20 subcategories. For example, the building contains exhibition space, atmosphere space, shopping space, and dining space, and each tag includes all the relevant valid words in the database.

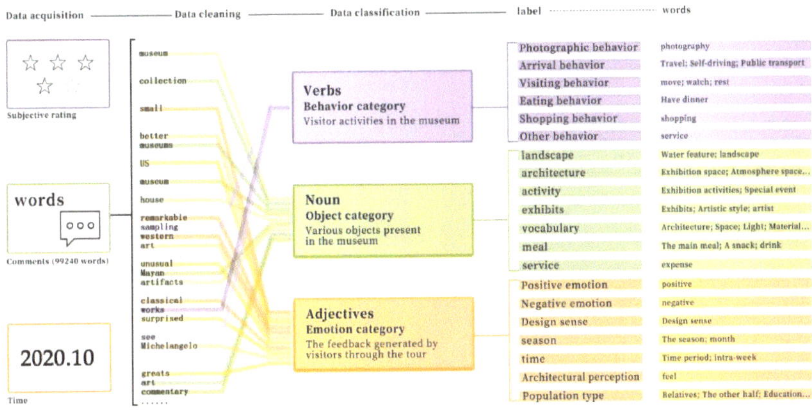

Figure 4. Data Processing.

The process of building a database refers to collecting and dividing data into word lists with the aid of programming technology. After data filtering, the effective data were divided into three categories, and the initial connection between the data and the use of the environment was established. Finally, each piece of data was linked to the specific circumstance in reality through coding.

2.3. Semantic Network Analysis Mechanism

The open-source data visualization software Gephi-0.10.1 was adopted in this study to analyze the connection status of the 44 tags and demonstrate the causal mechanism of exploring space use. The reviews were split into sentences (ending with ".", "?", "!", etc.) to explore the correlation mechanism between the different factors (Figure 5).

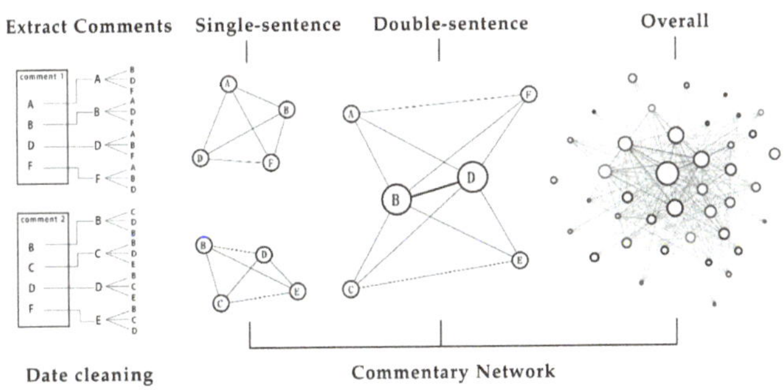

Figure 5. Gephi Visual Network Logic.

1. The size of the point: reflects the number of occurrences of the tag in the database. The larger the point, the more attention the tag has received from visitors;
2. Number of lines: reflects the influence dimension of the tag in the network. The more lines on the tag, the bigger the influence dimension of the point in the overall network;
3. Line width: reflects the intensity of the influence between two tags. For example, the more pairs of two words found in the same review, the thicker the edge between the two points, and the higher the mutual influence, and vice versa;
4. The position of the point: reflects the position of the tag in the whole network. The closer the point is to the center of the network, the more attention the tag receives and the bigger impact it has on other tags. On the contrary, tags at the edge of the network are relatively unnoticed by visitors and are not an important factor affecting their museum tour experience (Figure 6).

(a) (b) (c) (d)

Figure 6. Gephi Grammar. (**a**) The size of the point; (**b**) Number of lines; (**c**) Line width; (**d**) The position of the point.

3. Visualizing Data and Data Analysis

3.1. Overall Word Frequency Analysis

Overall word frequency analysis can reveal the focus of visitors' attention in the museum. This analysis serves to validate whether the museum's functions and positioning align with its intended goals while also facilitating the identification of areas ripe for future development.

At the category level, when the proportion of behavior is the highest, visitors' activities in the museum are mainly behaviors such as shopping and dining. When the proportion of object is the highest, visitors' activities are usually connected with objects, such as buildings and exhibits. When the proportion of emotion is the highest, visitors' activities in the museum usually center on the design, the season, or the extreme emotional evaluation of the building.

At the subcategory and tag level, taking the behavior category as an example (Figure 7), word frequency (word frequency/average word frequency within the category) was compared using 1 as the ratio limit, thus quantifying the intensity of the focus, and the elements with the highest and lowest concern were identified.

Behavior category	Average value	Photographic behavior	Arrival behavior				Visiting behavior				Eating behavior	Shopping behavior	Acceptance behavior
	189	0. 19	3. 17				4. 69				0. 37	0. 13	1. 46
		×	√				MAX				×	MIN	√
			Average value	travel	self-drive	Public transport	Average value	move	see	rest			
			200	2. 3	0. 66	0. 08	295	0. 38	2. 32	0. 29			
				MAX	×	MIN		×	MAX	MIN			

Figure 7. Overall Word Frequency Analysis (word frequency/average word frequency within the category).

The word frequency analysis was then visualized, and the attention to different data was represented by the bar graph of 20 subcategories under the three categories of behavior (purple), object (green), and emotion (yellow). The following conclusions can be drawn from Figure 8:

1. Under the behavior category, the behavior that received the most attention was "visiting behavior", followed by "arrival behavior". Among them, the elements that visitors were mostly concerned with were "tourism", "viewing", and "service", indicating a strong emphasis on the overall tour experience.
2. Under the object category, the object that received the most attention was "exhibits", followed by "glossary". Among them, "art", "exhibits", "cost", and "architecture" were the elements that attracted more attention from the visitors, underscoring their appeal within the museum context.
3. Within the space, the areas that received a high degree of attention from the visitors were "dining space" and "shopping space", reflecting the importance of amenities and facilities catering to visitors' needs and preferences.
4. The visitors paid a lot of attention to the architectural design content, such as "lighting", "hard decoration", and other elements. The visitors also cared about architectural details, on which more comments and feedback can be found. Such insights

emphasize the significance of architectural aesthetics in shaping visitors' perceptions and experiences within the museum environment.

Figure 8. Overall Word Frequency (number of occurrences of each word).

3.2. Temporal Connection (Change of Month) Analysis

A frequency connection was established between each item and the time node to analyze the word frequency changes, reflecting the differences in attention across different months and the correlation between the changes in elements with the change in time. The following conclusions can be drawn from Figure 9:

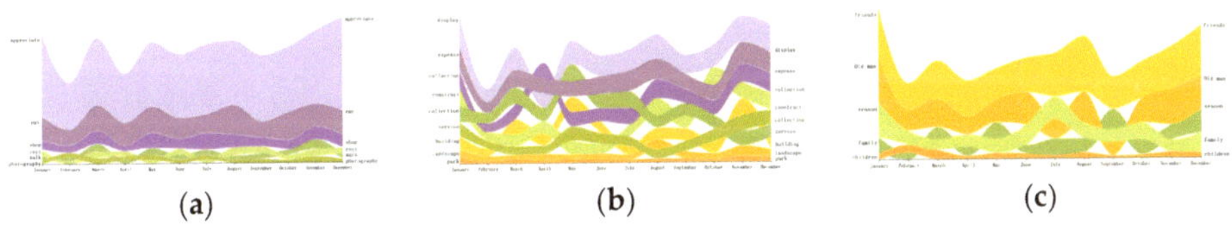

Figure 9. Word Frequency by Month. (**a**) Behavior in relation to time nodes; (**b**) Object in relation to time nodes; (**c**) Result data in relation to time nodes.

1. Under the behavior category, "having meals" is the element that attracted more attention, and was stable throughout the year and did not fluctuate greatly with the change of months.
2. Under the behavior category, a noteworthy correlation was observed between "rest" and "walking", where the prominence of one corresponded inversely with the prominence of the other, suggesting a trade-off in visitor activities.
3. Under the object category, "display" was the most stable element, maintaining consistent attention levels across varying time nodes.
4. Under the object category, "construction" received more attention than "architecture", except for a few time nodes.
5. The result data also reflect the relationship between group tours and seasonal changes. Only in the month of August, the number of family visitors as groups increased faster than that of individual visitors, highlighting the influence of seasonal dynamics on visitor behavior and preferences.

3.3. Spatial Connection Analysis

In addition to the aforementioned analysis, a direct correlation was established between each item and the architectural space to precisely quantify the rate of space utilization, thereby offering insight into visitors' inclination towards evaluating the spatial layout. By meticulously computing both the spatial area and the frequency of words employed in spatial commentary, a comprehensive understanding of visitors' perceptions of the space was attained. Furthermore, by leveraging the spatial scale of the subject under review, the spatial area corresponding to a predetermined number of reviews was meticulously delineated.

With all the above steps finished, a pivotal review-to-space area ratio was derived, enabling a rigorous quantitative examination. This ratio not only provides a metric for assessing the intensity of visitor engagement within specific spatial contexts but also furnishes valuable insights into the spatial dynamics that shape visitor experiences. The relationship between the number of reviews and the corresponding spatial area is illustrated in Figure 10, offering a visual representation of the nuanced interplay between visitor feedback and architectural space utilization.

1. Among the various areas assessed, "Landscape" occupied the largest overall area within the museum premises, while the "exhibition hall" stood out as the area garnering the highest number of reviews. This highlights the significance of both outdoor ambience and curated exhibition spaces in shaping visitors' experiences.
2. "Office" had the lowest reviews/area ratio, indicating relatively fewer reviews in proportion to its spatial extent. Conversely, "restaurant" had the highest reviews/area ratio, underscoring the heightened attention and engagement it elicits from visitors. Following closely were the "exhibition hall" and "service area", further emphasizing the importance of amenities and exhibition spaces within the museum environment.
3. Across all areas except for the "office", the reviews predominantly focused on "behavior" over "object" and "emotion". This suggests that visitors tend to prioritize their interactions and experiences within these spaces. Conversely, reviews of the "office" area primarily centered on "object", with a notable emphasis on the "layout" element. This deviation underscores the functional and design considerations associated with administrative spaces within the museum.

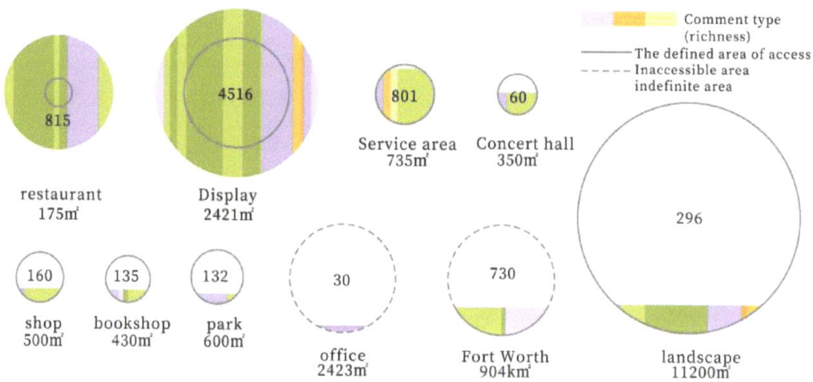

Figure 10. Space Area/Number of Reviews (space evaluation intensity).

3.4. Gephi Connection Mechanism Analysis

The ring diagram (Figure 11) presents the interplay among different categories of data, establishing a comprehensive relationship between behavior, object, and emotion. This supplementary analysis enriches the overall study by offering a nuanced perspective on visitor engagement within the museum environment. Aligned with the findings from the overarching word frequency map, the diagram underscores that visitors predominantly concentrated on eight key elements: "appreciation", "architecture", "construction", "exhibition", "exhibits", "collection", "ticket price", and "positive emotion". Notably, "positive emotion" emerged as the focal point, serving as the linchpin of the entire word frequency map. The remaining seven elements exhibited close associations with "positive emotion", underscoring their pivotal roles in shaping visitors' tour experiences. Furthermore, the diagram elucidates the symbiotic relationship among these elements. This holistic view not only illustrates the significance of emotional resonance in visitor engagement but also underscores the multifaceted nature of the factors influencing the overall museum experience. It can also be seen that:

1. Under the behavior category (purple), "watching" and "eating" were the elements that were most closely related to "positive emotion". This suggests the high quality of exhibitions and catering services offered at the Kimbell Art Museum.
2. Under the emotion category (yellow), excluding "positive emotion", "sense of design" was the most impressive element and received the most reviews. This indicates that visitors also place significant importance on the architectural design of the Kimbell Art Museum.
3. Under the object category (green), art exhibitions, architectural design and art museum services received a lot of attention, among which "exhibition", "collection", "construction", and "space" were the core factors that affected the tour experience in the museum. Despite its extensive landscaped area, the museum falls short in landscaping, as evidenced by the preceding points.

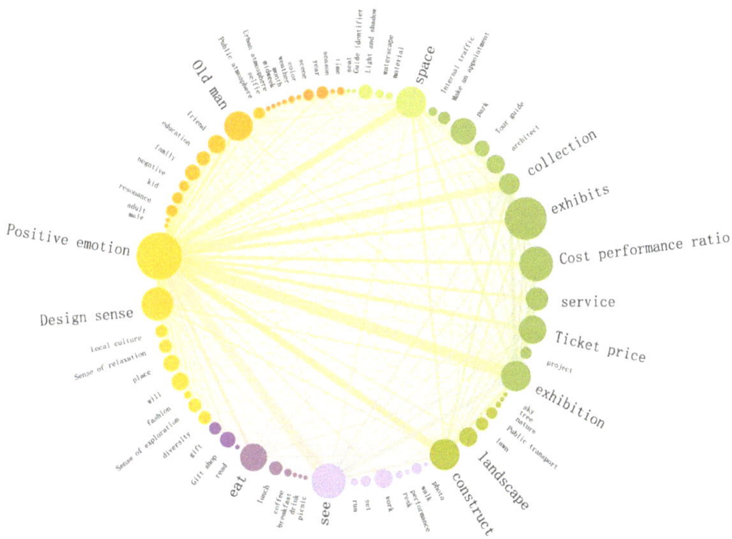

Figure 11. Gephi Ring Diagram (correlation between words).

The network diagram (Figure 12) provides insight into the group relationships that shape the core elements of the museum tour experience, offering a comprehensive view of the factors influencing visitor perceptions:

1. Visitors "positive emotion" mainly originated from the interconnections of "exhibition", "exhibits", and "collection" in the museum. These elements collectively contribute to fostering a sense of appreciation and enjoyment among visitors. The group of "construction", "sense of design", and "space" was the second source of positive emotion for visitors to the museum. The meticulous design and spatial arrangement within the museum environment enhance the visitors' overall experience. This further underscores the Kimbell Art Museum's excellence in exhibit curation and architectural design.
2. While "construction" garnered relatively less attention compared to "space", its association with "positive emotion" was notably stronger. This implies that the building

itself is seamlessly integrated into the visitor's exhibition experience, becoming a part of the exhibit.

3. The correlation between "landscape" and "walk" was not strong; visitors showed a low degree of willingness to comment on these two elements at one time. This observation implies that these aspects may not significantly impact visitors' overall tour experience.
4. "Weather", "season", and "month" were at the edge of the entire network, indicating that weather did not have much impact on the tour experience. And "Lawn", "sky", "trees", and "surrounding environment" were also at the edge of the network, indicating that outdoor landscape elements did not have much impact on the tour experience. In the context of the Kimbell Art Museum, the landscape area surpassed other functions significantly, suggesting a higher potential for visitor interaction. However, this indicates that the landscape design has not reached visitors' satisfaction levels.

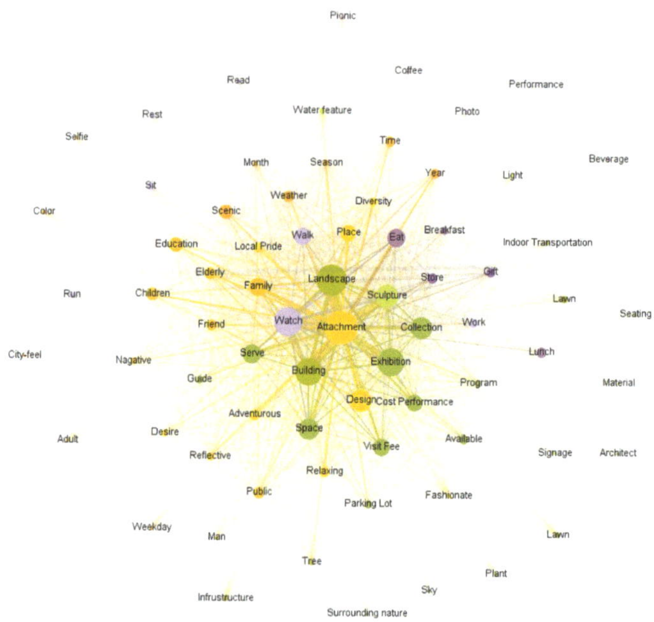

Figure 12. Gephi Network Diagram (focus of visitor reviews).

4. Conclusions

By combining the data analysis with the design features of the Kimbell Art Museum and conducting research from four perspectives of visitors' attention, temporal connection, space use, and overall connection [27], we can draw the following conclusions:

4.1. Spatial Layout

Kimbell Art Museum adopts a C-shaped spatial layout, and the ratio of exhibition area to rest area is 2:1 (Figure 13). According to the data analysis, in terms of visitors' attention, the words related to visiting behavior had the highest frequency of being mentioned, that is, visiting enjoyed the highest attention from the visitors. Among them, "appreciating" was the most common behavior. Mentions of "walking" and "rest" were lower, but they were almost even. In terms of temporal connection, the attention on walking and rest behavior influenced each other, maintaining a stable connection state of one high and the other low. It is known that exhibition occupies a dominant position in the behavior activities of the Kimbell Art Museum, and the attention on rest and walking was equal, which accords with

Chen Xiaotang (2016) in the post-occupancy evaluation of the museum: "A good rest space can help visitors recover their physical strength and improve their attention, so as to better finish the tour".

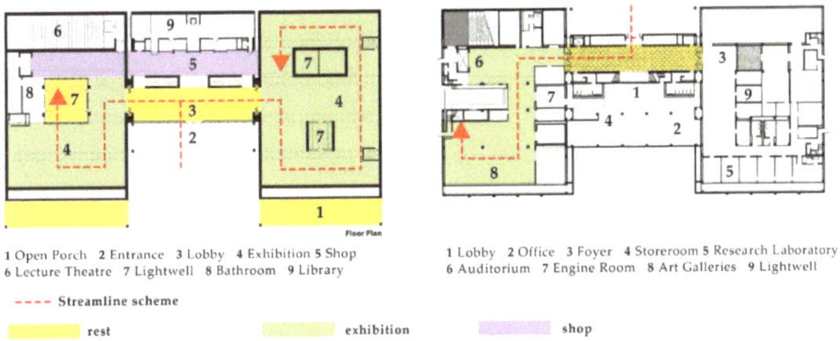

Figure 13. Floor Plan of the Kimbell Gallery (building layout and visitor routes).

4.2. Service Functions

The Kimbell Art Museum is equipped with service space, accounting for 17.37% of the total area of the venue (Figure 14). In terms of space evaluation intensity, "restaurant" (4.7 pieces/m^2) surpassed "exhibition" (1.9 pieces/m^2) to become the area with the highest space evaluation degree. In terms of temporal connection, "dining" and "shopping" both showed stable performance throughout the year, and the activity frequency did not fluctuate greatly with the change in the month. In terms of overall connection, "eating", second only to "seeing", was the most closely associated factor with "positive emotion" and is an important behavior in providing a quality sightseeing experience. It can be seen that the commercial attached space of the Kimbell Art Museum is well created, which improves the tour experience and has the potential for sustainable development [28].

Figure 14. Shopping and Dining Places.

4.3. Exhibition Environment Design

The Kimbell Gallery uses 16 unit vaults, two interior courtyards, and three light wells to bring in outdoor natural light (Figure 15) [29]. This design optimizes the viewing environment and seamlessly integrates with exhibits, contributing to its prominence among visitors, second only to functionality [30]. In terms of visitors' attention, showroom, which had 1561 comments, was the space with the highest number of comments. In addition, art exhibition, accounting for 44.52% of the attention, was the core visiting content of the museum. In terms of temporal connection, "exhibition" was the most stable among various factors in the object category. Unique activities and exhibitions avoid the off-season of visits brought about by seasonal changes. In terms of overall connection, "exhibition" in the object category was the core factor affecting the experience of art museums and was also the factor most closely associated with "positive emotion". It can be seen that the Kimbell Art Museum has a good exhibition environment, and visitors' attention is paid to proper places. According to the analysis, exhibition halls can be designed and opened around the lobby, park, catering area, and workshop to display exhibits of high attention. The exhibition area is used to presuppose people's tour paths to improve the overall attention of visitors to the non-exhibition space [31].

Figure 15. Exhibition Environment.

4.4. Architecture Appreciation

The Kimbell Art Museum is highly regarded in the field of architectural design. In terms of visitors' attention, visitors paid more attention to "construction" "(door, window, column, etc.) than "architecture" (light and shadow, material, etc.). Different from the analysis of light processing, spatial prototype, and form selection from a professional perspective, visitors were more interested in visible architectural elements such as structural style and decorative details [32]. There was little feedback on the overall appearance and design logic of the building. In terms of overall connection, the connection between "construction" and "positive emotion" was stronger than that of "space". However, in addition to being closely connected to "collection" and "exhibits", exhibition was mainly connected to "space", and "construction", "sense of design", and "space" constituted the second source of positive emotion for visitors in the museum. It is known that visitors are more sensitive to visible design elements in the process of visiting, and it is easier to obtain satisfaction through visible design elements. Therefore, formal and visible design should be properly retained in architectural details. Moreover, the frequency of "photography" behavior in the interior of buildings is low, and visitors are more inclined to perceive these buildings with their eyes. It can be considered that by combining with the building itself, an appropriate photography space can be created to guide visitors' attention to the building itself [33].

4.5. Landscape Design

The Kimbell Art Museum has a large landscape. (Figure 16) The construction area of the Kimbell Art Museum (9982 square meters) is similar to the landscape area (8600 square meters), but in terms of visitors' attention to the landscape, the visitors' attention was the lowest among those in the object category. In terms of temporal connection, visitors' evaluation of the landscape had an obvious peak in August, and August was the only month in which the rising trend of family visitors was higher than that of individual visitors. In terms of spatial evaluation intensity, the overall area of "landscape" was the largest, but the evaluation ratio was the lowest in the open space facing visitors. In terms of overall connection, only 3.5% of visitors' shared experiences of walking took place in the landscape. Among the 145 walking behaviors in the behavioral data, only 5 walking behaviors were directly connected with the landscape, and the landscape evaluations of "lawn", "sky", "trees", and "surrounding environment" were marginalized in the whole evaluation network. It can be seen that the spatial evaluation degree of the "landscape" area was very low, the temporal connection was strong, and it had almost no correlation with other behaviors such as exhibits and dining. The modernist landscape design effect of the Kimbell Art Museum failed to make good use of the advantages of the area, and its attraction to visitors is relatively weak. Food and beverage activities, commercial spaces, and exhibitions can be considered to break through the seasonal limitations and enhance the vitality of the landscape area.

Figure 16. Landscape of the Kimbell Art Museum.

5. Application Prospect

Big data mining technology and a semantic analysis concept are adopted in this study to establish an effective evaluation framework for the use of buildings and their landscapes (Figure 17), which has a broad application prospect.

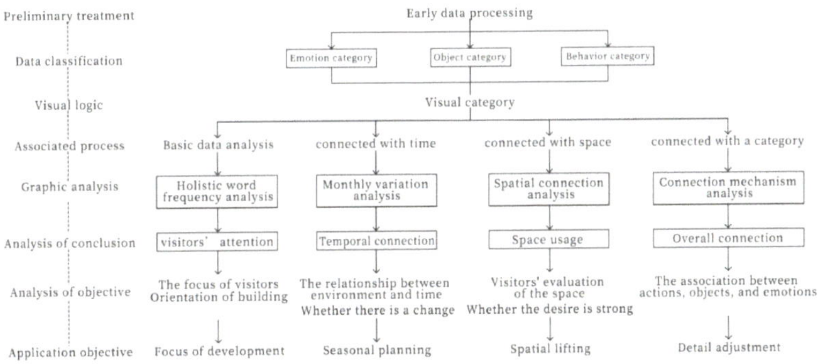

Figure 17. System of Analytical Methods for the Architecture Environment.

In terms of the pertinence of questions and the uniqueness of perspectives, the evaluation core of the traditional evaluation method has changed from the objective environment of buildings and preset questions of investigators to the spontaneous experience of visitors who have actually been to buildings. At the same time, it uses the rich diversity of big data to introduce a complex "multi-cause and multi-effect" investigation mechanism, which is different from the traditional evaluation method, which is dominated by accurate single evaluation. A mechanism is studied with the following characteristics: it has a change from structural to open and from top–down to bottom–top; it makes up for the shortcomings of soft experience and the spontaneity of visitors in traditional evaluation methods. It plays a specific role from a variety of design contents (objects) to a variety of design evaluation indexes (behavior and emotion) in massive data.

In terms of the scientificity of the method and the applicability of the object, the data source based on the subjective text evaluation on the network is more real and cumulative in both time and space [34]. It can cross geographical and time barriers and improve the efficiency of data collection and analysis. As computer technology can reduce labor time and speed up the research process, it can be applied to the built environment assessment of any public buildings with big data accumulation, especially libraries, shopping malls, exhibition centers, railway stations, etc., which have high exposure. Buildings such as the Kimbell Art Museum are public structures with inherent functionality, indoor spaces, and outdoor vistas. Visitor assessments pinpoint the connection between visitor activities and these components, enabling precise modifications. In different types of spaces, researchers can adjust the subitems in a macro category to form a new evaluation system to focus on spatial pain points and find usage problems to obtain systematic evaluation and design strategies.

By establishing the connection and continuous logic between the complex network data and the actual space use phenomenon, the semantic network technology quantitative analysis text, which has the advantages of large data volume, good accumulation, strong diversity, and high spontaneity, can be systematically applied to the post-occupancy evaluation of public buildings. It can improve the comprehensiveness and accuracy of the evaluation system and make the evaluation scope wider and deeper. The method of identifying problems from data and transforming them into design guidance is applied to provide more objective, detailed, and in-depth guiding value and theoretical support for the current demand analysis of public buildings and the future development trend.

Author Contributions: Investigation, X.Y.; Resources, D.L.; Writing—original draft, Z.Z. and M.W.; Writing—review & editing, B.Z. All authors have read and agreed to the published version of the manuscript.

Funding: This research received no external funding.

Data Availability Statement: The raw data supporting the conclusions of this article will be made available by the authors on request.

Conflicts of Interest: The authors declare no conflicts of interest.

References

1. Hutto, C.J.; Gilbert, E. VADER: A Parsimonious Rule-Based Model for Sentiment Analysis of Social Media Text. In Proceedings of the International AAAI Conference on Web and Social Media, Ann Arbor, MI, USA, 1–4 June 2014; AAAI Press: Palo Alto, CA, USA, 2014.
2. Simmons, R.F. Themes from 1972. In Proceedings of the 20th Annual Meeting on Association for Computational Linguistics, Toronto, ON, Canada, 16–18 June 1982; Association for Computational Linguistics: Toronto, ON, Canada, 1982; pp. 100–101.
3. Boiy, E.; Moens, M.-F. A Machine Learning Approach to Sentiment Analysis in Multilingual Web Texts. *Inf. Retr.* **2009**, *12*, 526–558. [CrossRef]
4. Zhu, X. *Research on the Subjective Evaluation Method of Built Environment*; Southeast University Press: Nanjing, China, 2005.
5. Vischer, J. Post-Occupancy Evaluation: A Multifaceted Tool for Building Improvement. *Learn. Our Build. A State Pract. Summ. Post-Occup. Eval.* **2002**, *6*, 23–34.
6. Xia, Z. From Post-occupancy Evaluation to Practice—Theory and Method. *World Archit.* **2014**, *8*, 18–19.
7. Wang, Z.; Zhuang, W. Research on the Post-occupancy Evaluation of the SD Method of Perceptual Evaluation Driven by Review Data—Taking Urban and Rural Historic Blocks as an Example. *New Archit.* **2019**, *4*, 38–42.
8. Wang, Z.; Zhuang, W. Research on Post renovation Evaluation of Old City Reconstruction Based on Review Data—Taking Beijing Bell and Drum Tower as an Example. *Huazhong Archit.* **2019**, *37*, 100–102.
9. Wang, T.; Yang, W. A Review of Research on Semantic Emotion Analysis Methods. *Comput. Eng. Appl.* **2021**, *57*, 11–24.
10. Corley, C.D.; Cook, D.J.; Mikler, A.R.; Singh, K.P. Text and Structural Data Mining of Influenza Mentions in Web and Social Media. *Int. J. Environ. Res. Public Health* **2010**, *7*, 596–615. [CrossRef] [PubMed]
11. Ramadhani, D.P.; Andry, A.; Febrianta, M.Y.; Damayanti, L.Z.A. Exploring Tourists' Behavioral Patterns in Bali's Top-Rated Destinations: Perception and Mobility. *J. Theor. Appl. Electron. Commer. Res.* **2024**, *19*, 743–773. [CrossRef]
12. Alaei, A.; Wang, Y.; Bui, V.; Stantic, B. Target-Oriented Data Annotation for Emotion and Sentiment Analysis in Tourism Related Social Media Data. *Future Internet* **2023**, *15*, 150. [CrossRef]
13. Song, J.; Chen, J.; Yang, X.; Zhu, Y. Research on Adaptive Reuse Strategy of Industrial Heritage Based on the Method of Social Network. *Land* **2024**, *13*, 383. [CrossRef]
14. Zhong, J.; Liu, W.; Wang, S.; Yang, H. Overview of Semantic Emotion Analysis Methods and Applications. *Data Anal. Knowl. Discov.* **2021**, *5*, 1–13.
15. Wu, D.; Yao, L. Sentiment Analysis of Online Comments through Semantic Networks and Rost Text Mining Software. In Proceedings of the 2021 IEEE Conference on Telecommunications, Optics and Computer Science (TOCS), Shenyang, China, 10–11 December 2021; pp. 869–872.
16. Becken, S.; Alaei, A.R.; Wang, Y. Benefits and pitfalls of using tweets to assess destination sentiment. *J. Hosp. Tour. Technol.* **2019**, *11*, 19–34. [CrossRef]
17. Bashynska, I.; Sarafanov, M.; Manikaeva, O. Research and Development of a Modern Deep Learning Model for Emotional Analysis Management of Text Data. *Appl. Sci.* **2024**, *14*, 1952. [CrossRef]
18. Chen, X. *Architects' Use of Post-occupancy Evaluation Methods and Practice in Museums*; South China University of Technology Press: Guangzhou, China, 2016.
19. Perrin, A. *Social Media Usage: 2005–2015*; Pew Research Center: Washington, DC, USA, 2015.
20. Mehta, V. Evaluating Public Space. *J. Urban Des.* **2014**, *19*, 53–88. [CrossRef]
21. Bitgood, S.C.; Loomis, R.J. Environmental Design and Evaluation in Museums. *Environ. Behav.* **1993**, *25*, 683–697. [CrossRef]
22. Žunić, A.; Bojana, Š.; Korina, B. Interaction between a Museum and a City in Europe. *Prostor* **2023**, *31*, 62–71.
23. Yao, Y. *Research on Subjective Evaluation Method of Post-Occupancy Built Environment Based on Network Reviews*; South China University of Technology Press: Guangzhou, China, 2020.

24. Lin, W.; Zhang, Q.; Wu, Y.J.; Chen, T.-C. Running a Sustainable Social Media Business: The Use of Deep Learning Methods in Online-Comment Short Texts. *Sustainability* **2023**, *15*, 9093. [CrossRef]
25. Giacomo Del, C.; Andreu, L.; Gallarza, M.G. Emotions and Visitors' Satisfaction at a Museum. *Int. J. Cult. Tour. Hosp. Res.* **2014**, *8*, 420–431.
26. Zhuang, W.; Mo, H. Overview of Basic Methods and Frontier Technologies of Post-occupany Evaluation of Buildings. *Time Archit.* **2019**, *4*, 46–51.
27. Nassef, M.; Mohammed, N.; Ibrahim, M. Investigating the Socio-Economic Sustainability within the Egyptian Museums over the Last Decade. *Sustainability* **2023**, *15*, 16746. [CrossRef]
28. Zhao, L. *Looking for the Lost Light and Shadow*; Dalian University of Technology Press: Dalian, China, 2003.
29. Kaçel, S.; Lau, B. Investigation of the Luminous Environment in Louis I. Kahn's Kimbell Art Museum. PLEA. 2013. Available online: http://web5.arch.cuhk.edu.hk/server1/staff1/edward/www/plea2018/plea/2013/Conference%20Papers/Session%20III/III.4%20Buildings%20-%20Daylight/investigation_of_the_luminous_environment_in_louis_i._kahn_s_kimbell_art_museum_.pdf (accessed on 6 May 2024).
30. Vermeeren, A.P.O.S.; Shih, H.-C.; van der Laan, R.; Calvi, L.; Yoon, J.; Keller, I. Designing Trajectories of Experiences: In Museums, around Museums, or Including Museums. In *Museum Experience Design: Crowds, Ecosystems and Novel Technologies*; Vermeeren, A., Calvi, L., Sabiescu, A., Eds.; Springer International Publishing: Cham, Germany, 2018; pp. 301–323.
31. Di Pietro, L.; Mugion, R.G.; Renzi, M.F.; Toni, M. An Audience-Centric Approach for Museums Sustainability. *Sustainability* **2014**, *6*, 5745–5762. [CrossRef]
32. Canbakal, A.; Nihan, A.Y.; Acar, H. Design Approaches to Museum Open Spaces with User Evaluations. *ICONARP Int. J. Archit. Plann.* **2023**, *11*, 397–427. [CrossRef]
33. Pop, I.L.; Borza, A. Factors Influencing Museum Sustainability and Indicators for Museum Sustainability Measurement. *Sustainability* **2016**, *8*, 101. [CrossRef]
34. Couldry, N. *Media, Society, World: Social Theory and Digital Media Practice*; Wiley & Sons: Hoboken, NJ, USA, 2012.

land

MDPI

Article

Assessing Inequality in Urban Green Spaces with Consideration for Physical Activity Promotion: Utilizing Spatial Analysis Techniques Supported by Multisource Data

Yunjing Hou, Yiming Liu, Yuxin Wu and Lei Wang *

College of Landscape Architecture, Northeast Forestry University, Harbin 150040, China; houyj@nefu.edu.cn (Y.H.); lymylxy@nefu.edu.cn (Y.L.); wuyuxin@nefu.edu.cn (Y.W.)
* Correspondence: wanglei@nefu.edu.cn

Abstract: Urban green spaces (UGSs) play a significant role in promoting public health by facilitating outdoor activities, but issues of spatial and socioeconomic inequality within UGSs have drawn increasing attention. However, current methods for assessing UGS inequality still face challenges such as data acquisition difficulties and low identification accuracy. Taking Harbin as a case study, this research employs various advanced technologies, including Python data scraping, drone imagery collection, and Amap API, to gather a diverse range of data on UGSs, including photos, high-resolution images, and AOI boundaries. Firstly, elements related to physical activity within UGSs are integrated into a supply adjustment index (SAI), based on which UGSs are classified into three categories. Then, a supply–demand improved two-step floating catchment area (SD2SFCA) method is employed to more accurately measure the accessibility of these three types of UGSs. Finally, using multiple linear regression analysis and Mann–Whitney U tests, socioeconomic inequalities in UGS accessibility are explored. The results indicate that (1) significant differentiation exists in the types of UGS services available in various urban areas, with a severe lack of small-scale, low-supply UGSs; (2) accessibility of all types of UGSs is significantly positively associated with housing prices, with higher-priced areas demonstrating notably higher accessibility compared to lower-priced ones; (3) children may be at a disadvantage in accessing UGSs with medium-supply levels. Future planning efforts need to enhance attention to vulnerable groups. This study underscores the importance of considering different types of UGSs in inequality assessments and proposes a method that could serve as a valuable tool for accurately assessing UGS inequality.

Keywords: urban green spaces; physical activity promotion; inequality; multiple data sources; improved 2SFCA method

Citation: Hou, Y.; Liu, Y.; Wu, Y.; Wang, L. Assessing Inequality in Urban Green Spaces with Consideration for Physical Activity Promotion: Utilizing Spatial Analysis Techniques Supported by Multisource Data. *Land* **2024**, *13*, 626. https://doi.org/10.3390/land13050626

Academic Editors: Xiwei Shen, Yang Song, Niall Kirkwood, Mary Padua, Bo Zhang and Rosalea Monacella

Received: 6 April 2024
Revised: 1 May 2024
Accepted: 4 May 2024
Published: 7 May 2024

1. Introduction

Rapid urbanization in China has brought various challenges to public health, despite improving people's living standards. For instance, high urbanization rates may lead to a decrease in residents' physical activity levels, indirectly contributing to the onset of various chronic diseases [1]. Urban green spaces (UGSs) serve as primary spaces for residents' outdoor physical activities, facilitating such activities through the provision of facilities and natural environments [2]. Therefore, ensuring that all residents have convenient access to UGSs is of paramount importance for comprehensively safeguarding their right to health.

Numerous studies have shown that there is inequality in the distribution of UGSs in urban areas. Research on this inequality primarily focuses on two dimensions: spatial and social. Spatial inequality of UGSs mainly refers to the mismatch between the distribution of UGSs in cities and the population [3,4]. Studies conducted in different countries and cities have revealed diverse spatial inequality patterns and causes [3,5–8]. Contradictions between construction costs and local finances are one of the main reasons for spatial inequality [9]. On the one hand, due to limited urban land resources for UGSs and profit-driven land allocation policies, UGS construction often lags behind other land uses such as commercial and residential areas [10]. Considering land prices, UGSs are often forced to be constructed in low-cost but sparsely populated suburbs [11], resulting in spatial mismatch with the population. On the other hand, local governments may view UGS construction as an important means to attract investment and achieve economic growth in urban planning, potentially leading to an uneven distribution of UGSs across different urban areas and further exacerbating issues such as gentrification [4,12,13]. From this perspective, how to optimize the spatial layout of UGSs at a low cost and high efficiency is an important challenge in addressing spatial inequality in UGSs.

The unreasonable distribution of UGSs may further result in unequal distribution among different social groups, namely, the socioeconomic inequality of UGSs [3,14]. Numerous studies have shown that there is inequality in access to UGSs among different income groups [12], races [14,15], and occupations [16]. Much of the research has focused on income-based UGS inequality, with lower-income groups often having fewer opportunities to access UGSs [7,12]. Particularly in developing countries like China, UGSs significantly influence housing prices in communities, making it difficult for lower-income groups to afford high-quality living environments and thus excluding them. Additionally, there is growing attention toward UGS inequality among different age groups. Due to physical limitations, children and the elderly are more vulnerable to environmental threats [17]. On one hand, issues like obesity, anxiety, and depression pose potential threats to children's mental and physical health, which could be mitigated by access to UGSs [18,19]. On the other hand, with urban areas facing an increasingly aging population, the elderly become an important user group of UGSs [20]. Access to UGSs is crucial for reducing the incidence of certain diseases and enhancing the well-being of the elderly [20,21].

In order to propose reasonable planning strategies, comprehensive assessments of inequality need to be conducted based on multiple spatial data sources using various models and analysis techniques. Recent studies have begun to utilize more advanced technologies to facilitate data acquisition and enhance the accuracy of results. Firstly, Python scripting proves to be a useful tool for efficiently gathering large volumes of data in a short time and has been widely employed in UGS-related research. For example, Liu et al. [22] conducted a cross-cultural comparison of UGS perceived quality using social media data collected through Python. Similarly, Zhang et al. [22] employed Python-based image data collection to assess UGS quality and measure accessibility. Secondly, real-time navigation and route planning have become valuable tools for obtaining accurate travel time data. For instance, Zhang et al. [23] utilized web service APIs from the Amap open platform to collect and analyze travel time data using Python scripts. Chen et al. [7] developed a method for measuring accessibility based on the Amap API, enabling a more precise assessment of UGS accessibility in Shanghai. Furthermore, unmanned aerial vehicle (UAV) observation technology has been employed in studies such as vegetation surveys and green roof observations [24,25], demonstrating significant potential for obtaining high-resolution images to accurately identify UGS structures.

Regarding assessment methods, the spatial inequality of UGSs is typically characterized using accessibility, which measures the opportunity for residents to access public service facilities. The two-step floating catchment area (2SFCA) method [26,27] is a commonly used accessibility measurement model, considering both facility supply and demand size as well as travel distance costs, and has been widely applied in recent research. Scholars have proposed several improved models to enhance the accuracy of results, including measurements of distance attenuation effects [3], improvements to fixed catchment areas [28,29], and enhancements to fixed transportation modes [30,31]. Additionally, other scholars have taken into account the competitive effects among multiple facilities. Prior to the 2SFCA method, they introduced the calculation of selection weights, resulting in the 3SFCA method [32]. Furthermore, Luo [33] considered the service capacity of facilities and distance costs. They introduced the Huff model [34] to further optimize the calculation method of selection weights, resulting in the H2SFCA method.

While considerable progress has been made in studying the inequality of UGSs, there are still some shortcomings in existing research. Firstly, there is limited research focusing on the assessment of accessibility to different types of UGSs. In fact, UGSs, based on differences in internal structure, have varying construction costs and may provide differentiated service functions. A precise assessment of accessibility to different types of UGSs can help to identify the types of services lacking in different areas, thereby achieving low-cost and high-efficiency layout optimization. Secondly, existing 2SFCA models still have several shortcomings: (1) the measurement of UGS supply scale only uses area as a measure, without considering the internal structural elements of UGSs, which may lead to inaccurate measurements of accessibility; (2) existing 3SFCA and H2SFCA models also overlook the impact of UGS structural elements on the calculation of resident selection weights; (3) the use of distance measurement ignores the travel speed of residents when using different modes of transportation; (4) typically, the centroid of a UGS is used as the supply point, ignoring the distance between the centroid of a larger-scale UGS and its entrance. Finally, there is limited research on the association between accessibility and urban spatial structure factors [8]. More in-depth research is needed to explain the causes of accessibility from multiple perspectives.

To address these shortcomings, this study first obtained basic data using advanced technologies such as Python scripting, UAV imagery acquisition, and the Amap API. Then, the 2SFCA model was improved from both supply and demand aspects, and spatial and socioeconomic inequalities of different types of UGSs were assessed separately (Figure 1). In conclusion, this study aims to provide feasible solutions for mitigating UGS inequality at a low cost and high efficiency through more accurate assessment methods of UGS inequality.

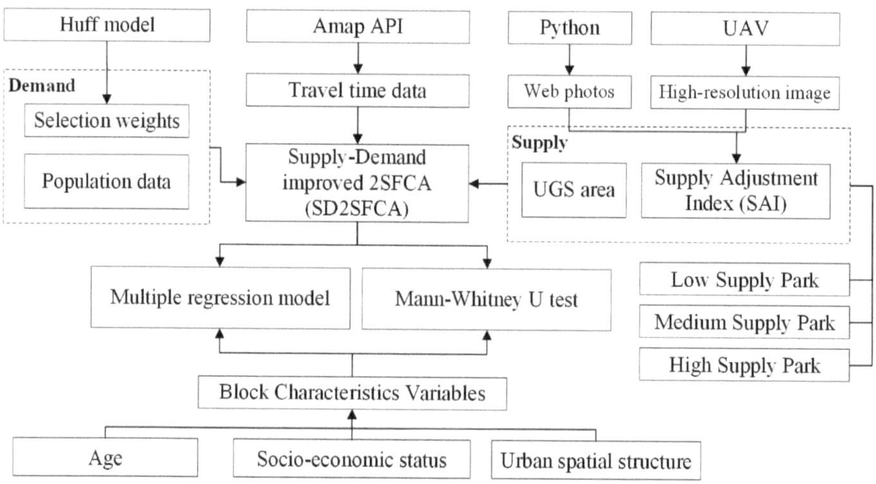

Figure 1. Research framework diagram.

2. Materials and Methods

2.1. Study Area

Harbin, located in northeastern China, is a significant large-scale industrial city in the country. This study selected the main urban area of Harbin as the research area (Figure 2). There are several reasons for this selection. Firstly, compared to surrounding areas, the main urban area has relatively complete urban planning and construction. It concentrates a significant number of UGS resources in the region with a high population density, making it highly meaningful for studying UGS accessibility. Second, existing UGSs in this area exhibit significant differences in scale and internal structure, aligning with the objectives of our study.

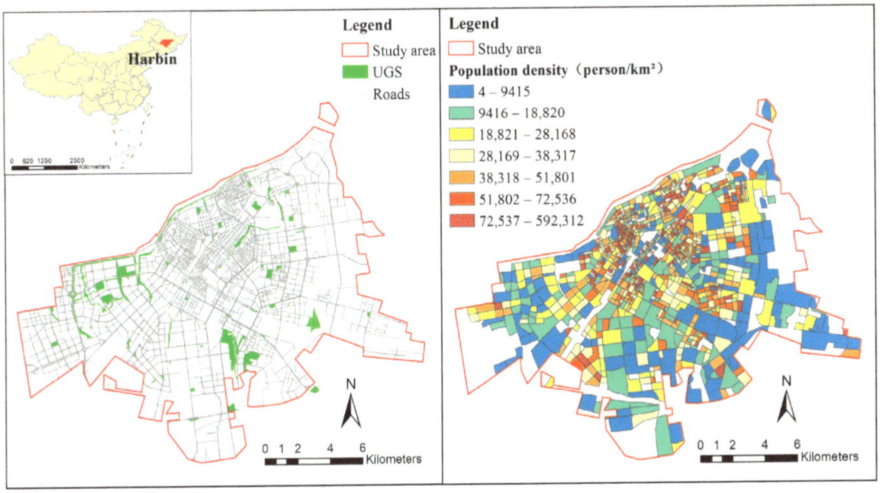

Figure 2. Study area.

2.2. Data Sources and Processing

2.2.1. UGS Boundary and Entrance Data

UGS boundary and entrance data were obtained from the Amap Open Platform (https://lbs.amap.com/, accessed on 2 April 2023). UGSs selected for this study met the following criteria: (1) had a certain area of hard surface available for activities, (2) had any type of vegetation, and (3) were free and open to the public. A total of 133 UGS areas were identified within the study area, including parks, greenways, and some greened plaza areas. We developed a Python program to collect the AOI boundary data of UGSs based on the Amap Open Platform API.

High-resolution satellite images of Harbin city were obtained from LocaSpaceViewer (LSV) (using Google Maps 2022 edition) and used as a base map. The entrances of each UGS were manually extracted on the ArcGIS platform. This was carried out because larger UGSs may have multiple entrances, and the distances between these entrances and the centroid can be considerable. This approach improves the accuracy of travel time calculations. For UGSs without clearly defined entrances (such as squares), the centroid was used as a representative supply point.

2.2.2. Web Photos and High-Resolution Aerial Imagery of UGSs

Web photos were sourced from Baidu Images and Dianping.com, accessed on 22 June 2023. A Python program was used to batch collect UGS photos, totaling 5425 images. For street corner parks, waterfront green spaces, and other areas without image sources, images were obtained from Baidu Street View.

Imagery data were obtained from high-resolution images captured by unmanned aerial vehicles (UAVs) in June 2023. After the UAVs autonomously flew and captured images according to programmed routes, the images were manually stitched together. The resulting image data will be used to assess the types and quantities of plants and water features within each UGS. In compliance with China's Interim Measures for the Administration of Unmanned Aerial Vehicles, flight permits were not obtained for certain residential, commercial, and industrial areas within the main urban area. For UGSs in these areas, manual identification and assessment were conducted based on Google satellite imagery and street view maps.

2.2.3. Population Data

Block-level analysis units were used in this study instead of administrative districts, facilitating the exploration of micro-level distribution patterns of UGS accessibility and more accurately identifying underserved areas. Basic population data were derived from the 7th National Population Census data (street-level) from 2020. Housing estate data for Harbin City (as of April 2023) were obtained from the Anjuke website (https://www.anjuke.com/, accessed on 6 April 2023) using Python. Utilizing housing estate data as auxiliary information, the population data at the subdistrict level were spatialized to the block units based on the proportion of total households within blocks to the total households within subdistricts.

2.2.4. Road Network and Travel Time Data

Road network data were sourced from the OpenStreetMap (OSM) platform. Walking was selected as the mode of travel in this study, as it is the most common mode of transportation for residents. Walking time was used instead of walking distance to more accurately reflect the real impedance between supply points and demand points. The Amap Web Service provides a routing API that offers walking, public transportation, driving query, and distance calculation interfaces in HTTP format. This method provides optional routes and travel times based on real-time traffic conditions, road networks, and modes of transportation [7]. Leveraging this functionality, walking time data from the starting point (block centroid) to the destination (UGS entrance) were obtained.

2.3. *A UGS Supply Adjustment Index (SAI) Considering Physical Activity Promotion Function*

2.3.1. Evaluation Framework of SAI

This study primarily considered the physical activity promotion function of UGSs and utilized a supply adjustment index (SAI) to adjust the calculation process of supply size. The evaluation index was derived from summarizing various urban green space quality evaluation tools [22,35–38] and previous literature on UGS impact on physical activity (Table 1). It mainly includes three indicator dimensions: facilities [2,39,40], natural environment [41,42], and safety [43,44], as well as nine assessment indicators such as walking trails, physical activity facilities, and recreational facilities. Among these, footpaths and sports facilities are essential prerequisites for promoting physical activity [2,39], while an adequate provision of leisure facilities, vegetation, and water features can indirectly attract residents to engage in physical activity within UGSs [2]. Additionally, sufficient safety conditions are an effective guarantee for physical activity [2,45].

Table 1. Evaluation indicator system of SAI based on multiple data sources.

Dimension	Element	Description	Scoring Method
Facility Conditions	Walkways	Pathways for slow walking in the park	Existence score
	Sports facilities	Fitness areas, dance squares, basketball courts, etc.	Category count
	Leisure facilities	Cafes, pergolas, umbrella seats, etc.	Category count
Natural Conditions	Vegetation	Lawns, dense forests, sparse forests, tree arrays, etc.	Category count
	Water features	Fountains, streams, artificial lakes, etc.	Category count
Safety Conditions	Barrier-free facilities	Ramps, tactile paving, etc.	Existence score
	Lighting facilities	Ground lights, street lights, etc.	Existence score
	Traffic	Separation of pedestrians and vehicles	Existence score
	Security measures	Surveillance cameras, security personnel, etc.	Existence score
Total Score	\	\	Sum of all scores

For each indicator, two scoring methods were adopted: existence score (1 point if the content exists, 0 points otherwise) and category count (1–3 points if 1–3 types exist, 4 points if 4 or more types exist, and 0 points otherwise). Three staff members simultaneously observed UGS photos and evaluated and scored them based on the evaluation indicators.

For each indicator of each UGS, the average score of all staff members' scores was taken as the final score for that indicator. After scoring, the scores of indicators in each dimension were summed to obtain the total quality score for each park. The ratio of the total quality score of each UGS to the total score of all indicators was used as the final SAI.

2.3.2. Classification of UGSs Based on SAI

Based on the SAI, this study categorizes UGSs into three types to reflect the supply level of elements related to physical activities within UGSs. (1) Low-supply park (LSP; $q \leq 0.3$): these UGS provide fewer types of services and are mainly used for residents' daily activities nearby. They have advantages such as low construction costs, flexible site selection, and convenient use. (2) Medium-supply park (MSP; $0.3 < q \leq 0.6$): these green spaces can offer more diverse functions, such as basketball courts, children's play facilities, fitness areas, etc., to meet the needs of different age and interest groups. They have higher construction and maintenance costs and can serve a larger range of resident activities. (3) High-supply park (HSP; $q > 0.6$): these green spaces have the richest functions and facilities. In addition to basic leisure functions, they may also include larger-scale activity spaces, such as large sports fields. They have the highest construction and maintenance costs and may attract a larger range of residents for activities.

It is important to emphasize that the three types of UGSs have different construction costs and focus on functions. However, this does not mean that UGSs providing fewer services are inferior to other UGSs. In fact, it is unrealistic to demand that all UGSs provide rich services in a city with limited resources. The key is to ensure a reasonable spatial distribution of various types of UGSs. At this level, measuring the accessibility of the three types of UGSs separately can help to more accurately identify the types of services lacking in each neighborhood, thereby achieving higher transformation benefits with lower costs as much as possible.

2.4. The Supply–Demand Improved 2SFCA (SD2SFCA) Method Considering Physical Activity Promotion

2.4.1. Improvement of Supply Scale

The traditional 2SFCA model uses area to represent the supply scale, which overlooks the internal components of UGSs and may result in inaccurate measurements of accessibility. In this study, the calculation method of the supply scale was adjusted based on the SAI of the UGS. The formula is as follows:

$$S_j^A = S_j * q_j$$

where S_j^A is the comprehensive supply scale of UGS j, S_j is the total area of j, and q_j is the supply adjustment index of j.

2.4.2. Improvement of Demand Scale

The traditional 2SFCA model uses population count to represent demand magnitude. Critics argue that this overlooks the competitive interaction between multiple facilities [32], as residents may only choose a facility that they find more satisfactory. When a resident's demand has been met by a particular facility, it should be subtracted from the overall demand. Accordingly, Wan et al. [32] introduced the calculation of choice weights before the 2SFCA method, resulting in the 3SFCA method. Luo [33] considered the service capacity of facilities and distance costs, introducing the Huff model [34] to further optimize the calculation of choice weights, resulting in the H2SFCA method. This method has been shown to reduce overestimation of accessibility and improve result accuracy [46,47]. However, these improvements still do not consider the influence of internal compositional elements of facility points. Therefore, this study introduces the proposed SAI into the Huff model to measure residents' choice weights for supply points. Additionally, the model in this study uses travel time based on road networks instead of traditional travel distance, aiding in a more accurate estimation of travel costs. The formula for calculating choice weights is as follows:

$$Prob_i = \frac{S_j^A t_{ij} G(t_{ij}, t_0)}{\sum_{k\in\{t_{kj}\le t_0\}} S_j^A t_{ij} G\left(t_{kj}, t_0\right)}$$

$$G(t_{ij}, t_0) = \begin{cases} \frac{e^{-\frac{1}{2}\times\left(\frac{t_{ij}}{t_0}\right)^2} - e^{-\frac{1}{2}}}{1-e^{-\frac{1}{2}}}, & t_{ij} \le t_0 \\ 0, & t_{ij} > t_0 \end{cases}$$

where $Prob_i$ is the probability of resident point i choosing park j, t_{ij} is the travel time from i to j, t_0 is the time threshold for the specified search range, and $G(t_{ij}, t_0)$ is the Gaussian decay function, with other parameters having the same meanings as above.

Next, centered at each supply point, the total population demand within the catchment area is calculated, so the supply–demand ratio of j is

$$R_j = \frac{S_j^A}{\sum_{k\in\{t_{kj}\le t_0\}} Prob_i G\left(t_{kj}, t_0\right) D_k}$$

where R_j is the ratio of the facility scale at supply point j to the population served within the search radius (t_0) and D_k represents the demand scale at demand point k, represented by the total population of k, with other parameters having the same meanings as above.

Finally, for each demand point i, search all supply points j within the time threshold range t_0 of i, summing R_j for each demand point, and adjusting the summation process using the selection probability and Gaussian function, resulting in the accessibility A_i at point i:

$$A_i = \sum_{j\in\{t_{ij}\le t_0\}} Prob_i G(t_{ij}, t_0) R_j$$

2.4.3. Improvement on Fixed Travel Time Threshold

Drawing upon previous research, residents may be willing to spend more time traveling to UGSs that offer additional services [28,47]. Conversely, for UGSs with fewer amenities, residents may incur shorter time costs. Therefore, a variable time threshold was used instead of the traditional fixed time threshold. The walking time thresholds for LSPs, MSPs, and HSPs were set to 10 min, 20 min, and 30 min, respectively, to more accurately reflect residents' travel behavior preferences.

2.5. Association between Block Characteristics and UGS Accessibility

After analyzing the spatial differences in accessibility, we further aimed to identify the specific associations between different block characteristics and UGS accessibility. Drawing on previous research [6,48,49], this study selects eight variables from three dimensions of block characteristics: population age structure, socioeconomic status, and built environment features (Table 2).

The selection of these variables is driven by three reasons. Firstly, we consider the proportion of elderly and adolescent populations in blocks to observe whether they face UGS inequality, as they are often considered high-demand groups for UGSs. Secondly, we examine the association between block socioeconomic status and accessibility using average housing prices. Thirdly, block population density, building density, and green space ratio represent land use characteristics, while the age of construction of blocks primarily reflects their relative age, and the distance to the city center reflects block location conditions. These built environment variables can to some extent reflect the level of urban development and may further influence UGS accessibility.

Initially, a multiple linear regression model was established, with the eight block characteristic factors as independent variables and the accessibility of various types of UGSs as the dependent variable, to examine potential influencing factors on accessibility from a global perspective. Subsequently, a focused analysis was conducted on the association between accessibility and socioeconomic status. Housing prices were divided into four levels (low, medium-low, medium-high, and high) based on quartiles, and the accessibility of UGSs and median values of block characteristic variables for each price level were calculated. The Mann–Whitney U test was then employed to compare the accessibility of UGSs at different price levels pairwise. Data analysis was conducted using SPSS 21.0 and Geoda 1.14.0 software.

Table 2. Potential influencing factors of accessibility.

Indicator	Variable	Explanation
Age	Proportion of old people (%) Proportion of children (%)	Proportion of population aged 65 and above within each block Proportion of population aged 16 and below within each block
Socioeconomic status	Housing price	Average housing price within each block
Built environment features	Population density Building density Greenery rate Block age (year) Distance to the city center	Total population quantity divided by block area Building footprint area divided by block area Total green space area within block divided by block area Median construction year of residential areas within each block Euclidean distance from block centroid to regional center

3. Results

3.1. Classification Results of UGS Based on SAI

Table 3 and Figure 3 present the classification results and spatial distribution of the 133 UGS within the area, respectively. Overall, UGS distribution is dense in the western part of the city, with a variety of types, while UGSs in the south and east are relatively sparse. MSPs are the most numerous, accounting for 48.87% of the total, concentrated in the northeast. LSPs have the fewest numbers, accounting for 24.06%, mostly distributed in the west and central areas. HSPs are concentrated in the west, followed by the southeast.

Table 3. Calculation results of SAI and classification results statistics.

	N	Minimum	Maximum	Mean	Sum	Std. Devitation
LSP	32	0.14	0.29	0.25	8.00	0.04
MSP	65	0.33	0.52	0.41	26.67	0.07
HSP	36	0.67	0.90	0.70	25.19	0.09

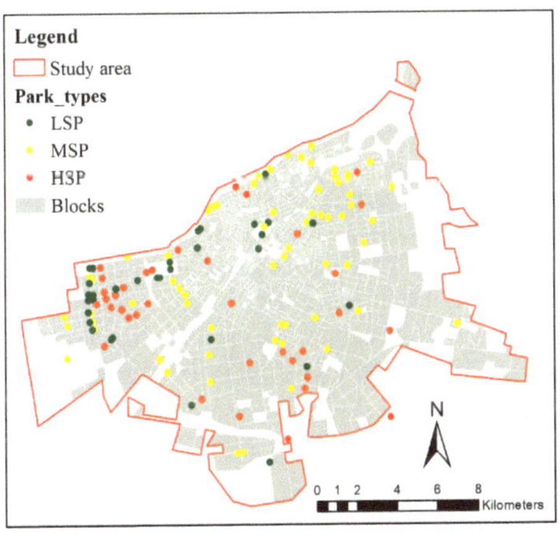

Figure 3. Spatial distribution of three types of UGSs.

3.2. Measurement Results of UGS Accessibility

3.2.1. Differences in Accessibility of Three Types of UGSs

Figure 4 illustrates the accessibility of overall UGSs (OP) as well as the accessibility results for three different types of UGSs. For the overall UGSs, high accessibility values are concentrated in the western, riverside, southeastern, and eastern areas within the second ring road of the city. Accessibility is generally lower in the southern and central parts of the city. Although these areas have a small number of UGS resources, their overall area and internal facilities are insufficient to match the higher population demands. Accessibility is poorest in the eastern outskirts, where obtaining any services is nearly impossible.

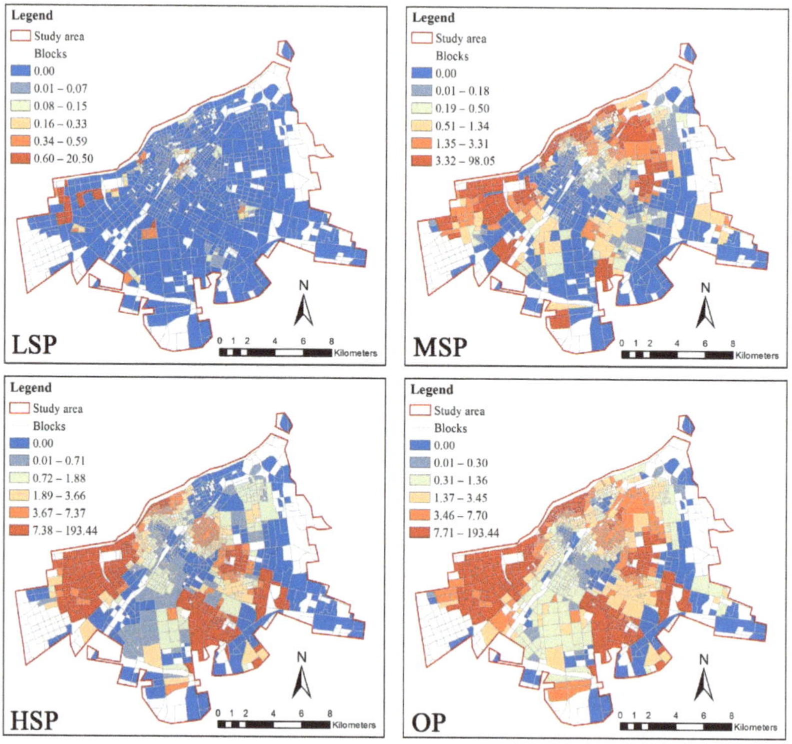

Figure 4. Spatial distribution of accessibility to three types of UGSs.

When considering the accessibility of the three types of UGSs separately, the results show that only about 11.72% of blocks can reach LSPs (Table 4), mainly distributed in the western and central parts of the city. The proportion of blocks served by MSPs is 62.18%, with accessibility generally decreasing from the suburbs toward the central urban areas, with high values concentrated in the western, riverside, and northeastern parts of the city. HSPs can serve 73.34% of blocks. Their accessibility distribution is more dispersed, with high values concentrated in the western, riverside areas, eastern, and southeastern parts.

It is evident that the number and distribution of LSPs within the region are severely unreasonable, and a significant portion of the high values in overall accessibility largely depend on HSPs. However, the number of HSPs is limited and not sufficient to serve a wider range of blocks. If only the overall accessibility of UGSs is evaluated, it will be difficult to identify the specific types of services that blocks truly lack.

Table 4. Descriptive statistics of UGS accessibility.

	Min	Max	Mean	Median	Standard Devitation	Serviced Blocks	Serviced Area/km^2	Serviced Population
LSP	0.00	20.50	0.11	0.00	1.04	127	12.14	390,967
MSP	0.00	98.05	1.60	0.18	4.82	674	95.52	2,553,738
HSP	0.00	193.44	5.25	1.33	13.01	795	110.42	2,965,519
OP	0.00	193.44	6.96	2.37	15.43	940	136.74	3,611,683

3.2.2. Classification of Areas with Inadequate Services

Based on the results of accessibility assessments, further classification of areas with inadequate services was conducted. Firstly, an overall classification was conducted, as shown in Figure 5a.

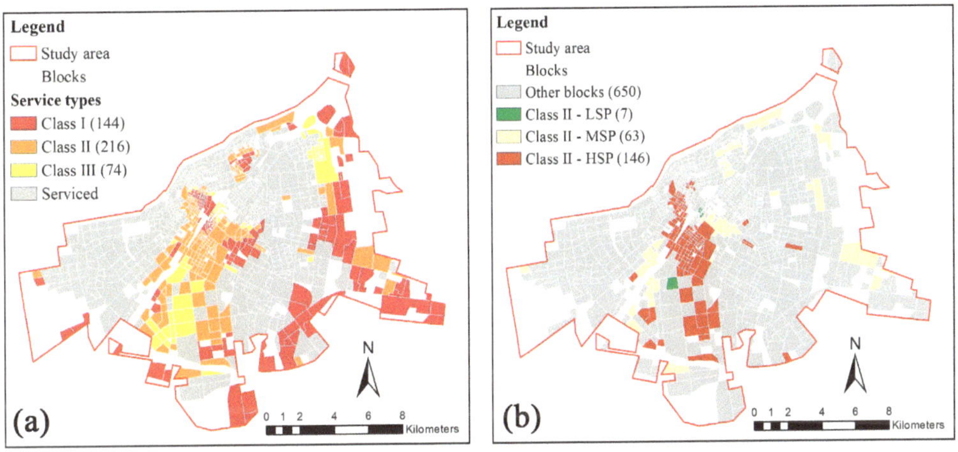

Figure 5. Classification of areas with inadequate services. (**a**) The overall classification of underserved areas; (**b**) Further classification of Class II.

Class I: Represents blocks with no access to any services, accounting for approximately 13.28% of the total. They are mainly distributed in the central, eastern, and southern suburban areas.

Class II: Represents blocks with access to only one class of UGS service, with relatively low accessibility, accounting for approximately 19.93% of the total. They are mainly distributed in the southern region. These blocks have relatively limited access to UGS classes, which may create significant service pressure and may not meet residents' diverse usage preferences. Subclassification of class II results is shown in Figure 5b. Most blocks in the south can only access HSPs with limited opportunities, and a very small number of blocks can only access LSPs. In contrast, a few blocks in the east, central, and north can only access MSPs.

Class III: Represents blocks with access to two or more classes of UGS services, but with relatively low accessibility. This is mainly due to the limited number of UGS classes, and the overall supply scale does not match the higher population demand. This class accounts for approximately 6.83% of the total and is mainly distributed in the southern and northeastern parts of the city.

3.3. Association between UGS Accessibility and Block Characteristics

3.3.1. Potential Influencing Factors of UGS Accessibility

Table 5 shows that the variance inflation factor (VIF) for each explanatory variable is less than 7.5, indicating the absence of multicollinearity between variables. For overall UGS accessibility, the proportion of elderly population, housing prices, block greenery rate, and distance to the city center are significantly positively associated with accessibility, while block population density and construction year are significantly negatively associated with accessibility.

Table 5. Multiple linear regression results.

Independent Variables	VIF	Standardized Coefficient			
		OP	LSP	MSP	HSP
Proportion of old people	1.352	0.124 ***	−0.031	0.150 ***	0.094 **
Proportion of children	1.257	−0.057	−0.019	−0.066 *	−0.042
Housing price	1.409	0.378 ***	0.089 *	0.399 ***	0.294 ***
Population density	1.224	−0.113 ***	−0.007	−0.083 **	−0.103 ***
Building density	1.470	0.010	−0.059	−0.015	0.022
Greenery rate	1.223	0.14 ***	−0.016	0.040	0.152 ***
Block age	1.462	−0.066 *	−0.019	−0.022 *	−0.069 *
Distance to the city center	1.378	0.283 ***	0.072 *	0.260 ***	0.234 ***
Adjusted R^2	-	0.208	0.017	0.168	0.149
p-value of the model	-	0.000	0.001	0.000	0.000

Note: * $p < 0.05$, ** $p < 0.01$, *** $p < 0.001$.

When considering different quality levels of UGSs, housing prices and distance to the city center have stable positive associations with all quality levels of UGSs. The block greenery rate has a significant positive association with HSP accessibility. Block age and population density have weak negative associations with MSP and HSP accessibility. The proportion of elderly population has a significant positive association with MSP and HSP accessibility, while the proportion of children population has a weak negative association only with MSPs. The block greenery rate has a significant positive association with HSP accessibility.

This indicates that children may be at a disadvantage in accessing adequate MSP services, while the elderly population may have a more matched accessibility to UGSs. Blocks with higher socioeconomic status, farther distance from the city center, newer construction age, and lower density are more likely to access UGSs. Blocks with higher greenery rates are also more likely to access HSPs.

3.3.2. Disparities in UGS Accessibility and Block Characteristics Based on Housing Price Levels

As shown in Table 6, compared to low-priced blocks, high-priced blocks have a lower proportion of elderly population and a higher proportion of children population. However, high-priced blocks exhibit a significantly lower population density and building density. The construction age of high-priced blocks is also generally later, and they tend to have higher greenery rates compared to low-priced blocks. Low-priced blocks may generally be situated farther from the city center. Additionally, the median MSP accessibility in low-priced blocks is slightly higher than in middle-low-priced blocks. This may be due to the fact that low-priced blocks are located in areas farther from the city center, where more UGSs are built, leveraging the surrounding natural resources.

The results of the Mann–Whitney U test in Table 7 indicate significant differences in the accessibility of MSPs and HSPs between blocks with high housing prices and those with other housing price levels. The median accessibility of UGSs in blocks with high housing prices is consistently the highest. Conversely, the accessibility of HSPs in blocks with low housing prices is significantly lower than that in the other three types of blocks. It is noteworthy that, although blocks with both high and low housing prices have relatively high median green coverage rates, the accessibility in blocks with low housing prices is significantly lower than that in blocks with high housing prices. This could be attributed to blocks with low housing prices having many underutilized potential UGSs, such as affiliated green spaces and protective green spaces.

Table 6. Block characteristics variables statistics based on housing price levels.

	(a) Low	(b) Low-Middle	(c) Middle-High	(d) High
Proportion of old people (%)	24.55	25.30	25.18	19.72
Proportion of children (%)	9.29	9.11	9.07	11.29
Population density (person/km^2)	27,726	42,947	33,893	25,941
Building density	0.49	0.52	0.53	0.38
Greenery rate	0.13	0.10	0.10	0.13
Block age (year)	23	25	24	18
Distance to the city center (m)	6836.77	4716.20	4179.25	5173.10

Table 7. Mann–Whitney U test results for UGS accessibility based on housing price levels.

	LSP	MSP	HSP	OP
(a) Low	0.00 (d **)	0.15 (d ***)	0.55 (b */c **/d ***)	1.65 (c */d ***)
(b) Low-middle	0.00 (d **)	0.10 (d ***)	0.85 (a */d ***)	2.21 (d ***)
(c) Middle-high	0.00	0.17 (d **)	0.86 (a **/d ***)	2.48 (a */d ***)
(d) High	0.00 (a/b **)	0.43 (a ***/b ***/c **)	1.30 (a/b/c ***)	4.47 (a/b/c ***)

Note: * $p < 0.05$, ** $p < 0.01$, *** $p < 0.001$. The letters before the housing price levels denote the group numbers. The groups with significant differences and their significance levels are indicated in parentheses after each median.

4. Discussion

4.1. Disparities in Types of UGS Services Obtained across Different Regions

The research findings indicate significant differentiation in the types of UGS services available in different urban areas. Firstly, the western part of the city facilitates access to various types of UGSs, which aligns with the emphasis on UGS development in newly developed urban areas. The service availability in the riverside areas is also adequate, consistent with previous studies [7,8,48], suggesting a universally positive role of proximity to water bodies in UGS development.

Secondly, most blocks in the eastern suburbs and southeastern parts can only access HSP services. However, due to the limited number of HSPs, their service coverage is relatively small. This suggests that the overall accessibility advantage in certain urban areas may not stem from a variety of green space resources but rather from the local advantages of a few large-scale UGSs. Additionally, most blocks cannot reach LSPs within the specified time threshold, and the accessibility level in the southern part of the city is generally poor, with some blocks only able to access HSP services with low accessibility opportunities. These phenomena may stem from a planning approach based on average indicators. Due to the larger scale and diverse service types of HSPs, they can quickly increase the overall green coverage and per capita green space area in the region to achieve the government's expected goals [48]. This creates a "superficial" perception of service adequacy and may lead planners to overlook other types of UGSs.

These findings highlight a contradiction: whether, with the same expected total area of UGSs, it is preferable to incrementally increase numerous small-scale UGSs with lower supply capabilities in a decentralized manner or to concentrate on fewer large-scale UGSs with higher supply capabilities in a centralized manner. Previous studies have addressed this issue [12,50], indicating that, compared to geographically concentrating resources and initiating several rounds of upscale development of large UGSs, focusing on smaller-scale interventions is more advantageous for UGS equity and can help to prevent gentrification. This study, from the perspective of different supply capabilities of UGSs, supports this viewpoint. Firstly, the construction and maintenance costs of HSPs are high, and they require large land areas as a foundation, which is unrealistic for high-density urban areas severely lacking in UGSs. Secondly, a single type of UGS may not fully meet the diverse needs of residents and may impose greater service pressures on HSPs. Studies have shown that residents prefer to engage in daily activities in smaller UGSs [51,52], possibly due to the disadvantages of larger UGSs such as crowding and difficulty in accessing activity areas. In contrast, LSPs and MSPs have several advantages, including greater flexibility, ease of encouraging residents to engage in physical activities [53], and contributing to strengthening community ties [54,55], among others. Therefore, in future planning, emphasis should be placed on incrementally supplementing more LSPs and MSPs in a decentralized manner to serve more blocks.

4.2. Association between Block Characteristics and UGS Accessibility

The research results indicate that house prices have a significant positive relationship with UGS accessibility. Except for low-supply parks (LSPs), which have a relatively limited accessible range, the median accessibility of various types of UGSs in high-priced blocks is the highest and significantly different from blocks of other price levels. This supports previous views [7,12], indicating significant socioeconomic inequality in UGS accessibility. This may be because high-income groups are generally willing to pay high costs for a better living environment [56] and have more opportunities to participate in decision-making processes related to their interests [57]. This may increase the housing prices in surrounding blocks [58], leading to significant segregation in accessing UGS services for low-income groups.

Similar to previous studies, this research also highlights differences in accessing UGSs between the elderly and children [16,22,49]. Areas with a higher proportion of children may exhibit lower accessibility to MSPs, while areas with a higher proportion of elderly residents may have higher accessibility to both MSPs and HSPs. This could be attributed to two factors. Firstly, families with children may choose to reside in older urban areas, which offer abundant commercial and educational resources [49], but lower UGS accessibility. Secondly, with the large-scale industrial transformation in China, the middle-aged and young unemployed population in old industrial areas are forced to migrate elsewhere, leading to an increase in aging population. However, measures such as industrial land replacement and renovation of old communities have also increased green spaces. Therefore, the elderly population may inadvertently benefit from access to urban parks [49].

Urban spatial structural factors partially explain the inequality of UGSs. Areas with high housing prices generally have later construction times and characteristics of low density and high green coverage. This partially explains the accessibility differences between new and old urban areas. New areas have better conditions for UGS construction, while old areas did not emphasize the importance of UGSs in the early stages and lacked comprehensive construction regulations, making it difficult to retrospectively increase UGSs in already built-up areas. However, this planning approach overlooks the lower population demand in new areas, leading to an increase in housing prices in these areas. Consequently, a large number of UGSs are overly concentrated in certain areas, exacerbating spatial inequality in UGSs and creating significant disadvantages for low-income groups in accessing UGSs.

4.3. Urban Planning and Policy Implications

Firstly, research on UGS spatial inequality should not only focus on areas lacking services but also consider the differences in UGS service types obtained in different areas. For example, the lower accessibility in the central urban area is due to the severe shortage of UGS supply, requiring prioritized expansion of the UGS total area. In contrast, for some blocks in the southern region, a more reasonable approach than continuing to increase HSPs is to add more LSPs and MSPs near these blocks and establish complete connections with existing HSPs. This can achieve greater benefits at lower costs.

Secondly, in terms of specific renovation strategies, attention should be paid to adopting stock planning schemes in areas where it is difficult to increase UGSs. There are three specific recommendations: (i) emphasize avoiding large-scale concentrated development and focus on small-scale interventions to supplement more LSPs and MSPs to bridge the gap; (ii) enhance facilities and services in existing LSPs to provide a more diverse range of functions to alleviate inequality [35]; (iii) advocate for open neighborhood planning to maximize the public benefits of closed non-park green spaces [59,60].

Finally, planners should prioritize addressing green inequality issues by actively intervening with policies and allocating funds to safeguard the health rights of low-income and vulnerable groups. At the same time, attention should be paid to increasing corresponding UGS service facilities in areas with a higher proportion of children and elderly populations to meet their specific needs.

5. Conclusions and Future Work

This study's main contributions are as follows: firstly, we utilized advanced data collection techniques, including Python data collection, drone image capture, and image stitching technology. This reduced the difficulty of obtaining data and improved the accuracy of the research results. Secondly, we considered elements related to physical activity in the measurement of UGS accessibility and improved the 2SFCA model from both supply and demand aspects, making the evaluation of accessibility more accurate. Thirdly, we separately evaluated the accessibility of different types of UGSs, which helps to formulate more precise planning strategies to ensure the maximization of benefits with limited construction costs. Fourthly, we evaluated the association between various urban spatial structure variables and accessibility, further explaining the reasons for UGS inequality based on previous research.

The conclusions of this study indicate the following: (1) there is significant differentiation in the types of UGS services available in different areas of the city. Government reliance on a single indicator for planning may lead to a severe lack of small-scale, low-cost UGSs, which hinders the full satisfaction of residents' diverse needs. (2) The accessibility of UGSs is significantly positively influenced by housing prices, with accessibility in high-priced areas notably higher than in low-priced areas, demonstrating clear socioeconomic inequality. (3) Children may be at a disadvantage in accessing UGSs, highlighting the need for future planning to pay greater attention to vulnerable groups.

Several limitations of this research framework must be acknowledged. Firstly, this study only used objective indicators to evaluate the sports service capacity of UGSs, without considering residents' subjective perceptions and needs in the park. In reality, residents' perceived quality of parks may vary [61] and be related to various factors [62]. Secondly, different modes of transportation were not considered for the accessibility evaluation of different types of UGSs. Lastly, due to the limitations of micro-data acquisition, the unequal differences in various vulnerable groups have not been further studied. Future research should be based on more field investigations, adopting a combination of subjective and objective data acquisition methods to further improve the reliability of the results.

Author Contributions: Conceptualization, Y.H. and Y.L.; methodology, Y.H. and Y.L.; project administration, Y.H. and L.W.; resources, Y.H. and L.W.; software, Y.L. and Y.W.; supervision, L.W.; visualization, Y.L. and Y.W.; writing—original draft, Y.L.; writing—review and editing, Y.H. All authors have read and agreed to the published version of the manuscript.

Funding: Project supported by the National Natural Science Foundation of China: Research on the Cold Island Mechanism of Seasonal Variations in Horizontal and Vertical Structures of Forests in Cold Cities (No. 42171246). Project supported by the China Postdoctoral Science Foundation: Research on the spatial mechanism of promoting physical activity in cold industrial community green spaces under the guidance of public health and the optimization model of "Sports-Green Integration" (No. 2020M670873).

Data Availability Statement: The source of all the data used in this study is provided in the manuscript.

Conflicts of Interest: The authors declare no conflicts of interest.

References

1. Lee, I.-M.; Shiroma, E.J.; Lobelo, F.; Puska, P.; Blair, S.N.; Katzmarzyk, P.T. Effect of physical inactivity on major non-communicable diseases worldwide: An analysis of burden of disease and life expectancy. *Lancet* **2012**, *380*, 219–229. [CrossRef]
2. McCormack, G.R.; Rock, M.; Toohey, A.M.; Hignell, D. Characteristics of urban parks associated with park use and physical activity: A review of qualitative research. *Health Place* **2010**, *16*, 712–726. [CrossRef]
3. Dai, D. Racial/ethnic and socioeconomic disparities in urban green space accessibility: Where to intervene? *Landsc. Urban Plan.* **2011**, *102*, 234–244. [CrossRef]
4. Kidokoro, T.; Sho, K.; Fukuda, R. Urban suburbia: Gentrification and spatial inequality in workers' communities in Tokyo. *Cities* **2023**, *136*, 104247. [CrossRef]
5. Hsiao, H. Spatial distribution of urban gardens on vacant land and rooftops: A case study of 'The Garden City Initiative' in Taipei City, Taiwan. *Urban Geogr.* **2021**, *43*, 1150–1175. [CrossRef]
6. Guo, S.; Song, C.; Pei, T.; Liu, Y.; Ma, T.; Du, Y.; Chen, J.; Fan, Z.; Tang, X.; Peng, Y.; et al. Accessibility to urban parks for elderly residents: Perspectives from mobile phone data. *Landsc. Urban Plan.* **2019**, *191*, 103642. [CrossRef]
7. Chen, Y.; Yue, W.; La Rosa, D. Which communities have better accessibility to green space? An investigation into environmental inequality using big data. *Landsc. Urban Plan.* **2020**, *204*, 103919. [CrossRef]
8. Xing, L.; Liu, Y.; Liu, X.; Wei, X.; Mao, Y. Spatio-temporal disparity between demand and supply of park green space service in urban area of Wuhan from 2000 to 2014. *Habitat Int.* **2018**, *71*, 49–59. [CrossRef]
9. Lu, Y.; Chen, R.; Chen, B.; Wu, J. Inclusive green environment for all? An investigation of spatial access equity of urban green space and associated socioeconomic drivers in China. *Landsc. Urban Plan.* **2024**, *241*, 104926. [CrossRef]
10. Chen, W.Y.; Hu, F.Z.Y. Producing nature for public: Land-based urbanization and provision of public green spaces in China. *Appl. Geogr.* **2015**, *58*, 32–40. [CrossRef]
11. Wu, L.; Kim, S.K. Exploring the equality of accessing urban green spaces: A comparative study of 341 Chinese cities. *Ecol. Indic.* **2021**, *121*, 107080. [CrossRef]
12. Wolch, J.R.; Byrne, J.; Newell, J.P. Urban green space, public health, and environmental justice: The challenge of making cities 'just green enough'. *Landsc. Urban Plan.* **2014**, *125*, 234–244. [CrossRef]
13. Wu, L.; Rowe, P.G. Green space progress or paradox: Identifying green space associated gentrification in Beijing. *Landsc. Urban Plan.* **2022**, *219*, 104321. [CrossRef]
14. Rigolon, A. A complex landscape of inequity in access to urban parks: A literature review. *Landsc. Urban Plan.* **2016**, *153*, 160–169. [CrossRef]
15. Nesbitt, L.; Meitner, M.J.; Girling, C.; Sheppard, S.R.J.; Lu, Y. Who has access to urban vegetation? A spatial analysis of distributional green equity in 10 US cities. *Landsc. Urban Plan.* **2019**, *181*, 51–79. [CrossRef]
16. Shen, Y.A.; Sun, F.; Che, Y.Y. Public green spaces and human wellbeing: Mapping the spatial inequity and mismatching status of public green space in the Central City of Shanghai. *Urban For. Urban Green.* **2017**, *27*, 59–68. (In English) [CrossRef]
17. Bolte, G.; Tamburlini, G.; Kohlhuber, M. Environmental Inequalities among Children in Europe—Evaluation of Scientific Evidence and Policy Implications. *Eur. J. Public Health* **2010**, *20*, 14–20. [CrossRef]
18. Maas, J.; Verheij, R.A.; de Vries, S.; Spreeuwenberg, P.; Schellevis, F.G.; Groenewegen, P.P. Morbidity is related to a green living environment. *J. Epidemiol. Community Health* **2009**, *63*, 967–973. [CrossRef] [PubMed]
19. Yigitcanlar, T.; Kamruzzaman, M.; Teimouri, R.; Degirmenci, K.; Alanjagh, F. Association between park visits and mental health in a developing country context: The case of Tabriz, Iran. *Landsc. Urban Plan.* **2020**, *199*, 103805. [CrossRef]
20. Xie, B.; An, Z.; Zheng, Y.; Li, Z. Healthy aging with parks: Association between park accessibility and the health status of older adults in urban China. *Sust. Cities Soc.* **2018**, *43*, 476–486. [CrossRef]
21. Pereira, G.; Foster, S.; Martin, K.; Christian, H.; Boruff, B.J.; Knuiman, M.; Giles-Corti, B. The association between neighborhood greenness and cardiovascular disease: An observational study. *BMC Public Health* **2012**, *12*, 466. [CrossRef] [PubMed]
22. Zhang, R.; Peng, S.; Sun, F.; Deng, L.; Che, Y. Assessing the social equity of urban parks: An improved index integrating multiple quality dimensions and service accessibility. *Cities* **2022**, *129*, 103839. (In English) [CrossRef]
23. Zhang, R.; Huang, L.; Wang, H. Accessibility Improvement and Renewal of Urban Park Green Space for the Elderly and the Disabled. *Forests* **2023**, *14*, 1801. [CrossRef]
24. Shao, H.; Song, P.; Mu, B.; Tian, G.; Chen, Q.; He, R.; Kim, G. Assessing city-scale green roof development potential using Unmanned Aerial Vehicle (UAV) imagery. *Urban For. Urban Green.* **2021**, *57*, 126954. [CrossRef]
25. Liang, H.; Li, W.; Zhang, Q.; Zhu, W.; Chen, D.; Liu, J.; Shu, T. Using unmanned aerial vehicle data to assess the three-dimension green quantity of urban green space: A case study in Shanghai, China. *Landsc. Urban Plan.* **2017**, *164*, 81–90. [CrossRef]
26. Radke, J.; Mu, L. Spatial Decompositions, Modeling and Mapping Service Regions to Predict Access to Social Programs. *Ann. GIS* **2000**, *6*, 105–112. [CrossRef]
27. Luo, W.; Wang, F. Measures of Spatial Accessibility to Healthcare in a GIS Environment: Synthesis and a Case Study in Chicago Region. *Env. Plann B Plann Des.* **2003**, *30*, 865–884. [CrossRef] [PubMed]
28. Dony, C.C.; Delmelle, E.M.; Delmelle, E.C. Re-conceptualizing accessibility to parks in multi-modal cities: A Variable-width Floating Catchment Area (VFCA) method. *Landsc. Urban Plan.* **2015**, *143*, 90–99. [CrossRef]
29. Luo, W.; Whippo, T. Variable catchment sizes for the two-step floating catchment area (2SFCA) method. *Health Place* **2012**, *18*, 789–795. [CrossRef] [PubMed]

30. Hu, S.; Song, W.; Li, C.; Lu, J. A multi-mode Gaussian-based two-step floating catchment area method for measuring accessibility of urban parks. *Cities* **2020**, *105*, 102815. [CrossRef]
31. Mao, L.; Nekorchuk, D. Measuring spatial accessibility to healthcare for populations with multiple transportation modes. *Health Place* **2013**, *24*, 115–122. [CrossRef] [PubMed]
32. Wan, N.; Zou, B.; Sternberg, T. A three-step floating catchment area method for analyzing spatial access to health services. *Int. J. Geogr. Inf. Sci.* **2012**, *26*, 1073–1089. [CrossRef]
33. Luo, J. Integrating the Huff Model and Floating Catchment Area Methods to Analyze Spatial Access to Healthcare Services. *Trans. GIS* **2014**, *18*, 436–448. [CrossRef]
34. Huff, D.L. A probabilistic analysis of shopping center trade areas. *Land Econ.* **1963**, *39*, 81–90. [CrossRef]
35. Wu, J.; Peng, Y.; Liu, P.; Weng, Y.; Lin, J. Is the green inequality overestimated? Quality reevaluation of green space accessibility. *Cities* **2022**, *130*, 103871. (In English) [CrossRef]
36. Knobel, P.; Dadvand, P.; Alonso, L.; Costa, L.; Español, M.; Maneja, R. Development of the urban green space quality assessment tool (RECITAL). *Urban For. Urban Green.* **2021**, *57*, 126895. (In English) [CrossRef]
37. Gidlow, C.; van Kempen, E.; Smith, G.; Triguero-Mas, M.; Kruize, H.; Gražulevičienė, R.; Ellis, N.; Hurst, G.; Masterson, D.; Cirach, M.; et al. Development of the natural environment scoring tool (NEST). *Urban For. Urban Green.* **2018**, *29*, 322–333. [CrossRef]
38. Bedimo-Rung, A.L.; Gustat, J.; Tompkins, B.J.; Rice, J.; Thomson, J. Development of a Direct Observation Instrument to Measure Environmental Characteristics of Parks for Physical Activity. *J. Phys. Act. Health* **2006**, *3*, S176–S189. [CrossRef]
39. Kaczynski, A.T.; Potwarka, L.R.; Saelens, B.E. Association of park size, distance, and features with physical activity in neighborhood parks. *Am. J. Public Health* **2008**, *98*, 1451–1456. (In English) [CrossRef] [PubMed]
40. Giles-Corti, B.; Broomhall, M.H.; Knuiman, M.; Collins, C.; Douglas, K.; Ng, K.; Lange, A.; Donovan, R.J. Increasing walking: How important is distance to, attractiveness, and size of public open space? *Am. J. Prev. Med.* **2005**, *28*, 169–176. [CrossRef]
41. Sugiyama, T.; Francis, J.; Middleton, N.J.; Owen, N.; Giles-Corti, B. Associations between recreational walking and attractiveness, size, and proximity of neighborhood open spaces. *Am. J. Public Health* **2010**, *100*, 1752–1757. [CrossRef] [PubMed]
42. Zhang, H.; Chen, B.; Sun, Z.; Bao, Z. Landscape perception and recreation needs in urban green space in Fuyang, Hangzhou, China. *Urban For. Urban Green.* **2013**, *12*, 44–52. [CrossRef]
43. Salmon, B.J. Where do children usually play? A qualitative study of parents' perceptions of influences on children's active free-play. *Health Place* **2006**, *12*, 383–393.
44. Ries, A.V.; Gittelsohn, J.; Voorhees, C.C.; Roche, K.M.; Clifton, K.J.; Astone, N.M. The environment and urban adolescents' use of recreational facilities for physical activity: A qualitative study. *Am. J. Health Promot.* **2008**, *23*, 43–50. [CrossRef] [PubMed]
45. Griffin, S.F.; Wilson, D.K.; Wilcox, S.; Buck, J.; Ainsworth, B.E. Physical Activity Influences in a Disadvantaged African American Community and the Communities' Proposed Solutions. *Health Promot. Pr.* **2008**, *9*, 180–190. [CrossRef] [PubMed]
46. Xing, L.; Liu, Y.; Wang, B.; Wang, Y.; Liu, H. An environmental justice study on spatial access to parks for youth by using an improved 2SFCA method in Wuhan, China. *Cities* **2020**, *96*, 102405. [CrossRef]
47. Zhang, J.; Cheng, Y.; Zhao, B. How to accurately identify the underserved areas of peri-urban parks? An integrated accessibility indicator. *Ecol. Indic.* **2021**, *122*, 107263. [CrossRef]
48. Liu, B.; Tian, Y.; Guo, M.; Tran, D.; Alwah, A.A.Q.; Xu, D. Evaluating the disparity between supply and demand of park green space using a multi-dimensional spatial equity evaluation framework. *Cities* **2022**, *121*, 103484. [CrossRef]
49. Xiao, Y.; Wang, Z.; Li, Z.; Tang, Z. An assessment of urban park access in Shanghai—Implications for the social equity in urban China. *Landsc. Urban Plan.* **2017**, *157*, 383–393. [CrossRef]
50. Schauman, S.; Salisbury, S. Restoring nature in the city: Puget Sound experiences. *Landsc. Urban Plan.* **1998**, *42*, 287–295. [CrossRef]
51. Kabisch, N.; Strohbach, M.; Haase, D.; Kronenberg, J. Urban green space availability in European cities. *Ecol. Indic.* **2016**, *70*, 586–596. [CrossRef]
52. Wendel, H.E.W.; Zarger, R.K.; Mihelcic, J.R. Accessibility and usability: Green space preferences, perceptions, and barriers in a rapidly urbanizing city in Latin America. *Landsc. Urban Plan.* **2012**, *107*, 272–282. [CrossRef]
53. Cohen, D.A.; Marsh, T.; Williamson, S.; Han, B.; Derose, K.P.; Golinelli, D.; McKenzie, T.L. The potential for pocket parks to increase physical activity. *Am. J. Health Promot.* **2014**, *28*, S19–S26. [CrossRef] [PubMed]
54. Zhang, H.; Han, M. Pocket parks in English and Chinese literature: A review. *Urban For. Urban Green.* **2021**, *61*, 127080. [CrossRef]
55. Liu, S.; Wang, X.H. Reexamine the value of urban pocket parks under the impact of the COVID-19. *Urban For. Urban Green.* **2021**, *64*, 127294. (In English) [CrossRef] [PubMed]
56. Xiao, Y.; Lu, Y.; Guo, Y.; Yuan, Y. Estimating the willingness to pay for green space services in Shanghai: Implications for social equity in urban China. *Urban For. Urban Green.* **2017**, *26*, 95–103. (In English) [CrossRef]
57. Shrestha, R.; Flacke, J.; Martinez, J.; Van Maarseveen, M. Environmental health related socio-spatial inequalities: Identifying "hotspots" of environmental burdens and social vulnerability. *Int. J. Environ. Res. Public Health* **2016**, *13*, 691. [CrossRef] [PubMed]
58. Jim, C.Y.; Chen, W.Y. External effects of neighbourhood parks and landscape elements on high-rise residential value. *Land Use Pol.* **2010**, *27*, 662–670. [CrossRef]
59. Ke, X.; Huang, D.; Zhou, T.; Men, H. Contribution of non-park green space to the equity of urban green space accessibility. *Ecol. Indic.* **2023**, *146*, 109855. [CrossRef]

60. Wen, Z.H.; Zhang, S.; Yang, Y.; Zheng, X.; Song, Z.; Zhou, Y.; Hao, J. How does enclosed private residential green space impact accessibility equity in urban regions? A case study in Shenzhen, China. *Urban For. Urban Green.* **2023**, *85*, 127968. (In English) [CrossRef]
61. Larson, K.L.; Brown, J.A.; Lee, K.J.; Pearsall, H. Park equity: Why subjective measures matter. *Urban For. Urban Green.* **2022**, *76*, 127733. (In English) [CrossRef]
62. Lau, K.K.L.; Yung, C.C.Y.; Tan, Z. Usage and perception of urban green space of older adults in the high-density city of Hong Kong. *Urban For. Urban Green.* **2021**, *64*, 127251. (In English) [CrossRef]

land

MDPI

Article

A Comparative Study of Perceptions of Destination Image Based on Content Mining: Fengjing Ancient Town and Zhaojialou Ancient Town as Examples

Jiahui Ding [1], Zheng Tao [1], Mingming Hou [2], Dan Chen [1] and Ling Wang [1,*]

1 School of Design, Shanghai Jiao Tong University, Shanghai 200240, China; dingjiahui@sjtu.edu.cn (J.D.); zheng.tao@sjtu.edu.cn (Z.T.); danchen.gator@sjtu.edu.cn (D.C.)
2 Department of Development and Planning, Shanghai Municipal Agricultural and Rural Commission, 100 Dagu Road, Shanghai 200003, China; mmhou@nyncw.shanghai.gov.cn
* Correspondence: wwlling@sjtu.edu.cn

Abstract: Ancient canal towns in Jiangnan have become important tourist destinations due to their unique water town scenery and historical value. Creating a unique tourist image boosts these ancient towns' competitive edge in tourism and contributes significantly to their preservation and growth. The vast amount of data from social media has become an essential source for uncovering tourism perceptions. This study takes two ancient towns in Shanghai, Zhaojialou and Fengjing, as case study areas. In order to explore and compare the destination images of the towns, in the perception of tourists and in official publicity, machine learning approaches like word embedding and K-means clustering are adopted to process the comments on Sina Weibo and publicity articles, and statistical analysis and correspondence analysis are used for comparative study. The results reveal the following: (1) Using k-means clustering, destination perceptions were categorized into 16 groups spanning three dimensions, "space, activity, and sentiment", with the most keywords in "activity" and the fewest in "sentiment". (2) The perception of tourists often differs significantly from the official promotional materials. Official promotions place a strong emphasis on shaping the image of ancient towns based on their historical resources, presenting a more general picture. Tourist perception, which is fragmented, highlights emerging elements and the experiential activities, along with the corresponding emotional experiences. (3) Comparing the two towns, Fengjing Ancient Town stands out, with more diverse tourist perceptions and richer emotional experiences. This underscores the effectiveness of tourism activities that use space as a media to evoke emotions, surpassing the impact of the spaces themselves.

Keywords: ancient town; destination image; UGC; DMO; machine learning

Citation: Ding, J.; Tao, Z.; Hou, M.; Chen, D.; Wang, L. A Comparative Study of Perceptions of Destination Image Based on Content Mining: Fengjing Ancient Town and Zhaojialou Ancient Town as Examples. *Land* **2023**, *12*, 1954. https://doi.org/10.3390/land12101954

Academic Editor: Heesup Han

Received: 20 September 2023
Revised: 19 October 2023
Accepted: 20 October 2023
Published: 23 October 2023

1. Introduction

In the 1980s, the emergence of tourism in Zhouzhuang marked the genesis of ancient town tourism in the country [1]. For all ancient towns, tourism development presents a vital opportunity to strike a balance between preserving historical culture and fostering socio-economic growth. Nevertheless, due to the shared historical backgrounds in the Jiangnan region, issues such as homogenization, superficiality, and theatricality have surfaced within the development models of ancient towns. This has led to a noticeable uniformity in the imagery of various Jiangnan ancient towns. However, these issues seem to be overly generalized. Each ancient town merits a more profound examination, as they may not be entirely similar to one another. Furthermore, it is essential to consider how different stakeholders perceive these towns. Research on destination images has emerged as an approach to delve into more nuanced and comprehensive depictions, facilitating a deeper understanding of ancient towns. This, in turn, paves the way for tailored development strategies for ancient towns to enhance their allure and competitiveness within the tourism industry.

A destination's image is directly correlated with marketing, branding, and tourists' willingness to choose a destination [2,3]. Research on destination image is instrumental in understanding tourist preferences, leading to a continuous enhancement of tourism product quality. This approach enables a destination to stand out from competitors in the competitive landscape, elevating its ecological niche within the market [4,5]. Destination images are often derived from two primary sources. One of these sources is the destination marketing organization (DMO), which utilizes marketing and advertising strategies to craft the image of the destination [6]. The second source is tourists, who convey their perceptions of the tourism destination through platforms such as social media, travelogues, or by communicating with others, resulting in a destination image based on user-generated content (UGC). While there are differences in how DMOs and UGC shape the destination image [7], both significantly impact the choices of tourists of destinations and decisions regarding travel behaviors [8]. The rapid advancement of information and communication technologies has shifted the focus of the tourism industry from traditional markets towards a digitally driven marketplace [3]. In addition to traditional printed materials, DMOs have started utilizing the internet as a means of promotion. Tourists increasingly share information and their tourism experiences on public platforms. UGC primarily originates from platforms such as Facebook, Instagram, Twitter, and Flickr [9], as well as platforms in mainland China like Sina Weibo, tourism websites, and Red. With the aid of social media platforms, both UGC- and DMO-crafted destination images have achieved broader dissemination, enhancing their impact on potential tourists.

The higher the overlap between the destination image shaped by the DMO and the image perceived by tourists, the more advantageous it is for the development of the destination [10]. This is because tourists expect DMO promotions to align with authentic travel experiences and are not receptive to "fake events", where the DMO promotes the destination image with purpose but not in a truthful manner [3,11]. While the significance of maintaining a consistent destination image between DMO and UGC has been emphasized, it is noteworthy that existing research on destination images often concentrates on individual stakeholders, rather than facilitating a comparative analysis that juxtaposes DMO and UGC [12–15]. Furthermore, research that systematically examines the disparities between DMO and UGC in both destinations remains scarce. To close the research gaps aforementioned, this study takes Fengjing and Zhaojialou ancient towns in Shanghai as case study areas and employs a content analysis of web text to address two research questions: (1) What is the level of alignment between tourists' perceptions of the ancient town destination image and the official promotional image? (2) Regarding different ancient towns, does the alignment between UGC and DMO destination images exhibit similarities? This study contributes to methodological advancements by employing machine learning techniques. By outlining textual content analysis, this research contributes to helping tourism marketers reflect on the balance between DMO and UGC and optimize the storytelling within their DMO content.

2. Literature Review

John Hunt, Edward Mayo, and Clare Gunn initiated research on destination image in the 1970s [16]. Early research methods for the perception of destination image, such as questionnaires, lacked reliable theoretical foundations, and the accuracy of data collection was often questioned. Additionally, comparability between different scales or questionnaires was lacking [17,18]. In the 1990s and beyond, more complex quantitative analysis methods emerged by means of the advanced state of destination image theory [17]. However, the strong preference for quantitative analysis led to a trend towards structured research, which in turn resulted in the overlooking of the unique characteristics of tourism destinations, and there was an incline in research perspectives—out of the 142 publications on destination image from 1972 to 2000, approximately 80% of the research focused on tourists, while studies involving other stakeholders of destinations, such as DMO and local residents, accounted for only 2% of the research [16]. Traditional content analysis methods in destination image research have relied on sources such as interviews and questionnaires with tourists, literary works, and audiovisual materials featuring destinations. These sources often have a limited amount of data, which makes it challenging to explore the evolution of destination images over longer time spans. As global tourism and the internet have developed, destination image research since the year 2000 has not only increased its focus on other stakeholders but has also transformed data source selection. In the era of the internet, online data content offers a rich and abundant source of information, covering longer periods. This has become a crucial aspect of contemporary destination image research. Internet data, particularly online text, has become a prevalent and valuable data source for studying destination images due to its accessibility and scale [10,19–21].

Zhang and Zhang (2009) extracted the top 60 high-frequency words from online reviews of the Chenshan Botanic Garden in Shanghai, and they found that the most frequently mentioned element across 16 categories was the sentimental experience [22]. When there is a need to process larger volumes of data, machine learning emerges as a powerful strategy for effectively handling and interpreting substantial datasets [3,23], potentially leading to the discovery of more intriguing results. Koblet and Purves (2020) collected over 7000 tourist reviews of England's Lake District National Park and processed the reviews using machine learning methods, and they found that many fresh experiences reported by users cannot be found in the underlying data [24]. Song et al. (2021) employed machine learning approaches to analyze over 20,000 tourist reviews of the Las Vegas Strip, which revealed that, despite the Las Vegas Strip boasting numerous world-renowned hotels, casinos, and resort properties that symbolize the area, most visitors expressed dissatisfaction with staying exclusively in their own hotel. Instead, they preferred to explore various locations on foot to experience a variety of attractions [25]. This result highlights a discrepancy between the conventional image of a destination and how tourists perceive it. Költringer and Dickinger (2015) utilized a web content mining approach to analyze three different sources of data, including UGC, DMO, and media, and this study concluded that UGC contains the most diverse information compared with DMO and media, and the content of UGC does not conform to the image DMO conveyed [10].

Research methods for studying the destination image of ancient towns primarily encompass traditional approaches such as interviews, questionnaires, visitor-employed photography (VEP), and textual analysis of tourist comments. The analyzed data volume remains limited, and the involvement of machine learning has not happened yet. Zhang et al. (2019) utilized a combination of questionnaires and photo-elicitation interviews to analyze the different perceptions of destination image in Tongli from tourists and residents, and their results showed that tourists' perceptions of architecture, streets, and lanes were higher than local residents', while the perception of historical and cultural elements was lower [26]. Tan et al. (2018) conducted a text classification and frequency analysis on Weibo data and concluded that tourists' perceptions of Zhaojialou were primarily focused on food [27]. Xu et al. (2017) analyzed photos taken by tourists and found that tourists exhibited a stronger perception of ancient bridges and residential buildings, which validated the effectiveness of the promotional efforts of DMO [28]. Dong and Xu (2017) compared tourist photos from Mafengwo with official pictures of Fengjing ancient town. They concluded that tourist photos placed more emphasis on activities and details, showing minimal influence from official imagery [29]. They argued that this finding contradicted Urry's "hermeneutic circle". In fact, Urry and Larsen (2011) acknowledged that the hermeneutic circle is an oversimplified explanation, and that tourist photos possess the capability to shape a new destination image, rather than replicating photos promoted by DMO [30]. Previous studies of Fengjing and Zhaojialou ancient towns lacked the involvement of machine learning, leading to limitations in data volume. While some research has compared the UGC and DMO of Fengjing, their analysis was primarily based on photographs, which often exclude experiential aspects beyond visual elements [30]. This may impose certain limitations. In contrast, the analysis of textual content could yield a different result. Therefore, this study takes a different approach by analyzing online comments and integrating machine learning. It aims to explore the perception of destination image in both ancient towns and investigate the discrepancies between tourist perceptions and DMO representations.

3. Methods

3.1. Case Study Area

In this study, Fengjing Ancient Town and Zhaojialou Ancient Town in Shanghai were selected as the case study areas (Figure 1). Although the two towns are located in the same city, Shanghai, there are differences in their geographic locations, historical and cultural backgrounds, and patterns of development. Fengjing Ancient Town retains the typical style of a water town, and is located in the southwest corner of Shanghai and the northwest corner of Jinshan District, which is the junction of Shanghai and Zhejiang; it is characterized by the regional culture of "Wu gen Yue Jiao" and was listed in the first batch of "China's historic and cultural villages" and the first batch of historic and cultural villages of Shanghai in 2005 [31]. Zhaojialou is located in Gexin Village of Pujiang Town, Minhang District, Shanghai, which is the origin of the cultivation culture in Shanghai. Zhaojialou retains many buildings from the Ming and Qing Dynasties and was recognized as one of "China's historic and cultural villages" in 2014.

Figure 1. Geographical location of Fengjing and Zhaojialou (source: authors).

This study comprises four sources of a dataset encompassing UGC and DMO data from Fengjing Ancient Town (FJ) and Zhaojialou Ancient Town (ZJL). Regarding UGC data, the study employed keyword searches for the two towns to retrieve Sina Weibo hashtag check-in comment data spanning from 2010 to 2020. For Fengjing ancient town, the original textual data (FJ-UGC) consisted of 10,619 comments, while Zhaojialou ancient town (ZJL-UGC) had 11,485. The study primarily employed a content analysis of the comments to establish tourists' perceptions of the destination image. Due to the absence of an official website for Zhaojialou, the study employed the Shanghai Cultural Tourism Promotion Website and three widely used tourism websites (Ctrip, Qunar, and Tuniu) to source official introduction articles for both ancient towns as DMO data. The DMO data for Fengjing (FJ-DMO) comprised five articles, and Zhaojialou (ZJL-DMO) consisted of four articles. All of the text materials consisted of Mandarin Chinese characters.

3.2. Data Processing

Text data were processed according to Figure 2.

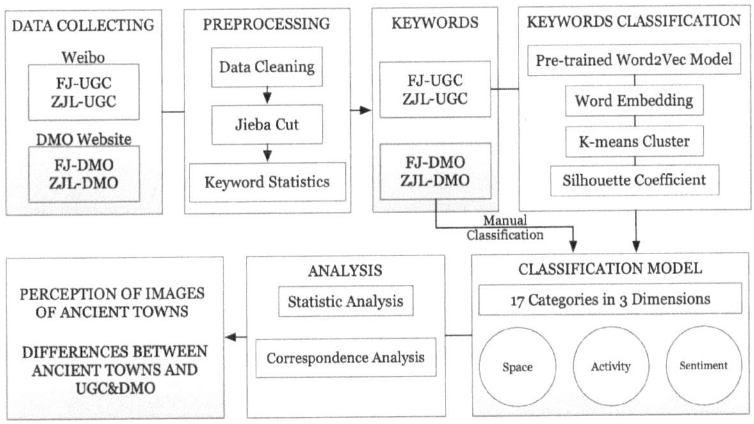

Figure 2. Route of data processing.

3.2.1. High-Frequency Keyword Extraction

The most frequently mentioned keywords in the comments, known as high-frequency keywords, represented the main points of tourists' attention and reflected their attitudes towards the destination [20]. Extracting high-frequency keywords involves three steps: data cleaning, tokenization, and statistics.

Blank comments and irrelevant content such as items only containing phrases like "share pictures", "I am at" in UGC data were removed. Additionally, duplicate comments were eliminated. As a result, the remaining comments of FJ-UGC were 9333, and ZJL-UGC had 10,716. Using regular expressions, only pure textual content without URLs and punctuation remained in both UGC and DMO data.

Jieba is a widely used Chinese text processing toolkit that is capable of segmenting text into words [23]. In this study, a user dictionary relevant to the ancient towns and a stop word dictionary including common pronouns, conjunctions, and unnecessary internet jargon were established. The UGC and DMO data, which had undergone data cleaning, were then subjected to tokenization using Jieba (github.com/fxsjy/jieba(accessed on 15 January 2022)). Word segmentation was performed multiple times due to Chinese words being composed of individual characters, which is different from English words. Many words identified during the initial segmentation, which needed to be excluded, were not present in the initial stop word dictionary. Additionally, some words are valuable and should be considered for inclusion in the user dictionary. The final segmentation results were established after refining the stop word dictionary and user dictionary.

Considering the disparity in data size between UGC and the DMO, distinct frequency thresholds were applied: keywords with a frequency greater than or equal to 5 for UGC and those with a frequency greater than or equal to 2 for DMO were selected for further analysis. Despite UGC data containing over 300,000 characters, significantly more than the 15,000 characters in DMO content, it is noteworthy that UGC offers limited valuable information, resulting in the exclusion of the majority of its content. DMO data are relatively stable due to the articles being published by different institutions almost conveying the same content, whereas UGC exhibits substantial fluctuations that require more data to obtain more representative results. Moreover, it is worth noting that the presence of introductions from multiple institutions for both towns is rather scarce. The study has diligently sought to compile a comprehensive sample of articles released by DMOs. Additionally, the analysis primarily focuses on comparing percentages rather than absolute values, as this approach provides a more comparable perspective. Table 1 presents the count of the keywords finally selected and the corresponding frequency from the four data sources.

Table 1. High frequency keywords.

Source	FJ-UGC	ZJL-UGC	FJ-DMO	ZJL-DMO
Number of keywords	641	714	167	141
Frequency	9555	13,819	524	748

3.2.2. Keyword Classification

Content analysis is not simply a word frequency count but a reasonable way of classifying the data [32]. This study harnesses clustering algorithms to categorize keywords. On the one hand, the substantial volume of UGC data lends itself well to the application of machine learning in clustering analysis. On the other hand, employing a unified classification facilitates a horizontal comparison between UGC and DMO data. Consequently, the study exclusively performed clustering on the UGC data, which was subsequently applied to the DMO data as well.

(1) Word embedding

To let computers "understand" human language, it is necessary to subject the keywords to word embedding processing, thereby converting them into word vectors [33]. The Word2Vec model is one of the most commonly used word embedding techniques [34,35]. After inserting the keywords into the Word2Vec model, corresponding word vectors were obtained. Following this, a clustering algorithm was applied to process these word vectors, effectively categorizing the unstructured textual data [20]. In this study, a pre-trained Word2Vec model released by Tencent AI Lab [36] was utilized to embed keywords and acquire their respective word vectors. Some of the keywords could not be embedded in the model due to the absence of corresponding vectors. These words were manually categorized after the clustering process.

(2) K-means clustering

The study used K-means clustering to cluster the word vectors of FJ-UGC and ZJL-UGC. The silhouette coefficient was utilized to evaluate the clustering effectiveness and determine the optimal number of clusters in the K-means clustering [37]. The research discovered that the ideal k values for FJ-UGC were 21 and 24, while for ZJL-UGC, the optimal k values were 8 and 15 (Table 2). To ensure a consistent classification standard for the keywords of both ancient towns, the study initially chose the closest numbers, 21 and 15, as the initial cluster numbers for the keywords of Fengjing and Zhaojialou. This number was later unified based on human judgment.

(3) Categories of keywords

The unification of the UGC keyword classification for both datasets was carried out through manual judgment, guided by the principle that "each category contains only a set of keywords with the same meaning or connotation" [38] and that "each category is distinct" [39]. Adjectives related to emotions were categorized into positive, neutral, and negative sentiments based on the Gooseeker platform's (gooseeker.com (accessed on 13 January 2022)) sentiment corpus. Based on this, the clustering results of the UGC data for both ancient towns were unified. Categories and keywords without actual significance were omitted. This led to the development of a comprehensive classification method that includes sixteen categories under three dimensions (Table 3).

Table 2. Silhouette score of k-means cluster.

K-means model of FJ-UGC							
Cluster number	10	15	20	21	24	25	30
Silhouette score	0.06210957	0.06513383	0.06720589	0.072402	0.072835	0.06117635	0.06356114
K-means model of ZJL-UGC							
Cluster number	5	8	10	15	20	25	30
Silhouette score	0.0690023	0.075108	0.05928351	0.075067	0.05729202	0.06224708	0.06160499

Table 3. Categories with descriptions.

Dimension	Category	Description
Space	Architectural Space	Enclosed spaces for various activities, such as shumai shops, restaurants, bars, etc.
	Gathering Space	Open spaces like markets, squares, ticket booths, and parking lots, where people gather or stay
	Street Space	Old streets, streets, alleys, lanes, streets, etc.
	Waterfront Space	Areas near rivers or formed by rivers, such as riverbanks, rivers, and both sides of the river
	Landmark Space	Clearly defined landmarks or buildings, like Zhihe Bridge, Heping Street, Wuyou Xian
Activity	Time	Specific time periods like months, days of the week, times of day, holidays, weekends
	External traffic	Modes of transportation such as buses, subways, driving
	Individual behavior	Day trips, sightseeing, rest, tourist activities.
	History and culture	Indicating the origin or customs of ancient towns, or art and things related to it
	Nature	Climate, weather, animals, plants, atmosphere, etc.
	Artificial	Human-related elements like small bridges, flowing water, houses, architecture, folk paintings, bridges
	Human	Family members, friends, tourists, local people
	Food	Dishes, cuisines, etc.
Sentiment	Positive	Sentiments such as good, like, delicious, happy
	Neutral	Descriptions such as uncrowded, few people, typical, and commercialized
	Negative	Sentiments like boring, not good, unfortunate, tired, disappointed, and regretful

3.3. Data Analysis

The statistical analysis calculated the proportion of word frequencies for the three dimensions and sixteen categories in each data source. Based on the results of the statistical analysis, a two-sample z-test was conducted using the "statsmodels" library in Python to ascertain the existence of significant differences between UGC and DMO within each category. Subsequently, a correspondence analysis was performed on the four sources of data in order to assess the differences in the perception of the destination image of the two ancient towns across different data sources. Correspondence analysis allows multiple datasets to be visualized in a two-dimensional plot, using the spatial arrangement of data points on the plot to assess the degree of correlation and differences between variables [10]. In this study, all correspondence analysis plots were generated utilizing the SPSS 26 software.

4. Results

4.1. Statistical Analysis

The results of the statistical analysis (Table 4) indicate a significant difference in the categories between DMO and UGC data (Figure 3). Among the top three categories in both sets of UGC data were "Food", "Positive", and "Individual Behavior". In ZJL-UGC,

"Food" had the highest proportion (28.74%), while in FJ-UGC, "Positive" was the most prominent (18.06%). In the DMO data for both towns, the top two categories were "History and culture" and "Artificial". However, in FJ-DMO, "Food" (12.60%) ranked third, which was very close in proportion to the UGC data for Fengjing. In ZJL-DMO, the third-ranked thematic word was "Architectural Space" (11.72%).

Table 4. Proportion of 16 categories.

Dimension	Category	Proportion of Keywords (%)			
		FJ-UGC	ZJL-UGC	FJ-DMO	ZJL-DMO
Space	Architectural space	1.75	1.58	2.29	11.72
	Gathering space	1.31	0.89	4.20	3.56
	Street space	0.65	1.10	2.67	2.09
	Waterfront space	0.71	0.34	2.48	1.26
	Landmark space	1.71	1.61	12.21	10.46
	Subtotal	6.12	5.52	23.85	29.08
Activity	Time	8.95	8.13	3.24	1.26
	External traffic	1.44	0.96	3.24	0.00
	Individual behavior	16.08	10.45	6.30	9.00
	History and culture	5.78	4.64	24.81	28.03
	Nature	9.05	7.01	0.38	0.42
	Artificial	4.47	2.79	14.50	20.50
	Human	8.85	10.10	4.58	2.51
	Food	13.59	28.74	12.60	3.35
	Subtotal	68.22	72.82	69.66	65.06
Sentiment	Positive	18.06	15.60	3.63	3.56
	Neutral	3.79	2.94	2.86	2.30
	Negative	3.81	3.12	0.00	0.00
	Subtotal	25.66	21.66	6.49	5.86
Total		100.00	100.00	100.00	100.00

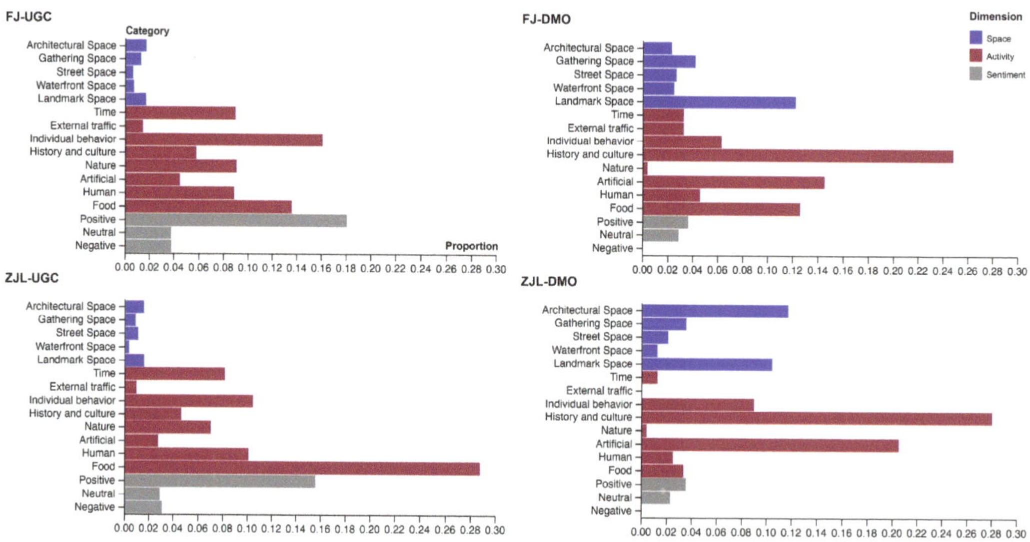

Figure 3. Statistical comparison chart of the proportion of keywords in 16 categories.

Based on the results of the two-sample z-test (Table 5), it is evident that in both Fengjing and Zhaojialou, there exist distinct differences between UGC and DMO. In Fengjing, all categories exhibited significant differences between UGC and DMO, except for "Architectural Space", "Food", and "Neutral". In Zhaojialou, only "Individual behavior" and "Neutral" did not display significant differences.

Table 5. Two-sample Z test of UGC and DMO proportion of two towns.

Category	Fengjing		Zhaojialou		Alpha
	Z-Score	*p*-Value	Z-Score	*p*-Value	
Architectural space	−0.9157	0.3598	−15.8542	0.0000	0.05
Gathering space	−5.3817	0.0000	−5.8204	0.0000	0.05
Street space	−5.1872	0.0000	−2.0148	0.0439	0.05
Waterfront space	−4.4471	0.0000	−3.2367	0.0012	0.05
Landmark space	−15.7748	0.0000	−13.9305	0.0000	0.05
Time	4.5214	0.0000	5.4801	0.0000	0.05
External traffic	−3.2653	0.0011	2.1467	0.0318	0.05
Individual behavior	6.0147	0.0000	1.0235	0.3061	0.05
History and culture	−16.8759	0.0000	−22.2088	0.0000	0.05
Nature	6.8923	0.0000	5.6330	0.0000	0.05
Artificial	−10.2577	0.0000	−21.0770	0.0000	0.05
Human	3.3908	0.0007	5.4764	0.0000	0.05
Food	0.6475	0.5173	12.1717	0.0000	0.05
Positive	8.5041	0.0000	7.2117	0.0000	0.05
Neutral	1.0885	0.2764	0.8133	0.4160	0.05
Negative	4.5514	0.0000	3.9207	0.0001	0.05

In order to compare the relative changes in proportions of UGC and DMO data across categories, the study calculated the relative change [(DMO-UGC)/UGC] in DMO compared to UGC (Figure 4). The results aligned with the findings from the two-sample z-test, indicating that all five categories within the space dimension showed increases, with the exception of Architectural Space in Fengjing. This suggests that the DMO primarily emphasized the promotion of the unique spatial aspects of the ancient towns. However, tourists appeared to be more focused on the emotional experiences elicited by these spaces, potentially leading to an oversight of the physical spaces themselves.

For Fengjing ancient town, "Landmark space" exhibited the highest increase in DMO compared to UGC (6.162 times), followed by "History and culture" (3.293 times) and "Street space" (3.094 times). On the other hand, all categories in the sentiment dimension in DMO showed a decrease. The most significant decrease was observed in "Negative", where the frequency of the negative keywords in DMO was 0. The second-largest decrease was "Nature" (−0.958).

For Zhaojialou, "Architectural space" (6.393 times) and "Artificial" (6.359 times) showed a high increase, followed by "Landmark Space" (5.511 times) and "History and Culture" (5.044 times). Similar to Fengjing, the "Negative Emotion" category in Zhaojialou's DMO data showed one of the largest decreases, which was also because the frequency of negative keywords was 0 in DMO. Additionally, a significant decrease was observed in the "Nature" category, which was similarly notable in Fengjing. Another category that showed a substantial decrease was "External traffic", as DMO keywords did not include terms related to reaching the ancient town. Furthermore, "Food" (−0.884) is another category where DMO differed significantly from UGC.

4.2. Correspondence Analysis

Statistical results were analyzed according to four sources, FJ-UGC, ZJL-UGC, FJ-DMO, and ZJL-DMO, as well as three dimensions: space, activity, and sentiment (Figure 5). The results aligned closely with statistical analysis. A stark contrast between UGC and DMO in both towns was evident, indicating that the perceived tourist destination image in the two towns differed significantly from the image portrayed by DMO. Notably, DMOs emphasized the space dimension, particularly in ZJL-DMO, reflecting the emphasis on spaces in promoting Zhaojialou. On the other hand, UGC data leaned more towards the sentiment dimension, particularly in FJ-UGC, indicating the rich emotional experiences visitors had in Fengjing. While the four sources did not exhibit significant differences in the activity dimension, it is worth noting that categories in activity dimensions dominated across all datasets. The following sections will provide a more detailed analysis of the specific categories within three dimensions.

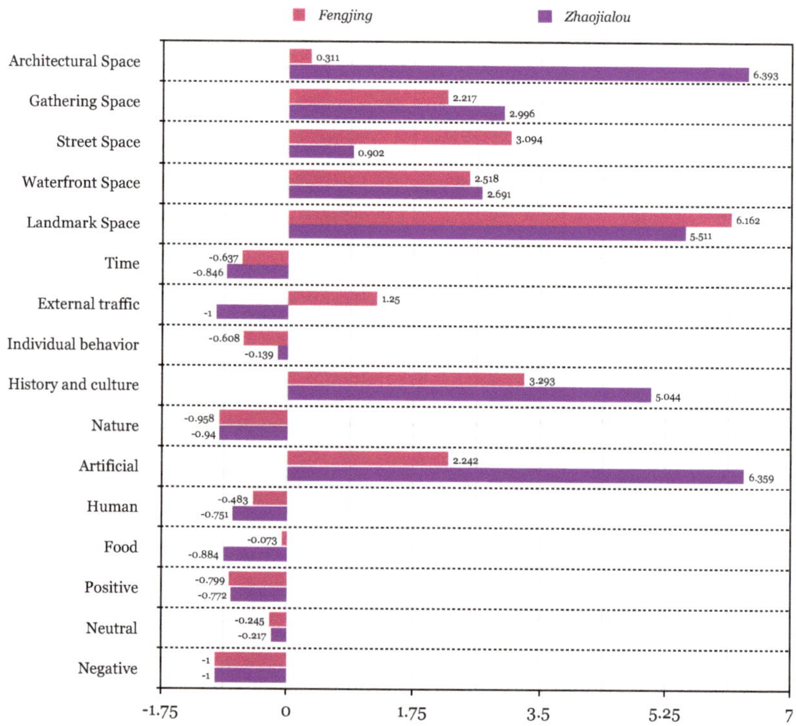

Figure 4. Relative change [(DMO − UGC)/UGC] in DMO compared to UGC.

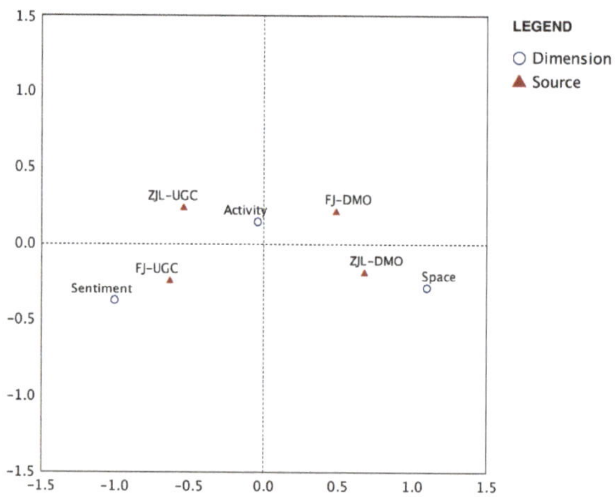

Figure 5. Correspondence analysis map of three dimensions.

4.2.1. Comparing the Space Dimension and Its Categories

The statistical results indicate that in the space dimension, DMO, in comparison to UGC, exhibited a significant increase in the proportion of categories. This increase was obvious in "Landmark space". For instance, in Fengjing, landmarks like Zhihe Bridge, Nandajie Street, and Heping Street are highlighted, while in Zhaojialou, Mei Garden, Li Garden, and Li Geng Hall are emphasized. The "Landmark Space" is positioned closer to the two DMOs in the correspondence analysis (Figures 6 and 7).

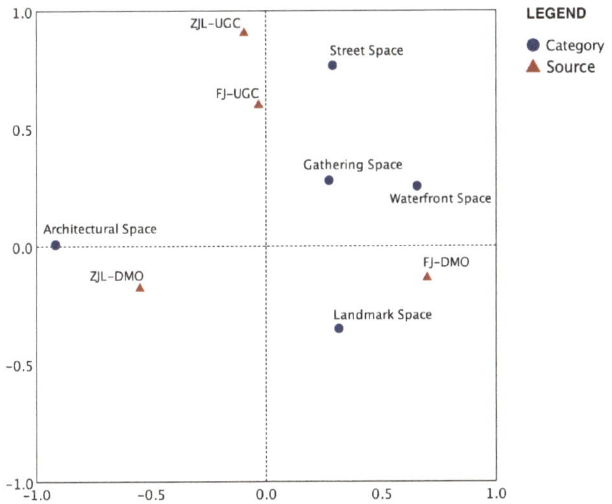

Figure 6. Correspondence analysis map of the space dimension.

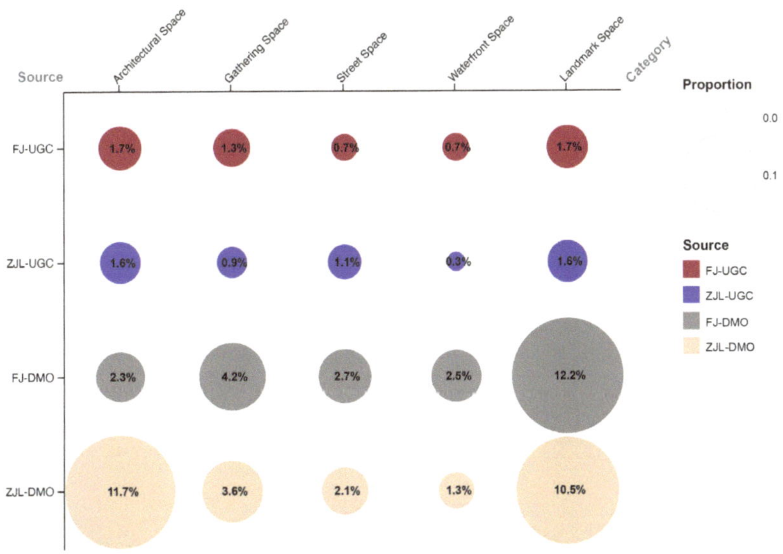

Figure 7. Categories in the space dimension.

ZJL-DMO emphasized spaces such as courtyards, halls, and side rooms, with "Architectural space" accounting for the highest proportion (11.72%) in ZJL-DMO, far more than ZJL-UGC (1.58%). The correspondence analysis map also demonstrates that "Architectural space" aligned with ZJL-DMO. On the other hand, "Landmark space" leaned more towards FJ-DMO, while "Street space" was closer to both sets of UGC. No significant differences were observed in the other two spatial categories, "Gathering Space" and "Waterfront Space".

The keywords of categories in space dimension from the four sources were further compared and significant differences were found in the "Landmark space" keywords between DMO and UGC for both towns. Unlike FJ-DMO, FJ-UGC mentioned more landmarks, including "Old Stage", "People's Commune", "Corridor", "Starbucks", and "Three Bridges Courtyard". Interestingly, keywords that reflect the conjunction of Wu and Yue's cultural geographic features in Fengjing, such as boundary stones and boundary rivers, were rarely mentioned in both FJ-DMO and FJ-UGC. This suggests a lack of emphasis on these important features in both promotional materials and tourist perceptions.

The most mentioned landmarks in ZJL-UGC were dining places like "Wu Youxian" "Lao Bayang Restaurant", and "Qingju Noodle", while the most mentioned landmarks in ZJL-DMO were attractions like "Li Garden", "Mei Garden", and "Li Geng Hall". Zhaojialou also exhibited differences in "Architectural space". ZJL-DMO focused more on architectural terms like "courtyards", "alcove", "depth", and "halls", which differed from the emphasis on dining establishments in ZJL-UGC. Although other types of spaces showed varying proportions in DMO and UGC data for both towns, the content of high-frequency keywords remained largely consistent.

4.2.2. Comparing the Activity Dimension and Its Categories

K-means clustering was utilized to categorize keywords related to tourist activities in the ancient towns into eight categories. These categories encompass "External Traffic", "Activity Time", and "Individual Behavior", along with five categories representing aspects perceived by tourists through their activities: "Nature", "Artificial", "Human", "History and culture", and "Food". While the proportions of these eight categories varied in UGC and DMO data, the corresponding analysis map (Figures 8 and 9) clearly illustrates that the two sets of UGC and DMO data sources were distinct and situated on opposite sides of axis Y, which indicates significant disparities between the activities tourists perceived and the official promotional activities depicted by DMO.

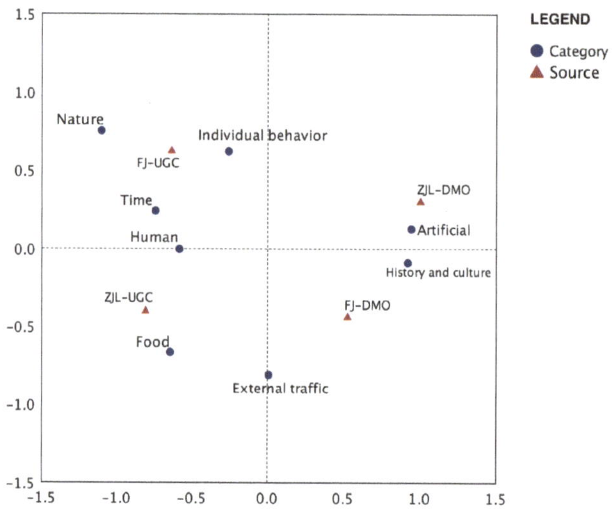

Figure 8. Correspondence analysis map of the activity dimension.

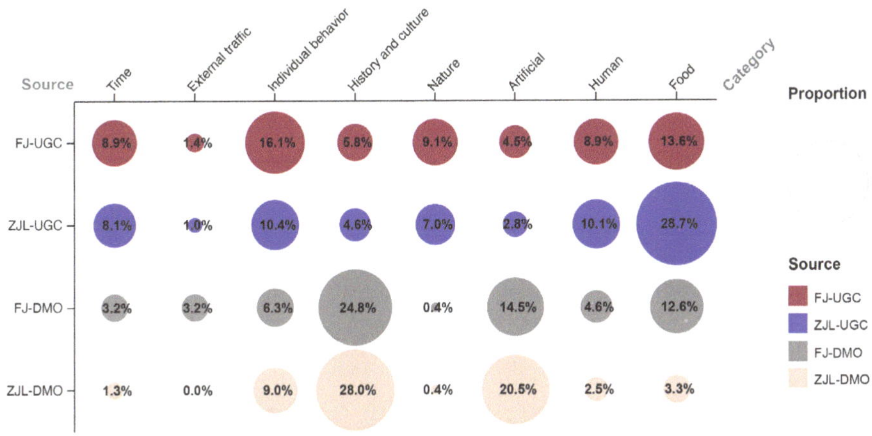

Figure 9. Categories of the activity dimension.

According to the corresponding analysis map, "History and culture" and "Artificial" were concentrated on the side of DMO sources, particularly in close proximity to ZJL-DMO. This reflects that DMOs, especially ZJL-DMO, perceive "Artificial" and "History and culture" as aspects that can showcase the developmental history, cultural essence, and material heritage of the ancient town, thereby highlighting its distinct features. However, in comparison to historical and cultural aspects, tourists were more focused on tourist behavior, food experiences, natural scenery, characters, and the timing of these activities (Figure 8). DMO barely mentioned elements like climate, atmosphere, plants, and animals. FJ-UGC exhibited a more noticeable focus on "Individual behavior", for example, rest, photography, excursions, and boat trips, as well as "Time", like specific days of a week or holidays. Additionally, FJ-UGC emphasized natural elements, including climate, atmosphere, and plants and animals. ZJL-UGC, on the other hand, prominently highlighted tourists' keen interest in "Food", in which keywords encompassed everyday items like Shanghai soup buns, as well as emerging trendy dishes. The distinctive traditional cuisine of Zhaojialou, known as the "Three Treasures", was also mentioned, albeit with less emphasis compared to non-traditional foods. In terms of "Human", the UGC of both towns primarily focused on close relationships, like family members and friends, with less focus on local residents or other tourists.

While FJ-UGC showed a weakness in the perception of "History and culture", the study uncovered keywords such as "love", "Nothing but thirty", and "reading" within it. These keywords correspond to Fengjing Ancient Town's distinct features, such as its Jiangnan wedding customs, role as a filming location for TV shows, and its historical significance as a hub for scholars. This underscores Fengjing's unique qualities, setting it apart from the typical "small bridges and flowing waters" image of other canal towns, which is needed for a further exploration and increased awareness among tourists. Furthermore, the widespread dissemination of movies and television, innovative wedding customs events, and the establishment of new attractions like "The Archway of Champion" have significantly contributed to enhancing perceptions of and emotional connections to the culture of Fengjing.

4.2.3. Comparing the Sentiment Dimension and Its Categories

A significant disparity between UGC and DMO in terms of the proportion of three categories of sentiment dimension across the two ancient towns was shown by the statistical analysis. The proportion of the sentiment dimension in UGC was much higher than in DMO: 25.66% and 21.66% in FJ-UGC and ZJL-UGC and 6.49% and 5.86% in FJ-DMO and ZJL-DMO (Table 4). "Positive" had a significant presence in both sets of UGC data, which indicates that tourists are highly satisfied with their experiences in the two ancient towns (Figure 10), particularly in FJ-UGC, where positive emotions (18.06%) counted as the largest among 16 categories (Figure 11). Proportions of "Neutral" were similar in both towns and the proportion of "Negative" was minimal. DMO data contained no keywords related to "Negative", while the proportions of "Positive" and "Neutral" were marginally different, which suggests that DMO narratives tend to be more objective in style and may lack emotional resonance with tourists.

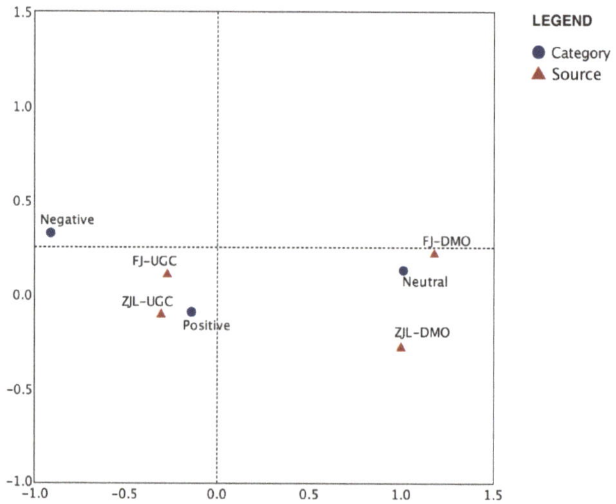

Figure 10. Correspondence analysis map of the sentiment dimension.

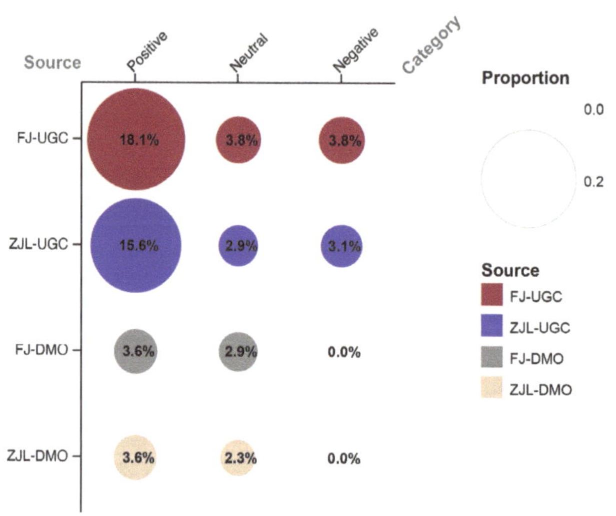

Figure 11. Categories of the sentiment dimension.

Among the three categories in both UGC datasets, the most significant difference lay in "Neutral". Fengjing gives tourists a sense of "quiet" and "casual", whereas Zhaojialou showcases a more bustling and commercialized tourism atmosphere with references to "crowd" and "commercialization". In terms of positive emotions, high-frequency keywords in both towns include phrases like "delicious", "happy", and "like". Perceptions of

Fengjing capture the feeling of "quaint", "tranquil", and "comfortable" that evoke senses of "pleasure", "leisure", and "happiness". In contrast, Zhaojialou brings forth "Positive" senses of "liveliness" and "enjoyment". The keyword "delicious" is highlighted 392 times, ranking first in "Positive", which echoes the outstanding presence of "Food" in the activity dimension of Zhaojialou. "Negative" perceptions of both towns share similar keywords such as "boring", "bad", "disappointing", and "regret", indicating that the tourist experience fell short of expectations. Additionally, some visitors to Fengjing felt "exhausted", while in Zhaojialou, some perceived it as "human gridlock" and "overcrowded".

5. Discussion

5.1. The Visitor-Perceived (UGC) Destination Image of Ancient Town Differs Significantly from That Officialy Promoted (DMO)

Just like previous studies have concluded, tourists focus more on sentiment [22]. Additionally, historical and cultural elements are barely shown on the list of tourists' perceptions [26], which is contrary to what DMO portrays—detailed origin, culture, and stories of the towns in a historical context. The results explain that tourists do not concentrate on historical, traditional, or cultural essence but on personal experience [40]. Due to their "immediate thought" or "immediate observation", promoted by social media, the authenticity of ancient towns is forgotten [41]. The destination image shaped by DMO can establish a brand [42]. A positive and strong local brand can create cultural value and evoke emotional attachment, thereby further boosting the sales of specific local products [43–45]. The water town Zhouzhuang, for instance, has successfully cultivated the brand of being the "Top Water Town in China", with its image of "bridges, rivers, and residential homes" being widely recognized and propagated [46]. In such a scenario, other water towns would find it challenging to compete with Zhouzhuang solely by using the spatial image of "bridges, rivers, and residential homes" to attract tourists. Fengjing aims to establish its brand as the "Millennium Conjunction of Wu and Yue", which holds significant cultural significance. However, its DMO seems to prioritize highlighting "Landmark space" while overlooking this crucial cultural symbol. Similarly, Zhaojialou Ancient Town positions itself as the "Origin of Shanghai's Cultivation Culture", but its DMO description consists mainly of specialized architectural keywords, possibly due to emphasizing its restoration project that began in 2008. It is important to note that not all tourists are professionals, from what was reported in UGC; their purpose for visiting ancient towns is primarily experiential since the large proportion of their comments did not reach the core part of ancient towns, which means the historical and cultural factors. DMO failed to resonate with tourists because their emphasis on historical and cultural aspects was not integrated into the experience of tourists, which further led to the gap between DMO and UGC. However, it is worth acknowledging that while UGC reflects immediate perceptions, they can be fragmented. In order to obtain a comprehensive understanding, these perceptions require further research and refinement.

5.2. Activities Evoke Sentiments Which in Turn Shape the Differentiated Perception of the Destination Image

The study revealed that the perception of the destination image of ancient towns follows a systematic process of "Space-Activity-Sentiment". As tourists engage in deeper activities, the frequency of sentiment keywords corresponding to more specific spaces gradually increases (Figure 12). The deeper their engagement in activities, the stronger their sentiment becomes. Similarly, the higher the alignment between space and activities, the more the tourists' experiences meet their expectations, leading to a stronger positive emotional response and a heightened perception of the space. Conversely, low alignment makes emotions lean toward the negative side, and the perception of the space could be diminished. For instance, consider the "Wu Yue" featured landmark in Fengjing, represented by the boundary monument and river. Since there are no spaces for visitors to stay and no corresponding activities provided in the vicinity, tourists often pass by quickly without forming emotional connections. As a result, the spaces have low perception.

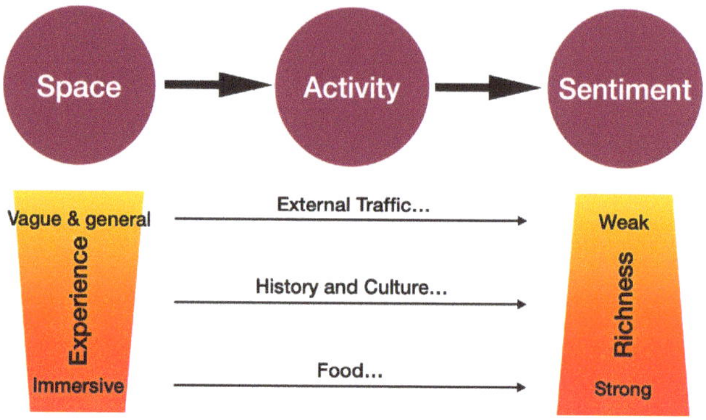

Figure 12. Relationship between three dimensions.

The perspective of the DMO of Zhaojialou sees culinary experiences as a small fraction (3.35%) of the town's unique offerings. However, in the tourists' perception, Zhaojialou Ancient Town is recognized as a destination famous for food (28.75%), which overlaps with the previous research [27]. This branding image does not have a direct connection with the town's historical and architectural attributes.

5.3. *The Culture of the Ancient Town Continues and Regenerates through Tourism*

The differences between DMO and UGC reveal that some non-traditional and emerging elements have captured greater interest among tourists. For instance, in Fengjing, tourists mentioned "Nothing but Thirty", Starbucks, "Three Bridges Courtyard", and "People's Commune". Additionally, "Wu You Xian", "Lao Ba Yang Restaurant", soup duns, and shumai were mentioned by tourists in Zhaojialou. These elements are not the traditional aspects of these ancient towns, yet they are prosperous. Media plays a crucial role in establishing and maintaining the "intense pleasure" that people associate with a destination, often through means such as movies, TV shows, or travelogs [47]. Fengjing Ancient Town, being a location featured in various forms of literature and art, has been created and given new cultural value alongside these works. This has sparked emotional resonance among audiences. When these audiences visit Fengjing in person, the scenes they encountered in media will rekindle their emotional attachment to the place. Similarly, the People's Commune Memorial Museum in Fengjing has the same emotionally evocative effect on visitors who have experienced the historical phase of the people's commune. Tourists are drawn to Zhaojialou to experience emerging cuisines famous on social media. Food production is inherently a part of the cultural landscape, and dining experiences during the tourism process can fulfill not only material needs but also cultural and social desires of visitors [48–50]. The initial formation of ancient towns was primarily for the exchange of goods among the surrounding rural residents, and temporary markets gradually evolved into permanent shops, eventually developing into commerce-centered communities [51]. Ancient towns provide a physical space where people engage in "exchange", leading to the development of their culture. A theory from heritage tourism has also claimed that experience needs to be managed, which means that heritage tourism has the potential to revitalize a destination, enhance its infrastructure and environment, and serve as both a revenue generator and a source of employment, and in this way, historically significant buildings, existing craft expertise and traditional performances will not disappear for lack of social, political and financial support [40,52,53]. In the cases of Fengjing and Zhaojialou, new elements suit the preferences of contemporary tourists and encourage them to come and "exchange". During this process, the cultural essence of the ancient towns is sustained, and new cultural elements may emerge as a result.

6. Conclusions

Theoretically, the results confirm the former research findings that a gap between tourists' perception (UGC) and the official promotion by Destination Marketing Organization (DMO) exists [10,54]. Tourists often share content in their travel log that deviates from the image crafted by destination marketing efforts. DMOs should adjust their promotional strategies based on tourist perceptions and the actual situation. They should go beyond focusing solely on presenting general but vague descriptions of ancient towns, instead emphasizing the incorporation of emotionally rich content, contributing to the development of a stronger destination image [10,55]. Lastly, engaging in appropriate interactions with UGC platforms fosters emotional connection and identification with tourists, thereby effectively guiding the formation of the destination image. Furthermore, while "Negative" offers direct and accurate insights into tourist preferences which can aid DMOs in adjusting planning strategies or policies [20], it is important to recognize that emotional expressions on online platforms can be overestimated and inclined to criticize destinations with negative emotions. DMOs should interpret the perceptions and preferences of tourists reflected in UGC content objectively and comprehensively.

Practically, the breakthrough of this study lies in its utilization of a machine learning approach to analyzing online data, especially for the ancient towns. The acquisition of big data frees the research on destination image from the constraints of traditional small-scale surveys, granting this kind of research increasing credibility and objectivity. The data-driven classification approach employed in this study effectively combines the content of comments, aligning with the principles of grounded theory [10]. For handling larger volumes of data, employing machine learning approaches tends to yield more comprehensive and objective results. The analysis of UGC also provides novel insights for the managers of Fengjing and Zhaojialou, motivating them to explore tailored strategies for improving the tourist experience in these historic towns. Moreover, it aids in the identification of previously overlooked areas, enabling the integration of their planning strategies with tourist preferences in future endeavors. This, in turn, promotes the balanced development and preservation of these ancient towns, ultimately offering valuable insights for the protection and development of other similar historic sites.

There are certain limitations in this study, such as that the process of extracting and analyzing keyword content could potentially overlook nuances within the context and language used. Moreover, interventions based on expert knowledge and experience might be necessary to accurately classify keywords. Additionally, given the relatively narrow scope of UGC content sources, this study might inadvertently exclude insights from tourists beyond those using Sina Weibo. Future research will strive to encompass a broader range of UGC platforms, encompassing diverse content forms such as travel logs and photographs, thereby capturing a more multifaceted panorama of tourist perceptions. Furthermore, in order to let the machine learning fit more on specific tasks, pre-trained language models which provide fine-tuning functions should be considered in the future.

Author Contributions: Conceptualization, J.D. and L.W.; methodology, J.D.; software, J.D.; validation, J.D., L.W., Z.T. and D.C.; formal analysis, J.D., M.H. and D.C. investigation, J.D. and L.W.; resources, L.W.; data curation, J.D., D.C. and Z.T.; writing—original draft preparation, J.D.; writing—review and editing, J.D. and L.W.; visualization, J.D.; supervision, L.W.; project administration, L.W.; funding acquisition, L.W. All authors have read and agreed to the published version of the manuscript.

Funding: This research was funded by the Shanghai Planning Office of Philosophy and Social Science, grant number 2019BCK001.

Data Availability Statement: Not applicable.

Conflicts of Interest: The authors declare no conflict of interest.

References

1. Zhang, D.; Qiu, F. A Summary of ancient town tourism studies at home and abroad. *Tour. Trib.* **2011**, *26*, 86–92.
2. Zhao, Z.; Zhu, M.; Hao, X. Share the Gaze: Representation of destination image on the Chinese social platform WeChat Moments. *J. Travel Tour. Mark.* **2018**, *35*, 726–739. [CrossRef]
3. Arefieva, V.; Egger, R.; Yu, J. A machine learning approach to cluster destination image on Instagram. *Tour. Manag.* **2021**, *85*, 104318. [CrossRef]
4. Vu, H.Q.; Li, G.; Law, R.; Ye, B.H. Exploring the travel behaviors of inbound tourists to Hong Kong using geotagged photos. *Tour. Manag.* **2015**, *46*, 222–232. [CrossRef]
5. Lew, A.; McKercher, B. Modeling tourist movements: A local destination analysis. *Ann. Tour. Res.* **2006**, *33*, 403–423. [CrossRef]
6. Picazo, P.; Moreno-Gil, S. Analysis of the projected image of tourism destinations on photographs: A literature review to prepare for the future. *J. Vacat. Mark.* **2019**, *25*, 3–24. [CrossRef]
7. Mata, I.L.; Fossgard, K.; Haukeland, J.V.; Landa Mata, I. Do visitors gaze and reproduce what destination managers wish to commercialise? Perceived and projected image in the UNESCO World Heritage area West Norwegian Fjords. *Int. J. Digit. Cult. Electron. Tour.* **2018**, *2*, 294–321. [CrossRef]
8. Leung, D.; Law, R.; van Hoof, H.; Buhalis, D. Social Media in Tourism and Hospitality: A Literature Review. *J. Travel Tour. Mark.* **2013**, *30*, 3–22. [CrossRef]
9. Nusair, K.; Hua, N.; Ozturk, A.; Butt, I. A theoretical framework of electronic word-of-mouth against the backdrop of social networking websites. *J. Travel Tour. Mark.* **2017**, *34*, 653–665. [CrossRef]
10. Költringer, C.; Dickinger, A. Analyzing destination branding and image from online sources: A web content mining approach. *J. Bus. Res.* **2015**, *68*, 1836–1843. [CrossRef]
11. Govers, R.; Go, F.M. Projected destination image online: Website content analysis of pictures and text. *Inf. Technol. Tour.* **2004**, *7*, 73–89. [CrossRef]
12. Egger, R.; Gumus, O.; Kaiumova, E.; Mükisch, R.; Surkic, V. Destination image of DMO and UGC on Instagram: A machine-learning approach. In Proceedings of the Information and Communication Technologies in Tourism 2022: Proceedings of the ENTER 2022 eTourism Conference, Online, 11–14 January 2022; pp. 343–355.
13. Fatanti, M.N.; Suyadnya, I.W. Beyond user gaze: How Instagram creates tourism destination brand? *Procedia-Soc. Behav. Sci.* **2015**, *211*, 1089–1095. [CrossRef]
14. Yu, C.-E.; Xie, S.Y.; Wen, J. Coloring the destination: The role of color psychology on Instagram. *Tour. Manag.* **2020**, *80*, 104110. [CrossRef]
15. Huerta-Álvarez, R.; Cambra-Fierro, J.J.; Fuentes-Blasco, M. The interplay between social media communication, brand equity and brand engagement in tourist destinations: An analysis in an emerging economy. *J. Destin. Mark. Manag.* **2020**, *16*, 100413. [CrossRef]
16. Pike, S. Destination image analysis—A review of 142 papers from 1973 to 2000. *Tour. Manag.* **2002**, *23*, 541–549. [CrossRef]
17. Zhong, L.; Wu, B.; Xu, X.; Xu, Y. Literature review of overseas research on destination perception. *Hum. Geogr.* **2013**, *28*, 13–19. [CrossRef]
18. Driscoll, A.; Lawson, R.; Niven, B. Measuring tourists' destination perceptions. *Ann. Tour. Res.* **1994**, *21*, 499–511. [CrossRef]
19. Wong, C.U.I.; Qi, S. Tracking the evolution of a destination's image by text-mining online reviews-the case of Macau. *Tour. Manag. Perspect.* **2017**, *23*, 19–29. [CrossRef]
20. Huai, S.; Van de Voorde, T. Which environmental features contribute to positive and negative perceptions of urban parks? A cross-cultural comparison using online reviews and Natural Language Processing methods. *Landsc. Urban Plan.* **2022**, *218*, 104307. [CrossRef]
21. Xiang, Z.; Du, Q.; Ma, Y.; Fan, W. A comparative analysis of major online review platforms: Implications for social media analytics in hospitality and tourism. *Tour. Manag.* **2017**, *58*, 51–65. [CrossRef]
22. Zhang, R.; Zhang, J. Perception of tourism image of Chenshan Botanical Garden in Shanghai based on web text analysis and IPA model. *Chin. Landsc. Archit.* **2019**, *35*, 83–87. [CrossRef]
23. Li, J.; Xu, L.; Tang, L.; Wang, S.; Li, L. Big data in tourism research: A literature review. *Tour. Manag.* **2018**, *68*, 301–323. [CrossRef]
24. Koblet, O.; Purves, R.S. From online texts to Landscape Character Assessment: Collecting and analysing first-person landscape perception computationally. *Landsc. Urban Plan.* **2020**, *197*, 103757. [CrossRef]
25. Song, Y.; Wang, R.; Fernandez, J.; Li, D. Investigating sense of place of the Las Vegas Strip using online reviews and machine learning approaches. *Landsc. Urban Plan.* **2021**, *205*, 103956. [CrossRef]
26. Zhang, L.; Yang, K.; Liu, B.; Liu, S. A study on the regional characteristics perception of ancient towns in the south of the Yangtze River based on different perspectives of tourists and residents: A case study of Tongli Ancient Town. *Chin. Landsc. Archit.* **2019**, *35*, 10–16.
27. Tan, J.; Hua, Z.; Shan, L. Research on image of rural tourism in Shanghai based on micro-blog content analysis. *Shanghai Urban Manag.* **2018**, *27*, 29–35.
28. Xu, X.; Xu, C.; Gao, J. Using VEP method to understand tourists' perception on ancient water towns in southern Yangtze River: A case study of Shanghai Fengjing Ancient Town. *Areal Res. Dev.* **2017**, *36*, 121–126.
29. Dong, Y.; Xu, X. Using web2.0 interaction platform to explore tourism image:a case study of Fengjing Ancient Town. *Tour. Forum* **2017**, *10*, 37–48. [CrossRef]

30. Urry, J.; Larsen, J. *The Tourist Gaze 3.0*; Sage: Newcastle upon Tyne, UK, 2011.
31. Zhou, S. The revival strategy of the "Boundary River Culture" town—Based on the study of Shanghai Fenging Ancient Town. *Archit. Cult.* **2019**, *7*, 236–238.
32. Stemler, S. An overview of content analysis. *Pract. Assess. Res. Eval.* **2000**, *7*, 17.
33. Levy, O.; Goldberg, Y.; Dagan, I. Improving Distributional Similarity with Lessons Learned from Word Embeddings. *Trans. Assoc. Comput. Linguist.* **2015**, *3*, 211–225. [CrossRef]
34. Nawangsari, R.P.; Kusumaningrum, R.; Wibowo, A. Word2Vec for Indonesian Sentiment Analysis towards Hotel Reviews: An Evaluation Study. *Procedia Comput. Sci.* **2019**, *157*, 360–366. [CrossRef]
35. Sun, Y.; Liang, C.; Chang, C.-C. Online social construction of Taiwan's rural image: Comparison between Taiwanese self-representation and Chinese perception. *Tour. Manag.* **2020**, *76*, 103968. [CrossRef]
36. Song, Y.; Shi, S.; Li, J.; Zhang, H. Directional Skip-Gram: Explicitly Distinguishing Left and Right Context for Word Embeddings. In Proceedings of the 2018 Conference of the North American Chapter of the Association for Computational Linguistics: Human Language Technologies, New Orleans, LA, USA, 1–6 June 2018; pp. 175–180.
37. Zhu, L.; Ma, B.; Zhao, X. Clustering validity analysis based on silhouette coefficient. *J. Comput. Appl.* **2010**, *30*, 139–141+198.
38. Weber, R.P. *Basic Content Analysis*; Sage: Newcastle upon Tyne, UK, 1990; Volume 49.
39. Crowley, B.P.; Delfico, J.F. *Content Analysis: A Methodology for Structuring and Analyzing Written Material*; United States General Accounting Office (GAO), Program Evaluation and Methodology Division: Washington, DC, USA, 1996.
40. Munar, A.M.; Ooi, C.-S. *What Social Media Tell Us about the Heritage Experience*; CBS: Singapore, 2012.
41. Munar, A.M. Digital exhibitionism: The age of exposure. *Cult. Unbound* **2010**, 2, 401–422. [CrossRef]
42. Anholt, S. Nation-brands of the twenty-first century. *J. Brand Manag.* **1998**, *5*, 395–406. [CrossRef]
43. Murphy, L.; Benckendorff, P.; Moscardo, G. Destination brand personality: Visitor perceptions of a regional tourism destination. *Tour. Anal.* **2007**, *12*, 419–432. [CrossRef]
44. Morgan, N.J.; Pritchard, A.; Piggott, R. Destination branding and the role of the stakeholders: The case of New Zealand. *J. Vacat. Mark.* **2003**, *9*, 285–299. [CrossRef]
45. Gonzalez, A.M.; Bello, L. The construct "lifestyle" in market segmentation: The behaviour of tourist consumers. *Eur. J. Mark.* **2002**, *36*, 51–85. [CrossRef]
46. Li, D.; Zhang, J.; Lu, S.; Zhong, J. A Study on tourism image positioning of ancient water towns in Suzhou City—Take some ancient water towns as examples. *Areal Res. Dev.* **2007**, *2*, 81–85.
47. Urry, J. *The Tourist Gaze*; Sage: Newcastle upon Tyne, UK, 2002.
48. Tieskens, K.F.; Van Zanten, B.T.; Schulp, C.J.E.; Verburg, P.H. Aesthetic appreciation of the cultural landscape through social media: An analysis of revealed preference in the Dutch river landscape. *Landsc. Urban Plan.* **2018**, *177*, 128–137. [CrossRef]
49. Guan, J. Review of gourmet abroad and tourism studies: Talking about the extensive phenomenon of the concept of gourmet travel. *Tour. Trib.* **2012**, *27*, 85–92.
50. Shen, X.; Chen, M.; Ge, M.; Padua, M.G. Examining the Conceptual Model of Potential Urban Development Patch (PUDP), VOCs, and Food Culture in Urban Ecology: A Case in Chengdu, China. *Atmosphere* **2022**, *13*, 1369. [CrossRef]
51. Fei, H. Development of Chinese cities and towns. *Chin. Rural Econ.* **1996**, 3, 3–5+10.
52. Roche, M. Mega-Events and Micro-Modernization: On the Sociology of the New Urban Tourism. *Br. J. Sociol.* **1992**, *43*, 563–600. [CrossRef]
53. Willim, R. Looking with new eyes at the old factory-On the rise of industrial cool. In *Experiencescapes. Tourism, Culture and The New Economy*; Copenhagen Business School Press: Frederiksberg, Denmark, 2005.
54. MacKay, K.J.; Fesenmaier, D.R. An Exploration of Cross-Cultural Destination Image Assessment. *J. Travel Res.* **2000**, *38*, 417–423. [CrossRef]
55. Mazanec, J.A. Tourism-receiving countries in connotative Google space. *J. Travel Res.* **2010**, *49*, 501–512. [CrossRef]

land

Article

Evaluating Public Satisfaction and Its Determinants in Chinese Sponge Cities Using Structural Equation Modeling

Rui Wang [1], Youyou Wu [1], Jiaqi Niu [2], Na Wang [3] and Hong Wu [4,*]

[1] School of Art and Design, Wuhan University of Technology, Wuhan 430070, China; ruiwang1012@whut.edu.cn (R.W.); 343546@whut.edu.cn (Y.W.)
[2] The Bartlett School of Architecture, University College London, UCL, Gower Street, London WC1E 6BT, UK; ucbqniu@ucl.ac.uk
[3] SWA Group, 712 Main St., Floor 6, Houston, TX 77002, USA; nwang@swagroup.com
[4] Department of Landscape Architecture, The Pennsylvania State University, University Park, State College, PA 16802, USA
* Correspondence: huw24@psu.edu

Abstract: Public satisfaction is an important indicator of the success of environmental policies and management practices. China's sponge city development (SCD) initiative was launched in 2014 and has received international attention for its technical advancements and environmental achievements. Public satisfaction, however, has not been fully investigated in cities transformed by SCD. This study uses public surveys and structural equation modeling to evaluate people's satisfaction with SCD in four pilot sponge cities, and how familiarity with SCD, perceived benefits, concerns about adverse effects, and trust in government influence satisfaction levels. The results show that people in the four cities were, on average, slightly satisfied with SCD. Familiarity, perceived benefits, and trust in government were significant determinants of public satisfaction. On the contrary, concerns about the adverse effects of SCD did not significantly influence people's satisfaction. Moreover, a mismatch was found between government-led evaluation outcomes and satisfaction measured here. This study highlights the importance of social and perceived values in shaping people's satisfaction with SCD and provides suggestions for management strategies for enhancing public satisfaction, ultimately supporting the long-term effectiveness of urban stormwater management programs.

Keywords: public satisfaction; perceptions; concern; trust; sponge city development; green stormwater infrastructure; structural equation modeling

Citation: Wang, R.; Wu, Y.; Niu, J.; Wang, N.; Wu, H. Evaluating Public Satisfaction and Its Determinants in Chinese Sponge Cities Using Structural Equation Modeling. *Land* **2024**, *13*, 1225. https://doi.org/10.3390/land13081225

Academic Editors: Xiwei Shen, Yang Song, Niall Kirkwood, Mary Padua, Bo Zhang and Rosalea Monacella

Received: 28 June 2024
Revised: 1 August 2024
Accepted: 5 August 2024
Published: 7 August 2024

1. Introduction

Public satisfaction with environmental conditions is closely related to people's quality of life, making it a critical aspect for governments to monitor and improve in order to enhance environmental policymaking and resident well-being [1,2]. Effective environmental management also requires the cooperation of individuals and communities in supporting the operation and maintenance of infrastructure to achieve sustainable outcomes [3]. Conversely, dissatisfaction with environmental quality and related policies can lead to diminished public trust in government and increased participation in environmental activism, such as protests and campaigns [4]. Therefore, evaluating public satisfaction with environmental governance is essential to addressing potential conflicts between the local communities and authorities on environmental issues [5].

Due to climate change and urbanization, flooding and stormwater pollution are significant concerns for cities worldwide. Various water management initiatives have been implemented in Western countries to address these issues, such as low-impact development (LID) in the USA, Sustainable Urban Drainage Systems (SuDSs) in the United Kingdom, and Blue-Green Cities in Australia [5]. These initiatives aim to manage urban flooding, enhance urban water ecosystems, and promote social well-being.

In China, urban flooding has severely threatened public safety and quality of life. For example, in 2023, flooding affected 2.59 million people across 20 provinces, resulting in direct economic losses of 3.25 billion RMB, with more than 7300 houses collapsed or damaged [6]. Urban central areas are particularly vulnerable to flooding and stormwater pollution due to outdated infrastructure and inadequate maintenance, exacerbating the decline in residents' quality of life [7]. Many old urban neighborhoods still use combined sewer systems that channel rainwater and wastewater into centralized treatment plants. In severe wet weather conditions, these systems often result in the overflow of untreated wastewater into nearby water bodies, raising serious concerns over water quality [8].

By contrast, a more novel approach to stormwater management involves using decentralized green stormwater infrastructure (GSI), such as rain gardens, bioswales, green roofs, and grass filters, to intercept, infiltrate, retain, and purify stormwater runoff at its source before it enters the underground drainage system [9]. GSI components are typically small-scale and require minimal space, making them ideal for neighborhood renovations, street renewals, and the development of urban parks and green spaces [10]. Due to their natural elements—such as soil, plants, and stones—GSI facilities can often add a natural look to the urban environment if properly designed [11]. Additionally, as part of urban green spaces, GSI can provide other benefits, such as improving air quality, regulating microclimates, and providing recreational and educational opportunities for residents [12]. However, there are also concerns about possible disservices of GSI developments, such as when GSI investments spur the process of gentrification, leading to displacement and structural change within communities [13].

With the pros and cons of GSI, China's central government launched the sponge city development (SCD) initiative in 2014 and designated 30 pilot cities in 2015 and 2016 to test the capability of GSI, among other approaches [5]. According to the technical guide for SCD [14], additional approaches include upgrading outdated underground pipe systems, renewing watershed management plans, and developing flood warning systems at city or regional levels [15]. SCD is intended to offer a comprehensive upgrade to urban water systems through various interventions at multiple scales to more effectively address urban water challenges [16]. In 2019, the Chinese central government conducted a comprehensive performance evaluation of pilot sponge cities, rating 6 (i.e., Pingxiang, Chizhou, Suining, Nanning, Zhenjiang, and Baicheng) out of 16 pilot cities as excellent [17]. However, the specific assessment criteria and procedures were unavailable to the public. Consequently, it is unclear whether public satisfaction was a criterion considered in the assessment and to what extent residents are satisfied with SCD initiatives.

There is a general lack of empirical studies revealing public satisfaction with SCD in the pilot cities. Therefore, comparing measured public satisfaction levels with the performance ratings given by the central government can offer valuable insights into the social implications of SCD and identify areas for improvement from the public's perspective. Some studies explored public satisfaction within a certain project type. For example, Gu et al. [18] specifically focused on residents' satisfaction with SCD in old urban neighborhood renewal projects. They found that dissatisfaction was primarily related to maintenance and the participation approach during the renewal, which echoed another study indicating that nearly half of the residents were unaware of sponge city transformations in their communities [19]. Several other studies identified some sources of dissatisfaction without systematically investigating overall satisfaction from the public's perspective. For example, Wang et al. [20] interviewed SCD experts and revealed residents' complaints about design plans in neighborhood transformations in Zhenjiang. Dai et al. [21] reported residents' dissatisfaction with construction waste being dumped in neighborhoods without prior notice in Wuhan. Hawken et al. [22] surveyed satisfaction with SCD based on expert perspectives, finding that long-term care and monitoring, as well as the quality of construction and initial maintenance, often received low satisfaction ratings.

Regarding the determinants of public satisfaction, most studies have focused on how environmental factors affect people's behavior and life satisfaction. For instance, numerous studies have examined the effects of air quality, water pollution, and flood hazards on people's life satisfaction (e.g., subjective well-being) [23–25]. Other studies have identified physical and social attributes, such as the variety of amenities, cleanliness, and maintenance, as predictors of satisfaction with urban environments [26,27]. In the context of stormwater management, perceived factors have been frequently examined for their associations with people's stormwater-related behaviors rather than their satisfaction with stormwater policies. For example, Pradhananga and Davenport [28] found that a sense of attachment to the community's physical and social environment, as well as environmental concerns, impacts civic engagement in stormwater management. Wang et al. [29] studied the effects of a composite variable of perceived value (e.g., perceived effects on drinking water conditions, ecological and living environment, and housing prices) on people's willingness to pay for SCD. Their results indicated that perceived value could predict stormwater behaviors. However, it remains unclear whether these perceived values directly affect satisfaction with SCD.

Additionally, public trust can play a significant role in assessing satisfaction with environmental governance. Venkataramanan et al. [30] conducted a systematic review and found that a lack of trust in governments, along with knowledge and familiarity with flood-related issues, influences people's flood-related behaviors. In China, the public has generally exhibited high trust in the government in recent decades due to economic prosperity and effective public communication [31,32]. However, recent national and local press [33] indicate that urban flooding has resurfaced in cities that underwent SCD since 2015. These recurring events could decrease trust in the government's commitment to urban water management. Therefore, it is critical to monitor the status of public trust and its impact on satisfaction with SCD.

Moreover, research has shown that residents' satisfaction with SCD in old neighborhood renewal projects was affected by demographic factors, including age, education, and income [18]. Homeownership is another critical variable in China that symbolizes the Chinese dream and affects people's subjective well-being or life satisfaction [34]. Previous research has revealed that homeowners tend to be more satisfied with their housing and living environment conditions than renters due to better access to education, a stronger sense of belonging, and greater involvement in social activities [35]. However, little is known about whether homeownership leads to differing levels of satisfaction with SCD.

Overall, existing research highlights a gap in the comprehensive evaluation of public satisfaction with SCD and a lack of investigation into how social and perceived factors influence this satisfaction. Additionally, there is limited understanding of the disparity between public satisfaction levels and the central government's official assessments, creating challenges for future SCD policy and construction enhancements. To address these empirical research gaps, this study focused on four pilot sponge cities as case study areas to examine both the current status and determinants of public satisfaction with SCD. Specifically, this study investigated the effects of four factors: trust in the government, perceived benefits of SCD, concerns about adverse effects of SCD, and familiarity with SCD. It also explored differences in satisfaction levels between homeowners and renters, as well as among residents of the four cities. Insights from this research can inform better sponge city policies and implementation practices that align with public satisfaction, thereby enhancing social acceptance and garnering public support for SCD as a vital approach to urban stormwater management.

The study aimed to answer the following research questions:

(1) What is the status of overall public satisfaction, perceived benefits provided by SCD, concerns about adverse effects of SCD, familiarity with SCD, and trust in the government?
(2) How do satisfaction levels with SCD vary between homeowners and renters and across different cities?
(3) How do perceived benefits of SCD, trust in government, concerns about adverse effects of SCD, and familiarity with SCD affect public satisfaction with SCD?

2. Materials and Methods

2.1. Study Area

This study focused on four pilot sponge cities: Jiaxing, Zhenjiang, Pingxiang, and Chizhou (Figure 1). All four cities are located in the subtropical monsoon climate zone in southeastern China, with annual precipitation over 1000 mm. These cities also have abundant water resources and complex water ecosystems formed by rivers, lakes, and wetlands. However, rapid urbanization and outdated infrastructure have led to common ecological problems such as pluvial flooding and non-point source pollution in recent decades. The four cities vary in overall economic conditions and sponge city construction investments by local governments. For instance, Zhenjiang's annual per capita GDP and government SCD investment (USD 1.34 billion) far exceeded the other three cities (USD 570–790 million) over the three-year pilot period. In addition, the SCD pilot implementation areas differ in size; Zhenjiang and Pingxiang have larger areas (29–32.98 square kilometers) than Jiaxing and Chizhou (18–18.5 square kilometers).

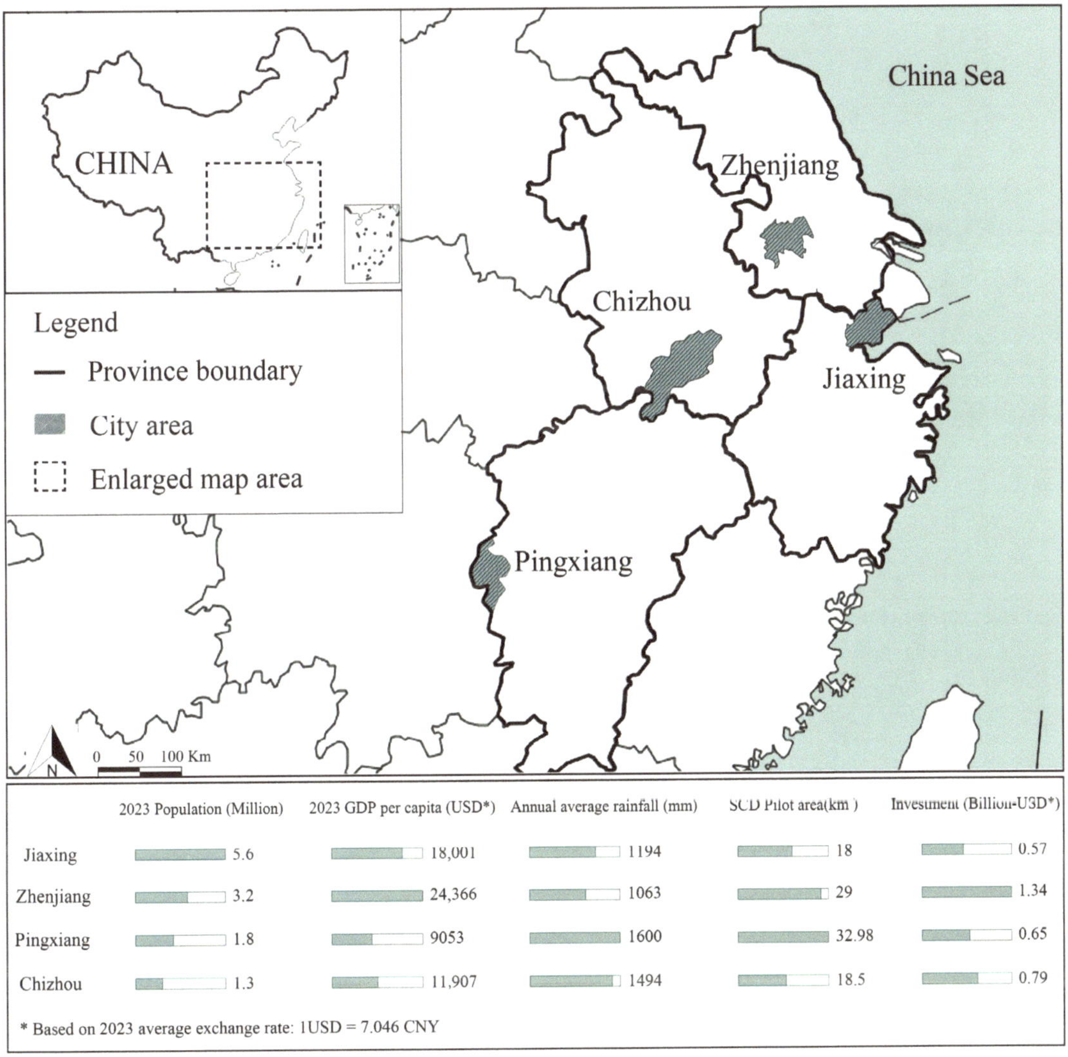

Figure 1. Geographic locations and background of studied pilot cities.

2.2. Data Collection

The sampling locations included residential areas and adjacent parks and plazas renovated or newly developed under SCD from 2014 to 2018. These sites feature various GSI elements, such as stormwater planters, rain gardens, sunken green spaces, and stormwater ponds (Figure 2), which significantly changed the visual characteristics of the landscape. A convenient sampling method was employed for this study. From October to December 2019, eight trained surveyors conducted intercept surveys using tablets to collect data via the Qualtrics platform. Surveyors randomly selected individuals at the sampling locations and asked if they had heard of SCD and their age. Those at least 18 years old who had heard of SCD were then introduced to the study's purpose and procedures. Verbal consent was obtained from the participants, who were informed that their participation was voluntary and that their personal information would remain confidential. Before starting the survey, participants received an introduction page detailing the study's purpose and the principal investigator's contact information. They were also informed that they could skip questions or withdraw from the survey at any time. The face-to-face nature of the survey allowed surveyors to provide additional information and address any questions participants might have. The surveyors also read the questions aloud for senior participants who had difficulty reading the texts on the tablets. At the end of the survey, participants received a small gift (a shopping bag) as a token of appreciation. All the survey materials were reviewed by the Institutional Review Board at the Pennsylvania State University (STUDY00013257).

Rain garden in Funan Residence, Zhenjiang

Stormwater planter in Yanyu Residence, Jiaxing

Bioswale in Jindian Residence, Pingxiang

Stormwater pond in Tiantang River Park, Chizhou

Figure 2. GSI facilities in surveyed cities (source: first author).

The questionnaire asked questions of six aspects: overall satisfaction with SCD, familiarity with SCD, perceived benefits of SCD, concerns about potential adverse effects of SCD, trust in governments' dedication to SCD, and demographic information. First, overall satisfaction with SCD was measured by asking, "How would you rate your level of satisfaction with the overall SCD in your city on a 5-point scale (1 = Completely dissatisfied; 5 = Completely satisfied)?" An "opt-out" option was provided for participants who preferred not to express an opinion on this topic. Second, familiarity with SCD was measured by asking, "How would you rate your familiarity with the concept of SCD on a 5-point scale (1 = Very unfamiliar; 5 = Very familiar)?" Next, the perceived benefits of SCD, concerns about adverse effects of SCD, and trust in the government were regarded as composite concepts and measured by multiple observed items, respectively. For the perceived benefits of SCD, participants were asked, "To what degree have you perceived the following benefits provided by SCD (1 = Not perceived at all, 5 = Well perceived)?" Seven specific benefits were assessed, including flood mitigation, public safety, water quality improvement, ecological benefits, recreational opportunities, aesthetics value, and job creation. Similarly, concerns about SCD's adverse effects were measured through five observed items identified from the literature. These concerns included "SCD costs too much", "Disturbance to daily life", "Poor performance and outcomes of SCD construction", "Occupies too much public space", and "Unappealing landscape aesthetics". Trust in governments' dedication to SCD was measured by asking the participants, "How much do you believe the government can effectively manage the following five aspects of SCD (1 = Do not believe at all, 5 = Strongly believe)?" The aspects included collaboration across government sectors, professionalism of SCD employees, sound governmental decision making, commitment to long-term maintenance, and valuing of public participation [36]. Finally, demographic characteristics such as age, gender, homeownership status, and city of residence were recorded to provide context for the survey responses.

2.3. Data Analysis

Data analysis was performed in SPSS Amos (version 27). First, descriptive statistics of all the measures were reported. Differences in people's perceived benefits, concerns about SCD's adverse effects, and trust in government by their homeownership status were analyzed using independent sample t-tests. Subsequently, a One-Way Analysis of Variance (ANOVA) test was performed to detect statistically significant differences in public satisfaction levels among the four selected cities.

Next, the structural equation modeling (SEM) method was employed [37] to assess the causal effects of the four chosen factors—perceived benefits of SCD, concerns about adverse effects of SCD, familiarity with SCD, and trust in the government—on public satisfaction. SEM is a statistical technique that allows researchers to analyze complex relationships between observed variables and underlying latent constructs [38]. By integrating principles from both factor analysis and regression analysis, SEM facilitates the identification of latent constructs from multiple observed variables and enables the simultaneous prediction of their effects on a dependent variable [39]. Given that individuals can perceive multiple benefits or concerns related to SCD and assess their trust in governments based on multiple aspects, a single observed variable for each construct is insufficient to capture the full range of these perceptions. Therefore, latent variables were developed to encompass the multiple dimensions of the studied constructs, and SEM was chosen for its capability to test the effects of multiple latent variables on a dependent variable.

Before building the SEM, a confirmatory factor analysis (CFA) was conducted to ensure the validity of the measurements for the three constructs of perceived benefits, concerns, and trust [40]. The SEM was then used to test the effects of these constructs on overall satisfaction, accounting for additional factors such as familiarity with SCD and demographic characteristics.

More specifically, Cronbach's alpha ($\alpha > 0.6$) was used to assess measurement validity in the CFA. To evaluate the goodness of fit for both the CFA model and the structural model, six fit indices were applied with their respective acceptable thresholds: composite reliability (CR > 0.6), Chi-square/degree of freedom (1 < CMIN/DF < 3), comparative fit index (CFI > 0.90), Tucker–Lewis index (TLI > 0.90), root mean square error of approximation (RMSEA < 0.08), and standardized root mean square residual (SRMR < 0.08) [41]. Factor loadings for each observed variable were assessed in magnitude using these cutoffs: statistically significant loadings greater than 0.30 were considered sufficient, and those greater than 0.60 were considered high [42]. Lastly, the analysis focused on the full sample without testing differences by city, considering the broader spectrum of the sample population. Preliminary results showed that demographic status did not significantly affect overall satisfaction, so these variables were removed from the formal analysis.

3. Results

3.1. Descriptive Statistics

Out of 607 participants who filled out the questionnaires, 528 answered most of the survey questions. Fourteen participants opted out of the question of overall satisfaction with SCD and were listwise excluded from the analysis. Therefore, the final dataset included 514 responses (Pingxiang = 92, Chizhou = 156, Zhenjiang = 155, and Jiaxing = 111). Less than half of the participants were male (40.9%), and the average age was 38.

Participants reported a slight-to-moderate level of overall satisfaction with SCD in their cities (Mean = 3.60) (Table 1). Their familiarity with SCD was neutral (Mean = 2.95). The average level of perceived SCD benefits was moderate to high (Aggregate mean = 4.29), with means of individual benefits ranging from 4.17 to 4.39 (Table 2). The average concern level was moderate (Aggregate mean = 3.96), with means of individual concerns ranging from 3.63 to 4.29. Lastly, the average trust level was moderate to high (Aggregate mean = 4.10), with means of individual trust items ranging from 4.01 to 4.30.

Table 1. Descriptive statistics of the mean values of studied variables.

Variables	Min	Max	Mean	SD	N	Alpha
Overall satisfaction with SCD [1]	1	5	3.60	0.931	514	
Perceived benefits of SCD [2]	1	5	4.29	0.796	514	0.95
Concerns about adverse effects of SCD [3]	1	5	3.96	0.707	514	0.82
Familiarity with SCD [4]	1	5	2.95	1.068	514	
Trust in governments' dedication to SCD [5]	1	5	4.10	0.884	514	0.94
Homeownership (Homeowner = 1)	0	1	0.80	0.398	514	
Gender (Male = 1)	1	2	1.59	0.492	514	
Age			38.29	15.608	500	

[1] Measured on five-point scale where 1 = Completely unsatisfied; 5 = Completely satisfied. [2] Mean value was aggregated from seven benefit items. [3] Mean value was aggregated from five concern items. [4] Measured on five-point scale where 1 = Completely unfamiliar; 5 = Completely familiar. [5] Mean value was aggregated from five trust items.

Table 2. Descriptive statistics of perceived benefits of SCD, concerns, and trust.

	Code	Observed Variable	Min	Max	Mean (SD)	N
Perceived Benefits	PB1	Mitigate urban flooding	1	5	4.31 (0.902)	514
	PB2	Improve public safety	1	5	4.28 (0.904)	513
	PB3	Mitigate stormwater pollution	1	5	4.37 (0.868)	513
	PB4	Increase recreational opportunities	1	5	4.17 (0.943)	513
	PB5	Enhance ecological values	1	5	4.34 (0.885)	514
	PB6	Promote green industry and create job opportunities	1	5	4.19 (0.925)	514
	PB7	Improve landscape aesthetics	1	5	4.39 (0.862)	514

Table 2. *Cont.*

	Code	Observed Variable	Min	Max	Mean (SD)	N
Concerns	CN1	SCD costs too much	1	5	3.63 (1.011)	514
	CN2	SCD disturbs my daily life (e.g., waste and noise)	1	5	4.06 (0.921)	514
	CN3	SCD does not effectively solve stormwater problems	1	5	4.29 (0.808)	514
	CN4	SCD occupies too much public space	1	5	3.79 (0.973)	514
	CN5	SCD negatively affects landscape aesthetic	1	5	4.04 (0.923)	514
Trust	TS1	Government sectors collaborate well	1	5	4.30 (0.888)	514
	TS2	SCD practitioners are professional	1	5	4.07 (0.987)	514
	TS3	Governments make sound decisions	1	5	4.01 (0.987)	514
	TS4	Governments commit to long-term maintenance	1	5	4.08 (0.999)	514
	TS5	Governments value public participation	1	5	4.02 (1.058)	514

Regarding satisfaction levels across cities, the one-way ANOVA showed statistically significant differences between two sponge city groups [$F(3, 510) = 12.149$, $p < 0.001$] (Table 3). Tukey post hoc tests revealed no statistically significant difference in satisfaction between Pingxiang and Chizhou ($p = 0.664$) or between Zhenjiang and Jiaxing ($p = 0.393$). However, the satisfaction levels in Pingxiang and Chizhou were both statistically significantly higher than in Zhenjiang and Jiaxing (Table 3).

Table 3. Comparison of public satisfaction across cities.

City	City	Mean Difference	Std. Error	*p*-Value
Pingxiang (Mean = 3.92)	Chizhou (Mean = 3.79)	0.14	0.119	0.664
	Zhenjiang (Mean = 3.31)	0.61	0.119	0.000 ***
	Jiaxing (Mean = 3.49)	0.44	0.127	0.004 **
Chizhou	Zhenjiang	0.48	0.102	0.000 ***
	Jiaxing	0.30	0.112	0.036 *
Zhenjiang	Jiaxing	−0.18	0.112	0.393

Note: Significance codes—* $p \leq 0.05$, ** $p \leq 0.01$, *** $p \leq 0.001$.

Homeowners and renters were indifferent in multiple aspects, including overall satisfaction with SCD, familiarity with SCD, trust in government, and concerns about SCD's adverse effects (Table 4). However, homeowners' perceived benefits of SCD were statistically significantly lower than those reported by renters ($p = 0.039$).

Table 4. Comparing mean ratings for all variables between homeowners and renters.

Variables	Homeowner Mean (SD)	Renter Mean (SD)	*p*-Value
Overall satisfaction with SCD	3.60 (0.954)	3.60 (0.838)	0.201
Familiarity with SCD	3.03 (1.070)	2.62 (1.010)	0.862
Trust in governments' dedication to SCD	4.09 (0.905)	4.14 (0.796)	0.208
Perceived benefits of SCD	4.26 (0.829)	4.42 (0.627)	0.039 *
Concerns about the adverse effects of SCD	3.95 (0.710)	3.99 (0.701)	0.653

Note: Significance codes—* $p \leq 0.05$; All *p*-values were interpreted using Levene's test for equality of variance.

3.2. Structural Equation Modeling

The CFA examined the validity of the measurements of perceived benefits of SCD, concerns about the adverse effects of SCD, and trust in government. As shown in Figure 3, the factor loadings ranged from the lowest of 0.54 for CN1 (SCD costs too much) to the highest of 0.94 for TS4 (governmental commitment to long-term maintenance). All factor loadings exceeded 0.3, so no observed variables were removed from the CFA model. Following the correlation of two pairs of error terms based on the modification indices in Amos, all the goodness of fit indices met the cutoff requirements: $X^2 = 331.918$, $p < 0.001$; CMIN/DF = 2.912, CFI = 0.969, TLI = 0.963, RMSEA = 0.061, SRMR = 0.0532), indicating a good model fit for the CFA model. Therefore, the three major latent constructs were validated through CFA and were subsequently integrated into the SEM in the next step.

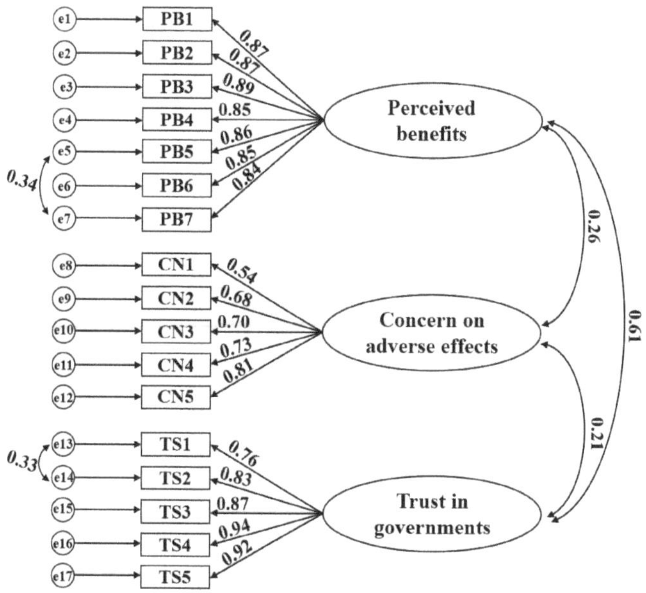

Figure 3. CFA of the three latent variables of perceived benefits, concerns, and trust ($X^2 = 331.918$, $p < 0.001$; CMIN/DF = 2.912, CFI = 0.969, TLI = 0.963, RMSEA = 0.061, SRMR = 0.0532).

Similarly, all goodness of fit indices of the full SEM model (Figure 4) met the required thresholds, indicating a good model fit ($X^2 = 399.916$, $p < 0.001$; CMIN/DF = 2.758, CFI = 0.965, TLI = 0.959, RMSEA = 0.059, SRMR = 0.0550). The model showed significant effects of perceived benefits, trust, and familiarity with SCD on overall satisfaction with SCD. More specifically, perceived benefits (weight = 0.425, $p < 0.001$) and trust (weight = 0.244, $p < 0.001$) exerted significant, positive, weak-to-moderate [42] effects on overall satisfaction. Familiarity (weight = 0.091, $p = 0.011$) exerted a significant but very weak effect on satisfaction. Concern, however, did not significantly affect overall satisfaction (weight = −0.018, $p = 0.658$).

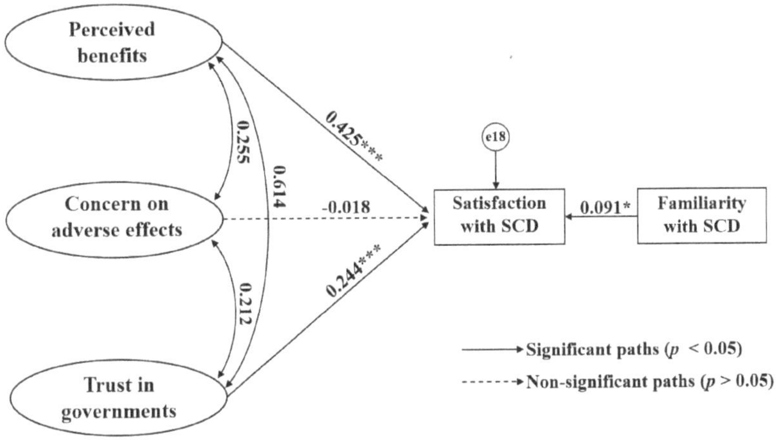

Figure 4. Structural equation model for the relationship of perceived benefits, concerns, trust, familiarity, and overall satisfaction with SCD (X^2 = 399.916, $p < 0.001$; CMIN/DF = 2.758, CFI = 0.965, TLI = 0.959, RMSEA = 0.059, SRMR = 0.0550). Significance codes: * $p \leq 0.05$, *** $p \leq 0.001$.

4. Discussion

Since 2014, the implementation of GSI and upgrades of the underground system under SCD policies have significantly transformed urban landscapes and altered social perceptions of stormwater management. The overall goal of this study was to elucidate public satisfaction with SCD and investigate its determinants. Overall, the results indicate positive feedback from the four surveyed pilot cities, including an average slight-to-moderate level of satisfaction, significant benefit perception, and high trust in government. In the subsequent discussion, we will first discuss the status of overall satisfaction and then explore the three major determinants of public satisfaction. The implications of these findings are summarized in the Conclusions section.

First, the slight-to-moderate average level of overall satisfaction (Aggregate mean = 3.60) indicates room for improving public satisfaction, especially in Zhenjiang (Mean = 3.31) and Jiaxing (Mean = 3.49) with the two lowest ratings. This satisfaction assessment from the public's perspective partially aligns with the central government's official evaluations, which rated Pingxiang, Chizhou, and Zhenjiang as excellent sponge cities, except for Jiaxing. Although the official evaluations of two of the three cities' overall performance align with public perceptions, the nuanced mismatch indicates that official metrics may not fully capture the public's experience or concerns, calling for more accessible and transparent criteria in official SCD evaluations. In the case of Zhenjiang, which has been frequently recognized in newsletters and reports for significant progress in enhancing its urban water environment, the potential overemphasis on environmental performance may have led to a neglect of social benefit provision, contributing to lower public satisfaction.

Second, the study results indicated that people perceived multiple benefits from SCD (Aggregate mean = 4.29, SD = 0.796), especially improvements in landscape aesthetics (Mean = 4.39, SD = 0.862) and water pollution reduction (Mean = 4.37, SD = 0.868). Indeed, China's SCD efforts have extended beyond the implementation of GSI, as each pilot city has formulated its own sponge programs to enhance benefit provision. For instance, Chizhou leveraged SCD funds to initiate river cleanup projects, renovate river banks, and add amenities to waterfront parks. In Pingxiang and Zhenjiang, urban parks have been upgraded with additional recreational facilities (e.g., jogging trails and seating benches) and educational opportunities (e.g., SCD exhibition halls) for urban residents (Figure 5). These enhancements suggest that the public can benefit from SCD in many other ways beyond the hydrological improvements in their living environment. These additional benefits are crucial for enhancing the public's quality of life and well-being.

Water quality has been improved with upgraded park space by sponge renovation project in Chizhou

Senior groups gathered at renovated river bank to chat and play card games in Yanliu yuan park in Chizhou

Pervious jogging trail added by sponge renovation project in Cuihu Park, Pingxiang

Sponge city themed exhibition hall in Sponge City Park, Zhenjiang

Figure 5. Infrastructures and benefits provided by sponge renovation projects (source: first author).

Subsequently, the SEM analysis revealed that people who perceived greater benefits (weight = 0.425, $p < 0.001$) were more satisfied with SCD. This indicates that enhancing people's perceived values of SCD is essential for increasing public satisfaction with sponge programs. This finding also aligns with previous research indicating that perceived values can positively influence people's willingness to pay for SCD [29]. Governments should provide the public with desired ecosystem benefits beyond the originally intended environmental benefits, such as recreation, aesthetics, and health [43]. Local governments tend to focus on maximizing the hydrological performance of sponge projects by converting extensive outdoor spaces into GSI. For example, some neighborhood sponge renovation projects in Zhenjiang replaced the original garden plants and trees with sunken areas featuring wild-looking plants [20]. This caused dissatisfaction among senior residents accustomed to the previous garden spaces. To avoid such issues, it is important to understand residents' preferences for the function and aesthetics of the community space before implementing GSI plans.

Third, the significant predictive effect of trust in governments (weight = 0.244, $p < 0.001$) demonstrated the importance of fostering trust between governments and the public. Previous research has shown that building social capital, such as through word-of-mouth within local communities, can enhance trust between governments and local stakeholders [4]. Increased trust not only boosts public satisfaction but also encourages public engagement and participation in stormwater management programs. Specifically, the results showed that two trust items, "TS3—Governments make sound decisions" and "TS5—Governments value public participation", received lower scores than the other trust items. This suggests a need for more transparent decision-making processes and platforms to improve communication and strengthen government-community relationships. Contrary to previous studies highlighting maintenance concerns [18], this study found relatively high trust in the government's commitment to long-term maintenance (Mean = 4.08, SD = 0.999). While our results reflect an overall positive relationship between local communities and SCD authorities when the survey was administrated, there is a risk of diminished trust if the outcomes of SCD fail to align with public expectations in the long run.

Fourth, our study found that people were not very familiar with SCD. Yet, the effect of familiarity on overall satisfaction was positive and significant, albeit at a low level (weight = 0.091, $p < 0.05$). The neutral-level familiarity echoed a previous study that revealed a lack of notice and low awareness of sponge renovation in their neighborhoods [19]. The positive impact of familiarity on satisfaction is also supported by a recent study that surveyed residents in Xi'an City, finding that residents of sponge communities were more satisfied with SCD than those of traditional communities [44]. Despite the weak effect, there remains potential to improve satisfaction by enhancing knowledge sharing through increased civic engagement and educational opportunities. These approaches were believed to boost perceived benefits and familiarity [45]. Since the sponge city concept and related GSI facilities remain relatively new, these results underscore the importance of raising awareness and educating the public about SCD and GSI.

Next, the lack of significant impact of concerns on satisfaction (weight = −0.018, $p = 0.658$) may be partially due to the mitigating effect of high trust in governments. Specifically, people tend to believe governments can effectively address the adverse effects of SCD, which reduces their attention to these issues in daily life. This high trust in governments aligns with previous studies on how trust influences flood risk perceptions and the positive effect of public participation in non-government organizations on the environment [46,47].

Additionally, the results indicated no significant differences between homeowners and renters regarding their overall satisfaction, trust in governments, concerns, or familiarity with SCD. This lack of differences contrasts with previous research suggesting that homeowners generally report higher satisfaction with their housing and living environment conditions than renters [35]. On the contrary, renters reported higher perceived benefits of SCD than homeowners ($p = 0.039$). Due to the transient nature of renting, renters might be more focused on and appreciative of improvements that enhance their quality of life in the short term, such as better stormwater management or enhanced green spaces. Homeowners, on the other hand, might be more focused on longer-term benefits or structural changes that impact property value and overall neighborhood quality, which may not align as closely with the specific benefits provided by SCD. Future research could further investigate the underlying reasons why homeownership affects benefit perceptions in SCD. Understanding these differences could help tailor SCD initiatives to better meet the diverse needs and expectations of various residential groups.

The significance of this study lies in its explicit exploration of public satisfaction with SCD and its determinants, addressing a notable gap in the literature. It bridges the gap between public perceptions and environmental policy, providing actionable insights that can enhance the design and implementation of sustainable urban solutions. By revealing how social perceptions and perceived benefits influence public satisfaction, the study deepens our understanding of the factors essential for effective stormwater management systems. The research is particularly relevant for countries with top–down governance systems, where integrating public perceptions into the development process has often been overlooked. Furthermore, by identifying differences between public and governmental perceptions, this study sets the groundwork for future research to refine SCD practices and better align environmental initiatives with community needs.

Finally, several limitations of this study should be acknowledged. First, SCD encompasses a range of construction approaches and management strategies across the pilot cities, which could lead to varying levels of public satisfaction depending on the specific aspects of SCD measured. This study only measured general satisfaction with SCD policies and construction as a whole. Future studies can be improved by evaluating multiple facets of SCD. For example, during the survey, some residents specifically expressed satisfaction with the national government's commitment to solving stormwater problems through SCD. However, they also voiced concerns about local governments' adherence to national policies and requirements. This suggests a possible difference in satisfaction with local vs. national authorities, which was not specifically measured in our study and presents a gap for future research. Second, the relatively small sample size in each surveyed city may limit the generalizability of the results to a broader population. Additionally, different cities' diverse local cultures may influence resident's needs and understanding of the living environment, leading to subjective satisfaction levels. Future studies can explore satisfaction across different cultural backgrounds and social contexts. Moreover, the survey was conducted in central areas of the cities, where many SCD projects were integrated with ongoing urban renewal efforts [7]. This integration could have created implementation challenges and conflicts between various regeneration policies, potentially skewing the perceived satisfaction with SCD. Future research can compare satisfaction levels in different city areas (e.g., central vs. suburban areas) to provide a more comprehensive understanding of public satisfaction with SCD.

5. Conclusions

Public satisfaction is an important indicator for evaluating the overall success of urban environmental policies and their implementation. Considering the impact of SCD on urban stormwater management and overall human well-being, it is crucial to address the current insufficient understanding of public satisfaction within the context of SCD. The lack of this information challenges local governments in refining existing GSI implementation strategies, potentially hampering the long-term management of urban water ecosystems. This study utilized public surveys and structural equation modeling to explore public satisfaction levels and their determinants. The findings reveal that individuals who perceive greater SCD benefits, hold higher trust in governments, and are more familiar with SCD are generally more satisfied. Additionally, renters perceived greater benefits from SCD than homeowners. Moreover, this study identified a minor discrepancy between the national government's performance evaluations of pilot sponge cities and public satisfaction levels.

Several important practical implications can be drawn from this study to improve people's overall satisfaction with SCD in southeast China and beyond. First, urban stormwater policymakers and program managers should recognize the significant role that local communities' social and perceived values play in shaping satisfaction with stormwater management programs. Investing in social and human capital is as crucial for improving technical and environmental effectiveness. Programs designed to build trust between governments and communities, such as civic engagement initiatives, training projects, and design workshops, can enhance familiarity with SCD and increase satisfaction. Second, since perceived benefits from SCD are positively associated with satisfaction, enhancing ecosystem benefit provision plans to deliver more of the desired benefits to local communities is essential. Aligning these benefits with community preferences can significantly improve the quality of life and overall satisfaction. Third, both local and national governments should enhance the SCD performance evaluation criteria by incorporating more social and cultural indicators. This will help ensure public satisfaction is considered in sponge city construction, addressing the existing mismatch between public and government perspectives and promoting the long-term social sustainability of SCD. Finally, given the subjective nature of public satisfaction based on the quality of SCD projects and local government services, developing local stormwater stewardship programs to periodically track and respond to residents' satisfaction can enhance government performance and the social and environmental impacts of SCD.

Author Contributions: Conceptualization, R.W. and H.W.; methodology, R.W. and H.W.; software, R.W. and Y.W.; validation, R.W. and H.W.; formal analysis, R.W.; resources, H.W.; writing—original draft preparation, R.W.; writing—review and editing, H.W., N.W. and J.N.; visualization, R.W. and Y.W.; supervision, H.W. All authors have read and agreed to the published version of the manuscript.

Funding: This research was supported by the Fundamental Research Funds for the Central Universities [No. 2024IVA066].

Data Availability Statement: The original contributions presented in this study are included in the article; further inquiries can be directed to the corresponding author.

Acknowledgments: The authors wish to thank all the interviewees for their participation. The authors also would like to express their gratitude to the editor and anonymous reviewers for their insightful and constructive comments.

Conflicts of Interest: Author Na Wang was employed by the company SWA Group. The remaining authors declare that the research was conducted in the absence of any commercial or financial relationships that could be construed as a potential conflict of interest.

References

1. Alkan, Y. User-focused alternative design model for Sarıçay channel and landscaping in terms of sports and health components: Çanakkale, Turkey. *Environ. Dev. Sustain.* **2020**, *23*, 7212–7229. [CrossRef]
2. Barclay, N.; Klotz, L. Role of community participation for green stormwater infrastructure development. *J. Environ. Manag.* **2019**, *251*, 109620. [CrossRef] [PubMed]
3. Bouckaert, G.; van de Walle, S. Comparing Measures of Citizen Trust and User Satisfaction as Indicators of 'Good Governance': Difficulties in Linking Trust and Satisfaction Indicators. *Int. Rev. Adm. Sci.* **2003**, *69*, 329–343. [CrossRef]
4. Brown, H.L.; Bos, D.G.; Walsh, C.J.; Fletcher, T.D.; RossRakesh, S. More than money: How multiple factors influence householder participation in at-source stormwater management. *J. Environ. Plan. Manag.* **2016**, *59*, 79–97. [CrossRef]
5. Chan, F.; Griffiths, J.; Higgitt, D.; Xu, S.; Zhu, F.; Tang, Y.-T.; Xu, Y.; Thorne, C. "Sponge City" in China—A breakthrough of planning and flood risk management in the urban context. *Land Use Policy* **2018**, *76*, 772–778. [CrossRef]
6. Ministry of Emergency Management of The People's Republic of China. Ministry of Emergency Management Released the National Natural Disaster Situation in May 2023. Available online: http://mp.weixin.qq.com/s?__biz=MjM5OTI3MjI1MQ==&mid=2651360134&idx=1&sn=c5d2a3fe3ea2bc01d89219062d96a57e&chksm=bcc221d28bb5a8c476583601d7178243cd232ca97a50ba0a86a99893835ee3f309a55459441f#rd (accessed on 5 July 2024).
7. Wang, R.; Wu, H.; Chiles, R.; Yang, Y. Sustainability outcomes and policy implications: Evaluating China's "old urban neighborhood renewal" experiment. *PLoS ONE* **2024**, *19*, e0301380. [CrossRef]
8. China National Radio. As the Rainy Season Approaches, Urban Waterlogging Reappears in Many Places. Can "Sponge Cities" Solve the Chronic Problem of Waterlogging? Available online: http://china.cnr.cn/gdgg/20220705/t20220705_525897763.shtml (accessed on 5 July 2022).
9. US EPA. What Is Green Infrastructure? [Overviews and Factsheets]. Available online: https://www.epa.gov/green-infrastructure/what-green-infrastructure (accessed on 30 September 2015).
10. Darnthamrongkul, W.; Mozingo, L.A. Toward sustainable stormwater management: Understanding public appreciation and recognition of urban Low Impact Development (LID) in the San Francisco Bay Area. *J. Environ. Manag.* **2021**, *300*, 113716. [CrossRef] [PubMed]
11. Echols, S.; Pennypacker, E. From Stormwater Management to Artful Rainwater Design. *Landsc. J.* **2008**, *27*, 268–290. [CrossRef]
12. Wang, R.; Brent, D.; Wu, H. Willingness to pay for ecosystem benefits of green stormwater infrastructure in Chinese sponge cities. *J. Clean. Prod.* **2022**, *371*, 133462. [CrossRef]
13. Ali, L.; Haase, A.; Heiland, S. Gentrification through Green Regeneration? Analyzing the Interaction between Inner-City Green Space Development and Neighbourhood Change in the Context of Regrowth: The Case of Lene-Voigt-Park in Leipzig, Eastern Germany. *Land* **2020**, *9*, 24. [CrossRef]
14. MHURD Ministry of Housing and Urban-Rural Development. Sponge City Technical Guideline-Low Impact Development. Available online: http://www.mohurd.gov.cn/wjfb/201411/%20t20141102_219465.html (accessed on 1 November 2014).
15. GOSC National Guidance on the Implementation of Sponge City Development. Available online: http://www.gov.cn/gongbao/content/2015/content_2953941.htm (accessed on 11 October 2015).
16. Ren, N.; Wang, Q.; Wang, Q.; Huang, H.; Wang, X. Upgrading to urban water system 3.0 through sponge city construction. *Front. Environ. Sci. Eng.* **2017**, *11*, 9. [CrossRef]
17. People's Government of Suining Municipal. New Grant of 0.12 Billion Issued to Suijing by the Central Government for Developing Sponge City Projects. Available online: https://www.suining.gov.cn/xinwen/show/f27f83bd66b6453fa98d3b2db1c8e664.html (accessed on 5 August 2019).
18. Gu, T.; Li, D.; Zhu, S.; Wang, Y. Does sponge-style old community renewal lead to a satisfying life for residents? An investigation in Zhenjiang, China. *Habitat Int.* **2019**, *90*, 102004. [CrossRef]

19. Gong, Y.; Fu, H.; Zhang, S.; Li, J.; Zhang, A.; Chen, Y.; Yin, D.; Liu, H. Research on Public Participation in Sponge City Construction. *China Water Wastewater* **2018**, *34*, 124963.
20. Wang, R.; Wu, H.; Chiles, R. Ecosystem Benefits Provision of Green Stormwater Infrastructure in Chinese Sponge Cities. *Environ. Manag.* **2022**, *69*, 558–575. [CrossRef] [PubMed]
21. Dai, L.; van Rijswick HF, M.W.; Driessen PP, J.; Keessen, A.M. Governance of the Sponge City Programme in China with Wuhan as a case study. *Int. J. Water Resour. Dev.* **2018**, *34*, 578–596. [CrossRef]
22. Hawken, S.; Sepasgozar SM, E.; Prodanovic, V.; Jing, J.; Bakelmun, A.; Avazpour, B.; Che, S.; Zhang, K. What makes a successful Sponge City project? Expert perceptions of critical factors in integrated urban water management in the Asia-Pacific. *Sustain. Cities Soc.* **2021**, *75*, 103317. [CrossRef]
23. Welsch, H. Environment and happiness: Valuation of air pollution using life satisfaction data. *Ecol. Econ.* **2006**, *58*, 801–813. [CrossRef]
24. Wu, W.; Chen, W.Y.; Yun, Y.; Wang, F.; Gong, Z. Urban greenness, mixed land-use, and life satisfaction: Evidence from residential locations and workplace settings in Beijing. *Landsc. Urban Plan.* **2022**, *224*, 104428. [CrossRef]
25. Zhang, B.; Song, Y.; Liu, D.; Zeng, Z.; Guo, S.; Yang, Q.; Wen, Y.; Wang, W.; Shen, X. Descriptive and Network Post-Occupancy Evaluation of the Urban Public Space through Social Media: A Case Study of Bryant Park, NY. *Land* **2023**, *12*, 1403. [CrossRef]
26. Mell, I. Examining the Role of Green Infrastructure as an Advocate for Regeneration. *Front. Sustain. Cities* **2022**, *4*, 731975. [CrossRef]
27. Liu, R.; Xiao, J. Factors Affecting Users' Satisfaction with Urban Parks through Online Comments Data: Evidence from Shenzhen, China. *Int. J. Environ. Res. Public Health* **2021**, *18*, 253. [CrossRef] [PubMed]
28. Pradhananga, A.K.; Davenport, M.A. Community attachment, beliefs and residents' civic engagement in stormwater management. *Landsc. Urban Plan.* **2017**, *168*, 1–8. [CrossRef]
29. Wang, Y.; Liu, X.; Huang, M.; Zuo, J.; Rameezdeen, R. Received vs. given: Willingness to pay for sponge city program from a perceived value perspective. *J. Clean. Prod.* **2020**, *256*, 120479. [CrossRef]
30. Venkataramanan, V.; Lopez, D.; McCuskey, D.J.; Kiefus, D.; McDonald, R.I.; Miller, W.M.; Packman, A.I.; Young, S.L. Knowledge, attitudes, intentions, and behavior related to green infrastructure for flood management: A systematic literature review. *Sci. Total Environ.* **2020**, *720*, 137606. [CrossRef] [PubMed]
31. Yang, Y.; Tang, M.; Zhou, W.; Huhe, N. The Effect of Media Use on Institutional Trust in China. *Probl. Post-Communism* **2014**, *61*, 45–56. [CrossRef]
32. Zhong, Y. Do Chinese People Trust Their Local Government, and Why?: An Empirical Study of Political Trust in Urban China. *Probl. Post-Communism* **2014**, *61*, 31–44. [CrossRef]
33. People.cn. Does Zhengzhou's Waterlogging Reveal a Failure of the Sponge City? See What Experts Say—Qiangguo News—People's Daily Online. Available online: http://www.people.com.cn/n1/2021/0723/c32306-32168268.html (accessed on 23 July 2021).
34. Wu, W.; Stephens, M.; Du, M.; Wang, B. Homeownership, family composition and subjective well-being. *Cities* **2019**, *84*, 46–55. [CrossRef]
35. Huang, Z.; Du, X.; Yu, X. Home ownership and residential satisfaction: Evidence from Hangzhou, China. *Habitat Int.* **2015**, *49*, 74–83. [CrossRef]
36. Tsang, S.; Burnett, M.; Hills, P.; Welford, R. Trust, public participation and environmental governance in Hong Kong. *Environ. Policy Gov.* **2009**, *19*, 99–114. [CrossRef]
37. Chin, W.W. The Partial Least Squares Approach to Structural Equation Modeling. *Mod. Methods Bus. Res.* **1998**, *295*, 295–336.
38. Hair, J. *Multivariate Data Analysis: A Global Perspective*, 7th ed.; Pearson Education: Upper Saddle River, NJ, USA, 2010; ISBN 9780135153093.
39. Kline, P. *An Easy Guide to Factor Analysis*, 1st ed.; Routledge: London, UK, 1994.
40. Ferraro, D.M.; Miller, Z.D.; Ferguson, L.A.; Taff, B.D.; Barber, J.R.; Newman, P.; Francis, C.D. The phantom chorus: Birdsong boosts human well-being in protected areas. *Proc. R. Soc. B Biol. Sci.* **2020**, *287*, 20201811. [CrossRef] [PubMed]
41. Wu, H.; Miller, Z.D.; Wang, R.; Zipp, K.Y.; Newman, P.; Shr, Y.-H.; Dems, C.L.; Taylor, A.; Kaye, M.W.; Smithwick EA, H. Public and manager perceptions about prescribed fire in the Mid-Atlantic, United States. *J. Environ. Manag.* **2022**, *322*, 116100. [CrossRef] [PubMed]
42. Hu, L.; Bentler, P.M. Cutoff criteria for fit indexes in covariance structure analysis: Conventional criteria versus new alternatives. *Struct. Equ. Model. A Multidiscip. J.* **1999**, *6*, 1–55. [CrossRef]
43. Yang, M.; Chen, D.; Shi, L.; He, J.; Liu, L.; Shang, X. How do construct a sponge city that can improve residents' satisfaction? Evidence from a suburb of Huizhou City, China. *Ecol. Indic.* **2022**, *142*, 109238. [CrossRef]
44. Luo, P.; Zheng, Y.; Wang, Y.; Zhang, S.; Yu, W.; Zhu, X.; Huo, A.; Wang, Z.; He, B.; Nover, D. Comparative Assessment of Sponge City Constructing in Public Awareness, Xi'an, China. *Sustainability* **2022**, *14*, 11653. [CrossRef]
45. Pierce, G.; Gmoser-Daskalakis, K.; Rippy, M.A.; Holden, P.A.; Grant, S.B.; Feldman, D.L.; Ambrose, R.F. Environmental Attitudes and Knowledge: Do They Matter for Support and Investment in Local Stormwater Infrastructure? *Soc. Nat. Resour.* **2021**, *34*, 885–905. [CrossRef]

46. Luechinger, S.; Raschky, P.A. Valuing flood disasters using the life satisfaction approach. *J. Public Econ.* **2009**, *93*, 620–633. [CrossRef]
47. Wang, Z.; Wang, H.; Huang, J.; Kang, J.; Han, D. Analysis of the Public Flood Risk Perception in a Flood-Prone City: The Case of Jingdezhen City in China. *Water* **2018**, *10*, 1577. [CrossRef]

land

MDPI

Article

Managing Landscape Urbanization and Assessing Biodiversity of Wildlife Habitats: A Study of Bobcats in San Jose, California

Yongli Zheng [1], Yuxi Wang [2,*], Xinyi Wang [3], Yuhan Wen [4] and Shuying Guo [5]

1 Department of Landscape Architecture, Heilongjiang Ecological Engineering College, Harbin 150080, China
2 The Bartlett School of Architecture, University College London, Gower Street, London WC1E 6BT, UK
3 The Bartlett School of Planning, University College London, Gower Street, London WC1E 6BT, UK
4 School of Architecture, Tianjin University, Nankai District, Tianjin 300072, China
5 Landscape Architecture and Spatial Planning Group, Wageningen University and Research, Droeven Forum, Daalse Steeg 2, 6708 PB Wageningen, The Netherlands
* Correspondence: ucbq167@ucl.ac.uk

Abstract: In the rapid process of urbanization, crucial habitats for mid-sized felids such as bobcats are increasingly compromised. This study employs Geographic Information System (GIS) tools and Machine Learning to investigate the subtle impacts of urbanization on bobcat habitats. Focused on the San Jose area, our extensive geospatial analysis has developed a complex ecological model for bobcat habitats. Our findings emphasize the significant influence of factors like vegetation cover, water body distribution, road traffic volume, and intersection density on the suitability of habitats for bobcats. Specifically, we discovered that while vegetation cover typically supports habitat suitability, its proximity to busy roads significantly undermines this advantage, indicating a need for strategic urban planning that incorporates wildlife mobility. By synthesizing natural and urban elements, we offer fresh insights into urban ecosystem management and propose specific conservation tactics: identifying optimal wildlife crossings, integrating corridors with urban infrastructure, and placing fencing and signage strategically to facilitate wildlife movement safely. These measures aim to reduce road-related threats and enhance the integrity of natural habitats, strengthening bobcat conservation efforts. More than its direct implications for bobcat conservation, this study offers actionable insights for urban wildlife conservation and introduces innovative methods for assessing and mitigating the broader ecological impacts of urbanization.

Keywords: bobcat; road crossings; suitability map; landscape urbanization; biodiversity; geospatial design

Citation: Zheng, Y.; Wang, Y.; Wang, X.; Wen, Y.; Guo, S. Managing Landscape Urbanization and Assessing Biodiversity of Wildlife Habitats: A Study of Bobcats in San Jose, California. *Land* **2024**, *13*, 152. https://doi.org/10.3390/land13020152

Academic Editor: Thomas Panagopoulos

Received: 5 December 2023
Revised: 19 January 2024
Accepted: 26 January 2024
Published: 28 January 2024

1. Introduction

Urbanization and the expansion of transportation networks significantly transform landscapes, often detrimentally affecting wildlife. In urban areas, this growth leads to the fragmentation and displacement of natural habitats, which, in turn, isolates wildlife populations and increases the risk of Wildlife–Vehicle Collisions (WVCs) in urban areas [1]. The reasons for transportation planners' often limited focus on avoiding critical habitats are complex. Traditionally, road construction prioritizes human-centric factors such as economic efficiency, traffic flow, and connecting urban centers, while environmental concerns, especially those related to wildlife habitats, are secondary [2,3]. This oversight stems partly from a lack of comprehensive environmental impact assessments and a limited understanding of habitat fragmentation's long-term ecological consequences [4]. Additionally, the urgency of infrastructural demands often leads to expedited planning processes that overlook environmental concerns [5], and a lack of collaboration between urban planners, ecologists, and conservationists exacerbates this issue [6], resulting in road networks that are efficient for humans but harmful to wildlife habitats.

These planning shortcomings have serious implications. According to one study, approximately 1 million wildlife are killed in vehicle collisions in the United States each year [7]. Roads cause not only immediate mortality but also lead to significant ecological disruptions, such as genetic isolation among wildlife populations [8]. As habitats are fragmented by roads, these populations become segmented, leading to reduced genetic diversity and increased vulnerability to various environmental threats [9].

Focusing on the bobcat (*Lynx rufus*) in the San Jose region, we see the direct consequences. San Jose's unique urban dynamics, typical of many rapidly developing urban centers in North America, provide a pertinent example of urban expansion impacting native wildlife habitats. Bobcats, native to this region and requiring extensive home ranges, are particularly vulnerable to urban encroachment and habitat fragmentation [10]. These creatures serve as ecological linchpins in the San Jose area, playing a crucial role in maintaining the balance of the local ecosystem. As apex predators, bobcats help regulate the populations of smaller mammals and rodents, preventing overpopulation and its associated ecological imbalances [11,12]. This predation is essential for controlling the spread of diseases and maintaining a healthy and diverse ecosystem. Furthermore, their presence is indicative of a thriving natural habitat, signifying a well-functioning ecosystem [13]. Therefore, the preservation of bobcats is not only essential for their own species' survival but is also integral to maintaining the overall health and balance of San Jose's urban environment, reflecting the interdependence of all species within this ecosystem.

To protect wildlife and enhance road safety, it is crucial to identify and implement suitable locations for wildlife crossings [14,15]. This challenge intertwines the welfare of both wildlife and humans, marking a pivotal moment in the development of San Jose's urban landscape [16]. The urgency of this task is highlighted by the high incidence of roadkill, which not only affects local ecosystems but also has global implications. Bobcats, as apex predators, are essential for maintaining the balance of the food chain, and their loss can lead to significant ecosystem disruptions. By strategically placing wildlife crossings, we can reduce these incidents, thus preserving the ecological integrity and biodiversity of the area [17]. Moreover, this approach addresses a broader concern: the impact of infrastructure development on wildlife [18]. Accurately evaluating and predicting wildlife crossing points is vital for mitigating habitat fragmentation and ensuring ecological connectivity, both locally and globally.

Moreover, the study of bobcat crossings is not confined to its immediate implications for this specific species. It stands as a prototypical example for addressing similar issues involving a multitude of wildlife species and urban infrastructure. Lessons learned from studying bobcats can be applied to inform the conservation efforts of other wildlife, ensuring that urban environments remain hospitable for diverse and resilient ecosystems [19]. In this way, the research on bobcat crossings in San Jose transcends its immediate context and contributes to a broader understanding of how urbanization impacts wildlife and ecosystems [19]. It underscores the urgent need for innovative solutions that balance the imperatives of urban development with the preservation of biodiversity and ecological integrity.

This article aims to illuminate the path towards harmonious coexistence between urban development and wildlife conservation, offering practical insights and solutions. We explore innovative urban planning approaches that integrate wildlife conservation, such as designing and implementing wildlife corridors and crossings specifically tailored for bobcats. By reducing wildlife–vehicle collisions, these solutions help preserve bobcat populations and maintain ecological balance. This study provides a blueprint for cities globally to navigate the challenges of urban expansion while safeguarding their natural habitats and the species that inhabit them.

2. Literature Review

2.1. Biodiversity Conservation

Bobcats are not merely inhabitants of the San Jose urban ecosystem, they serve as apex predators that maintain prey balance and contribute to the region's ecological balance [20]. As apex predators, bobcats traditionally regulate the populations of smaller mammals, which, in turn, affect vegetation and other wildlife species [21,22]. This cascading effect ripples through the food web, maintains biodiversity and checks species overpopulation. However, urban challenges in San Jose, this regulatory role of bobcats faces challenges. Despite their presence, many prey populations remain overabundant due to factors unique to urban settings. These include altered landscapes, availability of anthropogenic food sources, and reduced predation effectiveness in these modified habitats [23]. This phenomenon, often referred to in literature as the "predation paradox," indicates that the ecological role of apex predators like bobcats is more complex in urban areas than previously understood [24]. Furthermore, they contribute to the process of seed dispersal, an often-overlooked ecological service [25]. By consuming prey, bobcats inadvertently ingest seeds that can be carried to new areas, aiding in the distribution of plant species [26]. This not only promotes genetic diversity within plant populations but also bolsters the overall resilience of the ecosystem.

2.2. Genetic Connectivity

Urbanization has fragmented San Jose's bobcat habitats, leading to genetic isolation and potential biodiversity loss. Genetic diversity is crucial for disease resistance and adaptability to environmental change [9]. Establishing wildlife crossings is essential for maintaining genetic flow and bobcat population viability. Genetic diversity is the substrate for natural selection, enabling adaptation to environmental shifts [27]. Urban encroachment impedes connectivity and reduces the genetic variance of migrants, potentially diminishing the bobcats' adaptive capacity [28]. Furthermore, reduced genetic diversity can erode the adaptive potential of bobcat populations [29]. As environmental pressures mount, this genetic diversity is paramount for species survival [30]. Without it, populations become increasingly vulnerable to unforeseen challenges.

2.3. Road Safety

The intersection of bobcats and roadways in the San Jose region presents a significant challenge for both wildlife and human commuters. Wildlife–Vehicle Collisions (WVCs) present substantial challenges, encompassing property damage, personal injuries, and fatalities [31]. Property damage often involves significant costs for vehicle repair and impacts insurance premiums [32]. Injuries to drivers and passengers can range from minor to severe, requiring medical attention and possibly leading to long-term health consequences. Most critically, these collisions can result in fatalities, underscoring the serious risk they pose to the life of humans and wildlife [33]. The impacts of these collisions extend beyond the individual bobcats involved. Each incident contributes to the overall mortality rate of the population, potentially leading to long-term declines [34,35]. Additionally, WVCs can disrupt the ecological balance by removing apex predators from the ecosystem, which can lead to cascading effects on prey populations and vegetation dynamics [21].

In regions like California, where the intersection of wildlife and urban development is a growing concern, the implementation of well-designed wildlife crossings has shown promising results for species like bobcats [36]. For instance, in areas such as near Los Angeles and other parts of California, specifically designed overpasses and underpasses have been constructed to facilitate the safe movement of wildlife across busy roadways [37,38]. These structures have been strategically placed considering the natural movement patterns of wildlife, thereby minimizing their exposure to traffic and reducing the likelihood of collisions.

The results from these wildlife crossings have been encouraging. Monitoring studies have shown a significant decrease in wildlife–vehicle collisions in areas where these crossings are present [39]. Furthermore, they have contributed to maintaining genetic diversity within wildlife populations by allowing for greater movement and interaction among separate groups [38]. Therefore, the establishment of wildlife crossings is a practical measure that goes beyond conservation. It enhances public safety by reducing wildlife-related traffic incidents and plays a pivotal role in balancing urban development with the preservation of wildlife populations. The success of these crossings in California serves as a model for other regions facing similar challenges, ultimately paving the way for a safer coexistence between humans and wildlife in urban landscapes.

2.4. Urban Ecology

San Jose represents a microcosm of urban sprawl, mirroring the challenges that wildlife faces in rapidly urbanizing regions worldwide. The study of bobcat crossings stands as a prototypical example for addressing similar issues involving other wildlife species and urban infrastructure. It offers invaluable insights into the domain of urban ecology and wildlife conservation amidst the relentless tide of urbanization [40]. Understanding how bobcats adapt to this changing environment provides invaluable insights into the broader field of urban ecology. Their behavior, movements, and interactions with human infrastructure serve as a lens through which we can better comprehend the challenges and opportunities presented by urbanization [41].

2.5. Interdisciplinary Collaboration

This research necessitates a collaborative effort across various disciplines, uniting ecologists, urban planners, transport authorities, and conservationists to address the challenges at the interface of urban development and wildlife conservation. It emphasizes the intrinsic value of interdisciplinary cooperation in crafting solutions that consider both human and wildlife requirements while informing policy and planning decisions [42]. This interdisciplinary approach recognizes that the issues at hand are multifaceted and interconnected. Ecologists contribute insights into wildlife ecology and habitat needs, while urban planners provide expertise in infrastructure design [43]. Transport authorities bring vital data on traffic dynamics and engineering, and conservationists anchor the work in ethical and conservation policy [44]. Such cross-disciplinary collaboration yields comprehensive strategies that are sensitive to ecological and human demands [45]. This inclusive approach not only bolsters conservation efficacy but also enriches policy making with a broader grasp of the underlying complexities [46].

Moreover, incorporating varied perspectives leads to sustainable urban development strategies that embed wildlife conservation into urban planning, promoting a more harmonious and resilient cityscape [47]. This study, through an interdisciplinary lens, evaluates potential bobcat crossing sites within San Jose's urban matrix.

Two primary research questions guide this research: 1. What are the major factors impacting bobcats' living habitat? We assumed that the primary factors affecting bobcat habitat in San Jose, California are urban development, the amount of road traffic, and the presence of natural features such as vegetation and water sources. 2. How do these factors influence bobcats' ability to cross infrastructure in San Jose, California? We hypothesize in this regard that the above factors, particularly road traffic and urban infrastructure, will lead to habitat fragmentation and an increase in roadkill incidents, which will affect the movement patterns of bobcats and the connectivity of their populations. Through the integration of geospatial analysis and ecological modeling, this research aims to provide practical solutions that not only conserve bobcat populations but also contribute to the broader understanding of wildlife–urban interface management.

3. Materials and Methods

3.1. Study Site

The study delineates a 30 km radius (approximately 18.64 miles) from the Calero Reservoir as its focal area, located 20 km (about 12.43 miles) from downtown San Jose, CA. The Calero Reservoir, a crucial hydrological feature for the region, supports the surrounding ecosystems with ample water resources. Figure 1 illustrates several designated wildlife preserves within the study site, including the North Coyote Valley Conservation Area, Tulare Hill Ecological Preserve, and Basking Ridge Conservation Area—integral to the region's biological diversity. The area boasts a rich ecological mosaic of forests, grasslands, and watersheds, providing habitats conducive to a variety of wildlife species, including several protected ones. In recent years, a surge in roadkill events has been observed, particularly at the confluence of Highway 101 and Monterey Road (refer to Figure 1), with significant repercussions for human safety, bobcat populations, and the broader ecosystem integrity.

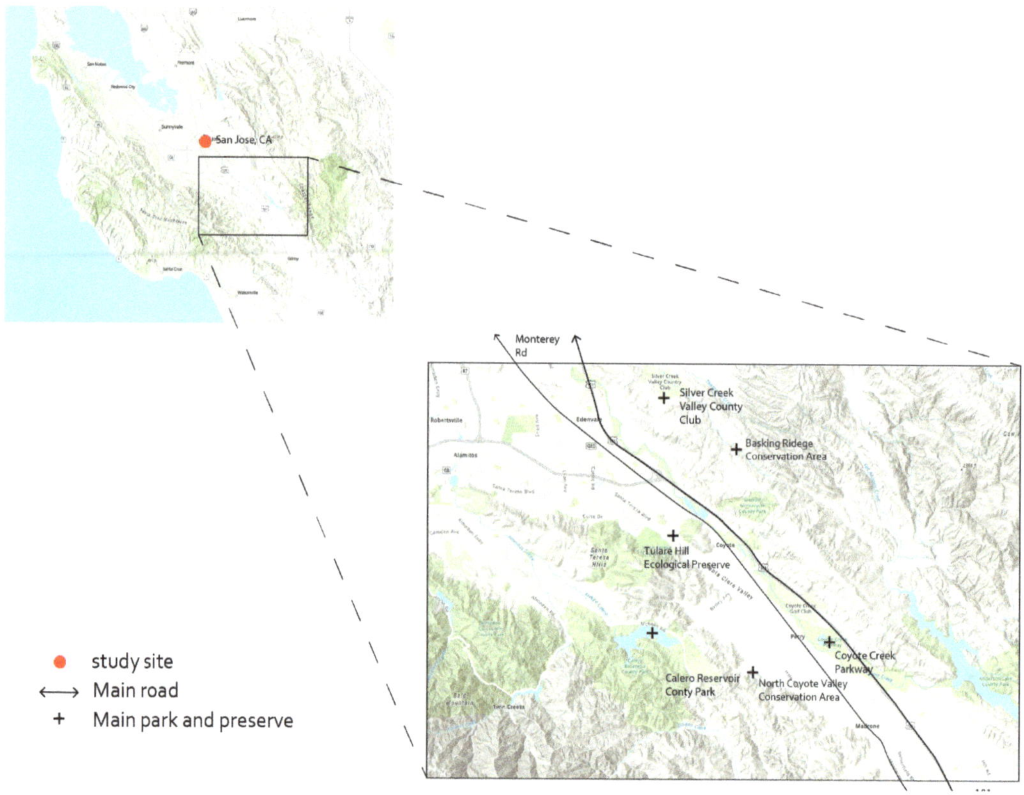

Figure 1. Bobcat Habitats and Infrastructure Crossways in the Calero Reservoir Vicinity, San Jose, California.

3.2. Data Sources and Scoring

This study conducted a literature review to identify key factors influencing the optimal locations for bobcat crossings. Articles published between 2019 and 2023 were sourced from the Web of Science database using keywords such as "Bobcat Crossing", "Habitat", "Living Condition", and "Infrastructure". The five-year period was selected to reflect recent developments in infrastructure and their effects on wildlife habitats. A dataset of 100–200 articles was curated to balance comprehensiveness with manageability, ensuring an in-depth analysis without sacrificing the breadth of the review. Within this dataset, 118 articles met the review criteria.

The research then extracted and quantified key factors from these articles, assessing word frequency in the abstracts to derive a weighted scoring system. These factors were categorized into six domains impacting bobcat habitats: "Vegetation Coverage", "Distance to Stream", "Vehicle Traffic Volume", "Bobcat Movement Pattern", and "Bobcat Roadkill Spots". This methodical approach informed the creation of a bespoke analytical framework, going beyond a mere synthesis of existing knowledge. Table 1 presents a comparative analysis of these factors, detailing associated keywords, frequency, and cumulative weight. The synthesis of factor analysis and weighting provides a novel contribution to the field by quantifying each factor's influence. Subsequent sections will delve deeper into these factors to assess their relevance to the siting of bobcat crossings.

Table 1. Key Factors for Bobcat Crossings.

Key Factors	Keywords	Frequency	Sum Weight
Vegetation Coverage	Open Water (Aquatic)	20	23.9%
	Developed Land	18	
	Riparian Forest and Shrub	15	
	Woodland	13	
	Grasslands	10	
	Wetland	8	
	Irrigated Agriculture	8	
Distance to Stream	Stream Length	25	19.6%
	Drainage Distance	14	
	Channel Network	12	
	Watershed	12	
	Stream Usage	10	
	Stream Segment	8	
	Riparian Zone	8	
Vehicle Traffic Volume	Average Daily Traffic (ADT)	15	15.7%
	Vehicle Flow	14	
	Road Usage	12	
	Vehicle Flow Analysis	12	
	Peak Hour Traffic	12	
Bobcat Movement Pattern	Bobcat Tracking	24	26.8%
	Bobcat GPS Telemetry	22	
	Bobcat Habitat Selection	20	
	Bobcat Migration	17	
	Bobcat GPS Collar Data	15	
	Bobcat Movement Ecology Studies	13	

Table 1. *Cont.*

Key Factors	Keywords	Frequency	Sum Weight
Bobcat Roadkill	Roadkill Hotspots	12	14.0%
	Wildlife Mortality Locations	11	
	Roadway Animal Incidents	10	
	Roadkill Mapping	10	
	Wildlife–Vehicle Conflict Zones	8	
	Animal–Vehicle Collision Sites	7	

Table 2 offers a comprehensive overview of the data employed in this project, incorporating information on vegetation coverage, water system characteristics, vehicle traffic volume, bobcat movement position points, and incidents of bobcat roadkill. All data were processed in the WGS 1984 coordinate reference system. The table provides granular details for each dataset, such as resolution, application, year of reference, and the originating source of the data.

Table 2. Data Sources.

Data	Reference Year	Resolution	Usage	Data Source
Vegetation Coverage Data	2023	Vector	Reclassification of Vegetation	The national Map. Gov
Water System Data	2023	Vector	Abstracting the Distance from Water	The national Map. Gov
Vehicle Traffic Volume	2023	Vector	Visualizing the Traffic Data	The national Map. Gov
Bobcat Movement Position Point Data	2023	Vector	Visualizing the Bobcat Movement Pattern	MOVEBANK
Bobcat Roadkill Data	2023	Vector	Extracting Space Distribution of Bobcat Roadkill	MOVEBANK

3.3. Data Analysis Criteria of Habitat Suitability Modeling (HSM)

This study employed Habitat Suitability Modeling (HSM) to develop a model for habitat suitability. HSM is considered the practical application of the ecological niche concept, utilizing environmental variables to predict the presence or abundance of species throughout the research area [48]. The research also harnessed the logic of the Landscape Planner's Toolkit, integrating Geographic Information Systems (GISs) to collect and process visual data such as Data Evaluation Parameters (DEPs) [49].

We categorized each factor into five equidistant values based on their impact on bobcat crossings, ranging from S1-Highest suitability (5 pts) to S5-Lowest suitability (1 pt); this division of Habitat Suitability Index (HSI) values into five equidistant levels facilitates easier comparison and contrast among the suitability of different habitats [50]. Our evaluation of the five values and their impact on bobcat crossings drew upon key insights extracted from the literature review, a method whose efficacy in assessing habitat suitability has been well documented.

HSM development integrates species occurrence with environmental data, employing statistical algorithms to analyze the relationship between these environmental factors and species occurrences, thereby predicting the probability of suitable habitats across the landscape [51]. In our analysis, the analysis incorporated key factors from Section 3.2 to ascertain the most crucial species occurrence data and environmental data. Subsequently, assigning habitat suitability using the Ordered Weighted Averaging (OWA) method [52]. The weighted key factors were then transformed into spatial distribution maps using OWA values, enabling spatial analysis within GIS. This approach has a strong precedent in current applications [53].

For the impact factors that cannot find the evidence for suitability score, this research applied the "Equal Interval" classification method [54]. In "Equal Interval" classification, each class occupies an equal interval along the number line. They are found by determining the range of the data [55]. The range is then divided by the number of classes, which gives the common difference. The class limits are established by starting at the lowest value and adding the common difference to obtain the upper limits of the first class, adding the common difference to this to obtain the limit of the second class, until the upper limit of the data is reached [56].

Finally, we overlaid all factors with different weights based on the word frequency, in ArcGIS Pro, then generated potential crossing locations where bobcats are likely to attempt to crossroads. Weighted overlay is a technique in GIS-based multi-criteria decision analysis that assigns weights and combines various thematic layers to create a comprehensive suitability map, widely used in habitat suitability modeling, land use planning, and site selection studies [57]. These areas represent critical points for wildlife connectivity. The formula of the weighted overlay method is shown below:

$$weightedoverlay = \sum_{i=1}^{n} C_i \times W_n \tag{1}$$

where C_i is each criterion (i) that has been reclassified, and W_n is the number of data (n) that were weighted.

3.3.1. Kernel Density

This study employs Kernel Density Analysis (KDA) as a pivotal ecological analytical method to evaluate bobcat habitats. Within the extensive realms of geospatial analysis, geography, and ecology, Kernel Density Analysis has proven to be a highly prevalent model, extensively applied for detecting ecological corridors [58–60], identifying biodiversity hotspots [61,62], and assessing potential conflict areas [63,64]. The methodology involves calculating the density of point features surrounding each output grid cell, contingent upon specified bandwidth and a chosen kernel function [65]. In this project, the KDA mechanism can be demonstrated as follows (2):

$$K_{(x)} = \frac{1}{nh2}\sum\nolimits_{i=1}^{n} K(\frac{x - x_i}{h}) \tag{2}$$

where for each habitat category, (n) represents the total number of locations that are suitable as bobcat's habitats, ($x - x_i$) represent the individual samples of habitat locations, (h) is the bandwidth (search radius of each habitat), and (K) is the kernel function, typically a Gaussian function.

In this project, Kernel Density Analysis (KDA) is initially employed to transform bobcat observation points into a continuous density surface [66]. Leveraging the spatial features of the density values derived from this continuous surface, polygons representing bobcat active regions are extracted, potentially indicative of their hunting grounds, habitats, or migration routes, drawing upon insights from prior reviews. The choice of bandwidth and kernel function among various parameters significantly impacts the accuracy of corridor estimation. With a very large bandwidth, the detection result fails to reflect the variability in data, whilst a very narrow bandwidth leads to noises [67]. Kernel function impacts how an individual sample is spread and weighted across the density surface. Herein, this study employs 1609.344 (1 mile) as bandwidth and Gaussian function as kernel function for its mathematical simplicity and statistical interpretability.

When considering the impact of roads on wildlife, by analyzing the activity density of bobcats and the location of roads, it is possible to determine where road infrastructure has the greatest impact on bobcats. This information is crucial for wildlife management and conservation planning, as it can guide how to design and place wildlife corridors or other mitigation measures.

3.3.2. Euclidean Distance

Euclidean distance serves as a metric to quantify the spatial relationship with bobcat habitats. Typically employed in prior studies to assess habitat isolation [68,69] and ecological importance evaluation [70,71], this metric calculates the shortest straight-line distance between points, providing a straightforward measure of spatial separation [65]. Equation (3) is the mathematical representation:

$$d = \sqrt{(x_2 - x_1)^2 + (y_2 - y_1)^2} \tag{3}$$

where (d) represents the distance between two points, which are points within study range and habitats, respectively, with (x_1, y_1) and (x_2, y_2) as the coordinates of these points. Herein, the calculation involves determining the distances from each pixel in the study area to the nearest bobcat-active points, considering key resources like water sources and food supply as indicated by prior reviews. Subsequently, varied weight values are assigned to these resources on the map, contributing to the creation of a weighted suitability raster based on pre-calculated Euclidean distances.

This method serves as a valuable tool for unraveling the spatial relationship between bobcats and their environment. It offers a straightforward means of assessing the spatial characteristics of animal habitats and identifying potential suitable areas. The analysis proves advantageous in swiftly pinpointing regions where bobcats can readily access resources, while also highlighting areas that may pose challenges due to difficulties in resource acquisition [35]. Through the overlay and calculation of Euclidean distances, researchers gain insights into animal behavioral patterns and ecological needs. This understanding, in turn, informs the development of effective wildlife management and conservation strategies.

4. Results

4.1. Vegetation Coverage (Sum Weight: 23.9%)

This study employed the Kernel Density method to process Vegetation Coverage data. Kernel Density Analysis uses kernel functions (such as the Gaussian kernel) to interpolate data points, forming a continuous density surface, thus accurately capturing variations in vegetation coverage on a small spatial scale. Bobcats prefer complex natural vegetation structures [17] and their presence is often associated with open water bodies, developed open spaces, grasslands/herbaceous and shrub/shrub habitats [72]. Forest-type land has a significant impact on the relative abundance of bobcats [73]. Although current research has analyzed the habitat types preferred by bobcats and their characteristics, there is not yet a clear and unified consensus on the preference order of these habitats. Therefore, in this study, we employed a method based on word frequency analysis to count habitat-related keywords mentioned in the literature under the "Vegetation Coverage" category, which are relevant to suitable locations for bobcat crossings. This method aims to quantify the relative importance of different land types for bobcats. Specifically, we classified these keywords into five levels using the Equal Interval method, assigning suitability scores from 5 (highest importance) to 1 (lowest importance) based on their significance, see Table 3.

Table 3. Suitability Ratings for Vegetation Coverage.

Keywords for Vegetation Coverage	Suitability Level	Current Value (Based on Word Frequency from Table 1)	Reclassified Value
Open Water (Aquatic)	S5—Highest Suitability	18.6–20	5 pts
Developed Land	S4—Higher Suitability	15.0–18.6	4 pts
Riparian Forest and Shrub	S3—Moderate Suitability	12.8–15.2	3 pts
Grassland; Woodland	S2—Lower Suitability	10.0–12.8	2 pts
Wetland; Irrigated Agriculture	S1—Lowest Suitability	8–10.0	1 pt

4.2. Distance to Stream (Weight: 19.6%)

In this study, the Euclidean Distance Analysis method was used to assess the Distance to Stream, a crucial factor in determining the suitability of habitats for bobcats. This method calculates the straight-line distance from each point within the study area to the nearest water stream, providing a clear quantification of the spatial relationship between water sources and bobcat habitats. Proximity to water sources is crucial for bobcats as these areas typically offer a richer food supply, including prey species attracted to water bodies. Additionally, water sources often support a variety of vegetation, providing necessary cover for bobcats for hunting and protection.

In urban environments, bobcats have been found to prefer areas around creeks and water channels [74], likely due to these areas offering a more favorable microclimate and more secluded environment compared to the more exposed urban landscapes [75]. Furthermore, water sources are essential for the hydration of bobcats, especially in urban areas where natural water sources may be scarce. Therefore, proximity to water is a significant factor in their movement and territory selection [76].

Consequently, areas closer to water sources are deemed more suitable for bobcat habitats. However, current research lacks specific studies on the precise distance between suitable bobcat habitats and water sources. To address this, our study applied the Equal Interval method to divide the distance from each point in the study area to the nearest stream into five levels, reflecting varying degrees of habitat suitability based on proximity to water sources. As shown in Table 3, this scoring system was developed based on a comprehensive review of bobcat behavioral patterns and ecological needs, assigning values from 5 (closest) to 1 (farthest) based on the reclassification of these levels according to their distance, as shown in Table 4.

Table 4. Suitability Ratings for Distance to Stream.

Distance to Stream	Suitability Level	Current Value (Based on Distance to Stream)	Reclassified Value
Very Close	S5—Highest Suitability	0–50 m	5 pts
Close	S4—Higher Suitability	50–100 m	4 pts
Moderate	S3—Moderate Suitability	100–150 m	3 pts
Far	S2—Lower Suitability	150–200 m	2 pts
Very Far	S1—Lowest Suitability	Above 250 m	1 pt

4.3. Traffic Volume (Weight: 15.7%)

This study utilized Kernel Density to analyze Traffic Volume data. Kernel Density reveals the spatial distribution characteristics of traffic by identifying areas of high traffic density.

Research has shown that bobcats exhibit avoidance behavior towards road areas and show the least overlap in their range within road-related habitats [77]. Bobcats are also particularly sensitive to traffic noise and vehicle lights [78]. Thus, areas with high traffic volume are not preferred habitats for bobcats. Using the Equal Interval method, this study classified the relative distances to high traffic volume areas. Based on proximity to these areas, from farthest to nearest, we assigned suitability scores from 5 to 1 to these locations, as shown in Table 5.

Table 5. Suitability Ratings for Traffic Volume.

Traffic Volume	Suitability Level	Average ADT per Day (Quantity of Vehicles)	Reclassified Value
Very low	S5—Highest Suitability	0–50,000	5 pts
Low	S4—Higher Suitability	50,000–110,000	4 pts
Moderate	S3—Moderate Suitability	110,000–370,000	3 pts
High	S2—Lower Suitability	370,000–920,000	2 pts
Very High	S1—Lowest Suitability	Above 920,000	1 pt

4.4. Bobcat Movement Patterns (Weight: 26.8%)

This study utilized the Kernel Density method to analyze data on Bobcat Movement Patterns. Kernel Density Analysis allowed us to visualize key areas and pathways of bobcat activity, such as frequently crossed locations or primary activity zones, aiding in understanding the spatial distribution of bobcats' behavioral habits and ecological needs. Using bobcat movement monitoring data from MOVEBANK in 2023, we produced maps of bobcat activity ranges and understood the intensity of activities within these areas. The Equal Interval method was applied to classify bobcat activity density into five different levels. For detailed categorization, refer to Table 5; these were assigned scores from 5 (highest density) to 1 (lowest density) based on the density gradient, see Table 6.

Table 6. Suitability Ratings for Movement Patterns.

The Density of Movement Patterns	Suitability Level	Crossing Times per Square Miles in 2023	Reclassified Value
Very High	S5—Highest Suitability	25,600–113,000	5 pts
High	S4—Higher Suitability	8460–25,600	4 pts
Moderate	S3—Moderate Suitability	2220–8460	3 pts
Low	S2—Lower Suitability	500–2220	2 pts
Very Low	S1—Lowest Suitability	0–500	1 pt

4.5. Bobcat Roadkill (Weight: 14.0%)

This study employs the Kernel Density method to analyze Bobcat Roadkill events. Kernel Density Analysis reveals the spatial clustering characteristics of bobcat roadkill incidents. By generating continuous density maps, this method clearly depicts the concentration trends of roadkill events in specific areas. The risk of road mortality for bobcats is significantly influenced by traffic levels [79], and roadkill incidents impact the activity and population of bobcats [80]. Increased traffic leads to an increased likelihood of vehicle encounters with wildlife [81]. As traffic volumes increase, the likelihood of bobcats crossing the roadway and being struck by vehicles as a result increases. Busy roads can disrupt roadways and can affect the habitats of mammalian predators (especially felines) by forcing these animals to cross roads more frequently to access different parts of their territories [82]. This increase in crossing behavior increases the risk of road traffic accidents. Therefore, based on bobcat roadkill data from MOVEBANK in 2023, our study applied the Equal Interval method to categorize the distances in high bobcat roadkill density areas into five levels, forming a continuous distance gradient. Each level is assigned a suitability score from 1 to 5 based on the relative distance from high-density roadkill areas. For detailed categorization, refer to Table 6; this grading method, by quantifying the frequency of bobcat roadkill events, reveals the spatial distribution patterns of habitat suitability affected by this factor, see Table 7.

Table 7. Suitability Ratings for Bobcat Roadkill.

The Density of Bobcat Roadkill	Suitability Level	Roadkill per Square Miles in 2023	Reclassified Value
Very low	S5—Highest Suitability	0.001–0.011	5 pts
Low	S4—Higher Suitability	0.012–0.053	4 pts
Moderate	S3—Moderate Suitability	0.054–0.214	3 pts
High	S2—Lower Suitability	0.215–0.836	2 pts
Very High	S1—Lowest Suitability	0.837–3.234	1 pt

4.6. Habitat Suitability Modeling (HSM)

The vector data obtained from Section 4.2 to 4.6 were processed by multiplying each factor's score by its corresponding weight and then summing them to obtain a composite score for each location. This reflects the contribution of each factor to land suitability. A visualized composite score map was then generated, summarizing the weighted sums [83]. Highly suitable habitats are represented in green, while the least suitable are in red.

In the San Jose region, our Habitat Suitability Modeling (HSM) integrated various ecological factors to reveal that the most conducive environments for bobcats, depicted as verdant zones on the composite maps, are predominantly located on the outskirts of urban development. These areas, in proximity to conservation zones such as Basking Ridge and Metcalf Motorcycle County Park, benefit from reduced human encroachment and roadkill risk due to their distance from major highways. Conversely, areas adjacent to and south of these highways, portrayed in shades of orange and yellow, manifest lower habitat suitability, with the least suitable regions, marked in red, located within densely urbanized areas. This pattern corroborates previous findings by Riley et al. [13], underscoring the detrimental effects of urban land use and adjacent road infrastructure on bobcat habitat viability.

Cross-layer analysis of Figure 2 elucidates the differential impact of various factors on habitat suitability. While urban regions and their associated roads demonstrate reduced suitability, this is contrasted by the lower scores in areas adjacent to the North Coyote Valley Conservation Area, despite their remote location from urban centers. Here, the widespread and uniform distribution of rivers correlates with increased habitat suitability, enhancing as one moves away from tributaries. Traffic dynamics further complicate suitability; the northwest regions near urban centers, with higher traffic volumes, signify lower suitability, aligning with infrastructure hotspots, particularly at intersections. However, the lowest roadkill map scores are concentrated along major thoroughfares, such as Highway 101 and Monterey Road, challenging conventional wisdom which posits that increased traffic and more lanes amplify mortality rates for bobcats [35].

An unexpected revelation from our study is the spatial congruence between areas of high habitat suitability and regions with frequent roadkill incidents, an intuitive contradiction that suggests ideal habitats might also present heightened risks for traffic-related fatalities. The habitat suitability map (Figure 2) positions primary bobcat territories at the city's periphery, correlating with areas of dense vegetation typical of conservation lands, parks, golf courses, and undeveloped tracts. The optimal habitats, situated between South San Jose and Metcalf Motorcycle Park, are dissected by major roadways. Although these roads provide a degree of suitability, they coincide significantly with roadkill hotspots when juxtaposed with individual maps. This observation compels a further discourse and consideration of roadkill intersections, especially where road planning and construction bisect suitable bobcat environments, potentially leading to a high frequency of roadkill events. The synthesis of these findings suggests a need for a multifaceted approach to urban development in San Jose, to better accommodate the needs of local wildlife, including bobcats. Specific management actions and recommendations will be detailed in the discussion section.

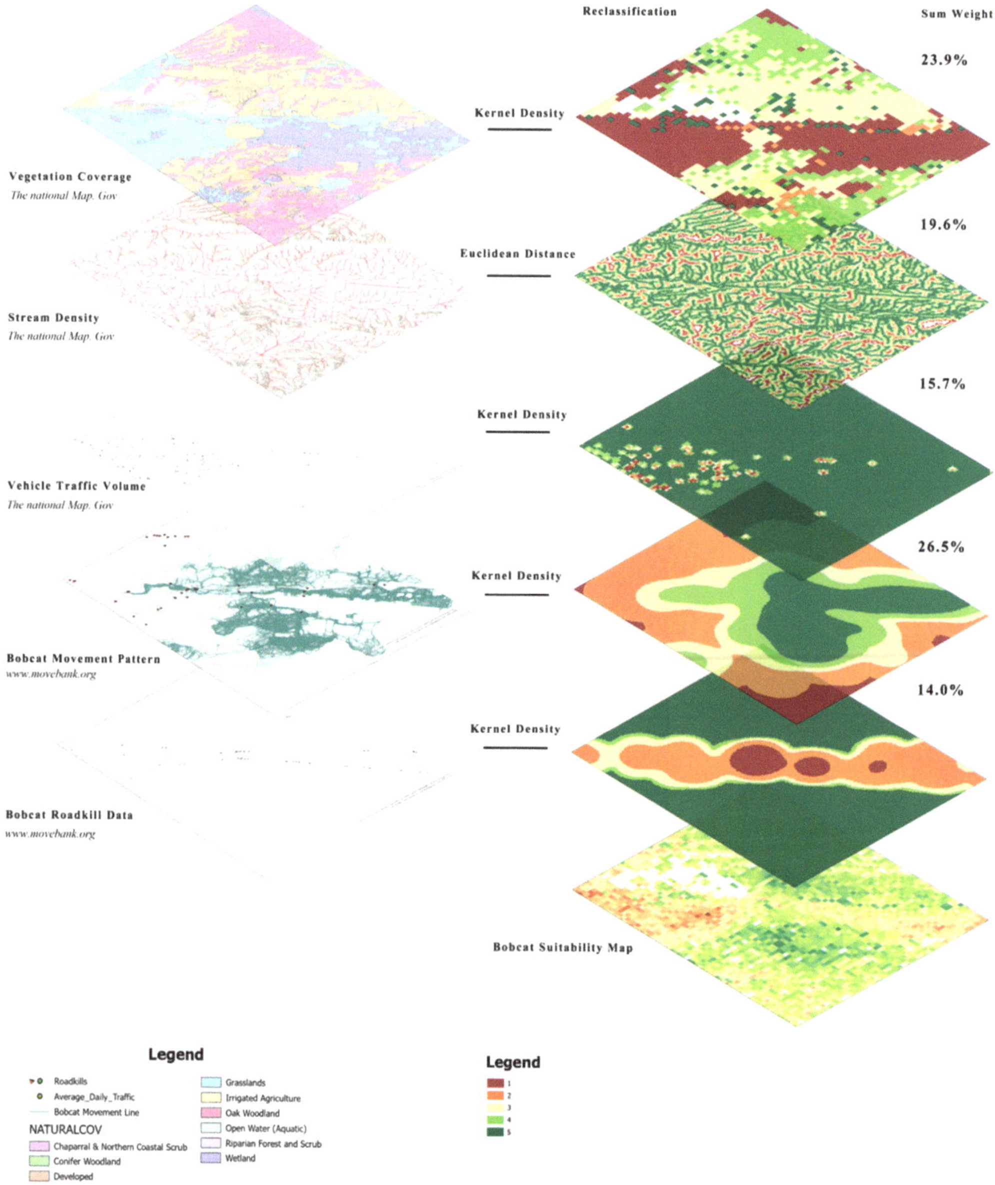

Figure 2. Comprehensive Habitat Suitability Analysis for Bobcats in San Jose, California: Integrating Vegetation Coverage, Stream Density, Traffic Volume, Bobcat Movement Patterns and roadkill dates, 2023.

5. Discussion

In this study, we explored the impact of human activities on bobcat habitat selection and suitability. Our analysis of the San Jose region, encompassing conservation areas such as Basking Ridge Conservation Area, Tulare Hill Ecological Preserve, and North Coyote Valley Conservation Area, indicated that major roads, including Highways 101 and 85, and urban land use, including the land between Highways 101 and 85, the silver leaf, and the town along the Monterey Road, called the Los Paseos significantly negatively impacted bobcat habitat suitability. These have led to habitat fragmentation and reduced connectivity within these suburban and natural landscape. Major roadways notably disrupt ecological continuity, severely disrupting the ecological environment and intensifying roadkill incidents, thereby restricting bobcat movement. Ruell et al. [28] also highlighted the negative impacts of roads on habitat connectivity.

Comparing the different layers, from Figure 2, we find that the interactions between various factors may influence suitability by affecting bobcat behavior. In our analysis, comparing the influence of roadkill incidents and traffic volume on habitat suitability, we found that while the general patterns of habitat suitability associated with roadkill incidents broadly align with those of traffic volume, the areas of highest impact for each do not overlap. Specifically, these factors differentially affect bobcat behaviors such as movement patterns, territorial range, and crossing frequency, with peak areas for roadkill incidents suggesting a higher risk to bobcats in certain regions [84], while traffic volume peak areas may correspond to areas with restricted bobcat movements due to noise and continuous vehicle presence. The occurrence of roadkill incidents is not concentrated at the peaks of traffic flow but aligns more with the suitability in the movement patterns of bobcats. Firstly, it is evident that the cause of roadkill incidents is not just heavy traffic flow, but the result of the interaction between traffic flow and bobcat movement patterns. Additionally, comparing the bobcat movement maps with layers of vegetation, streams, land use, etc., we can see some patterns similar to bobcat activities. Areas rich in vegetation and far from urban areas show higher suitability and activity density. This also reflects to some extent that natural environmental factors like vegetation are attractors of bobcat activity. This observation also underscores the significance of natural environmental factors as determinants of bobcat activity, evident from the bobcat movement maps. The patterns depicted in these maps show a clear correlation between areas of dense vegetation and increased bobcat activity [85]. Particularly, these areas, typically remote from urban settings, are not only characterized by higher habitat suitability but also by a greater density of bobcat activities. For example, areas such as Metcalf Motorcycle County Park and the portion southwest of the two preserves and the area between South San Jose have more dark green patches on the Suitability Map, demonstrating better suitability for bobcats. This is also an area where bobcat activity patterns are more intense (Figure 3). At the same time, however, the fragmental distribution of vegetation has led to more frequent travelling of bobcats on roads in this area, as seen in the illustration of bobcat movement patterns, even on roads, where their movement trajectories cross roads and connect the two reserves (Figure 2). This suggests that, aside from traffic flow, the natural landscape features, such as vegetation cover, significantly influence bobcat movements, attract their presence, and contribute to habitat preference, offering essential resources and sheltered breeding locations [85]. Vegetation cover is an important factor in bobcat habitat selection, providing them with necessary food sources and secluded breeding grounds [86].

Moreover, our analysis in Figure 3 revealed a dichotomy in the attractiveness of vegetation cover for bobcats. While areas rich in vegetation are highly appealing to bobcats, their proximity to transportation infrastructure increases the risk of roadkill. This finding is supported by Schmidt et al. [87], who indicated that bobcats near roads have a higher risk of roadkill. Cain [88] also observed an increased likelihood of roadkill for medium-sized carnivores in highly suitable habitats along roadways.

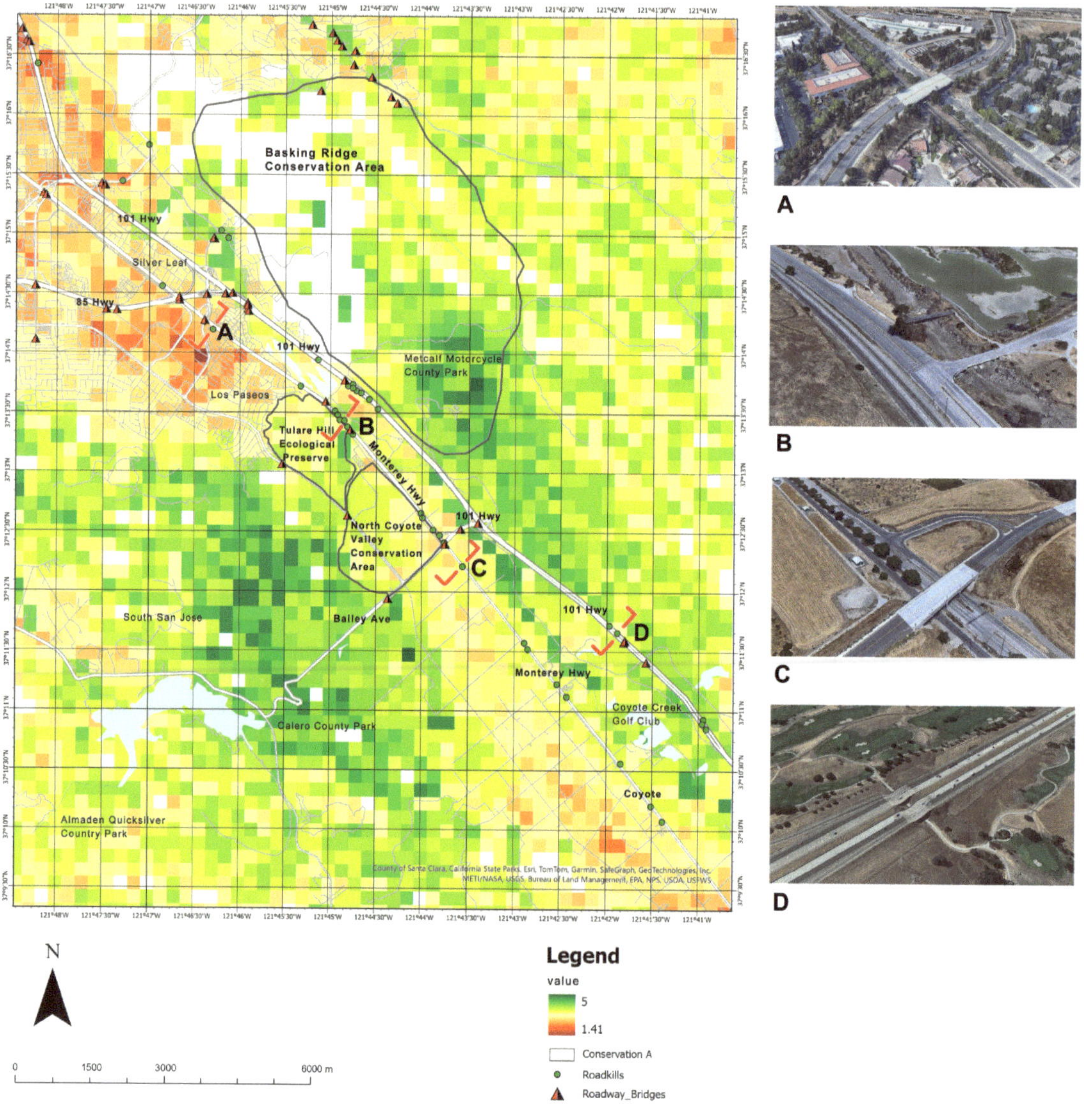

Figure 3. Overlay analysis of habitat suitability maps with infrastructure and conservation areas (**A**–**D**) are representative road sections that we selected for detailed analysis (Photos Source: Google Earth, 2023).

Overall, our research emphasizes that although areas with rich vegetation offer higher suitability for bobcats, road traffic infrastructure and roadkill phenomena significantly diminish this suitability. Some studies have also shown that predators, such as bobcats, tend to cross primary and secondary roads within their range of habitat. An increase in the density of primary and secondary roads results in more bobcats crossing and a greater risk of roadkill injury [84], due to their movement pattern. This is consistent with Serieys et al. [17], who found that bobcats prefer densely vegetated coniferous forests as habitats. However, abundant vegetation does not always equate to high suitability scores. Overlapping analysis (Figure 2) showed that roadkill incidents primarily occur on main roads, highly overlapping with areas frequently traversed by bobcats. This might be due to vegetation density providing cover but also potentially attracting bobcat activity, thereby increasing the risk of roadkill [89]. Our study indicates that in addition to considering the individual impact of various factors in habitat suitability assessments, the interplay and potential negative effects of these factors must also be considered. Therefore, future research should aim to adjust and optimize the suitability assessment system to more comprehensively reflect the complex relationships between environmental factors. Especially in assessing habitat suitability, it is crucial to consider not only the independent effects of environmental factors but also how these factors interact and affect the behavior and survival of wildlife. Moreover, our study underscores the need for conservation strategies, particularly in urban and suburban areas. When planning and managing natural reserves, the potential negative impacts of roads and urban land use on wildlife habitats should be fully considered. For instance, increasing ecological corridors and wildlife crossings can reduce roadkill incidents and enhance habitat connectivity. Additionally, raising public awareness about wildlife conservation, particularly among residents in urban and suburban areas, can help reduce disturbances to wildlife habitats. To delve deeper into bobcat habitat suitability analysis, we selected four areas (Figure 3) for case studies, comparing, learning, and applying our findings by overlaying suitability maps and roadkill points to analyze key points in depth.

Figure 3 shows that Segment A, being close to the town, has the lowest suitability score. Despite some vegetation, it is primarily urban greenery and close to houses. Segment B, near a water source, has vegetation only on the northeast side and is traversed by a flyover across the main road, resulting in good suitability. Segment C, with open views and belt-like vegetation on both sides, also crossed by a flyover, has slightly lower suitability compared to B. Segment D, farthest from the urban area, with broad views and some vegetation, also shows good suitability in the image. In conclusion, it is found that the suitability of different sections varies due to the different environmental landscapes in their surroundings. However, the same feature is that the habitat suitability is lower in the crossings where there is no advantage of natural resources such as vegetation and water. Whereas the abundance of favorable natural conditions and appropriate tunnel treatments will increase the suitability of the intersection location. In addition, the closer to the urban landscape, the lower suitability even though there are high quality vegetation and corridor bridge. For details, the comparison of the four sections reveals that areas with vegetation on both sides or near a water source have higher suitability, as clearly demonstrated by B and C. Additionally, in similar road environments, the comparison between B and C indicates that more abundant and favorable natural factors lead to higher suitability. This suggests that increasing vegetation around roads and in poorer corridors could be considered to reduce roadkill. Therefore, following this, we will conduct case studies and research based on the characteristics of different intersections (Figure 4), exploring measures to enhance suitability in the area.

Section A is characterized by a complex interplay of transportation networks and natural environments. In such areas, the challenge lies in bobcats potentially wandering into high-risk roadkill zones due to the blurred boundaries between roads, human settlements, and natural habitats. Based on Section 1-1 in Figure 4, railways may become their pathways, with animals using these linear structures for movement [90]. The common approach to mitigate threats from trains to mammals is to install fences along railway lines, preventing wildlife from entering the railway area. Fences, typically made of metal or wood and sufficiently high to deter target species, are proposed by [91]. However, fencing can restrict wildlife migration and habitat use, leading to habitat fragmentation and disruption of population genetic flow [57]. They also involve high construction and maintenance costs, requiring regular monitoring and repair. To address this, Spanowicz et al. [92] collected roadkill data in Canada and Brazil and designed an adaptive fencing plan, prioritizing high-risk areas to increase connectivity and reduce costs. Additionally, the authors of [93] developed a warning device that alerts animals as trains approach, enhancing their attention and facilitating learning. Moreover, for Section A, where roads are flanked by rich vegetation and central medians covered with dense plants, management measures including regular trimming and clearing roadside vegetation have been proven effective in reducing roadkill incidents for various carnivores [94–96]. Different species respond variably to the height and density of vegetation in road medians, influencing their road-crossing behavior [96]. However, more research is needed on the impact of road medians on bobcat crossing behaviors to achieve integrated vegetation management around these areas. For such sections, physical barriers like higher fences in high-risk areas along railways and roads can prevent bobcats from entering these zones. Additionally, comprehensive measures like managing road vegetation can effectively reduce animal roadkill incidents. Roadway A is characterized by its complex transportation network intertwined with the natural environment. The challenge with this type of area is that bobcats may stray into areas of high roadkill risk because the boundaries between the road, habitat, and natural environment are not obvious. Railroads may become their walking paths because typically there is sparse vegetation on both sides of the railroad, and animals may use these linear structures as a pathway for movement [90]. For such sections, for both sides of the railroad and roadway, physical separation, such as higher fences, can be added to prevent bobcats from straying into the area. And planting more native vegetation to increase cover to reduce the tendency of bobcats to cross railroads.

Section B is the area where roadkill incidents occur most frequently (Figure 3). It bisects a region highly suitable for bobcats, separating the reserve from a water source on the other side. There is a direct positive correlation between the risk of road mortality for carnivores and the presence of water bodies [97]. The proximity to water sources (Section 2-2 in Figure 4) may lead bobcats to prefer these areas as crossing points, being both drinking spots and hunting grounds, and increasing potential conflicts with human activities. For such fragmented and divided habitats, in addition to establishing wildlife crossings, it is also advisable to set up wildlife monitoring systems near these critical water sources, intervening when necessary. Through spatial overlay analysis of factors affecting bobcat survival and habitat suitability, we recognize the importance of establishing wildlife corridors along highway edges near highly suitable habitats. Further assessments can utilize spatial capture–recapture models to estimate bobcat densities in urban environments [74]. This will help to precisely identify bobcat activity hotspots. Moreover, studies like Reed et al. [11] show that combining expert experience models with empirical models simulating landscape connectivity can effectively identify crucial ecological corridors for medium-sized carnivores like bobcats. Data collected on wildlife activity and roadkill statistics are crucial in determining optimal locations for road crossing structures. Research indicates that wildlife underpasses not built based on such data have significantly lower usage efficiency compared to those planned and constructed with relevant data [98]. Section B is the area where roadkill occurs most frequently. Section B divides the area of high bobcat suitability and separates the reserve from the water source on the other side. The proximity to water sources may cause bobcats to prefer crossing at watering points and hunting areas. In addition, these areas may also be places where human activities are more concentrated, which increases the likelihood of bobcat conflicts with human. For such sections where suitable habitat is divided and fragmented, in addition to establishing animal passages, wildlife monitoring systems can be set up near these critical water sources and humans intervene when necessary.

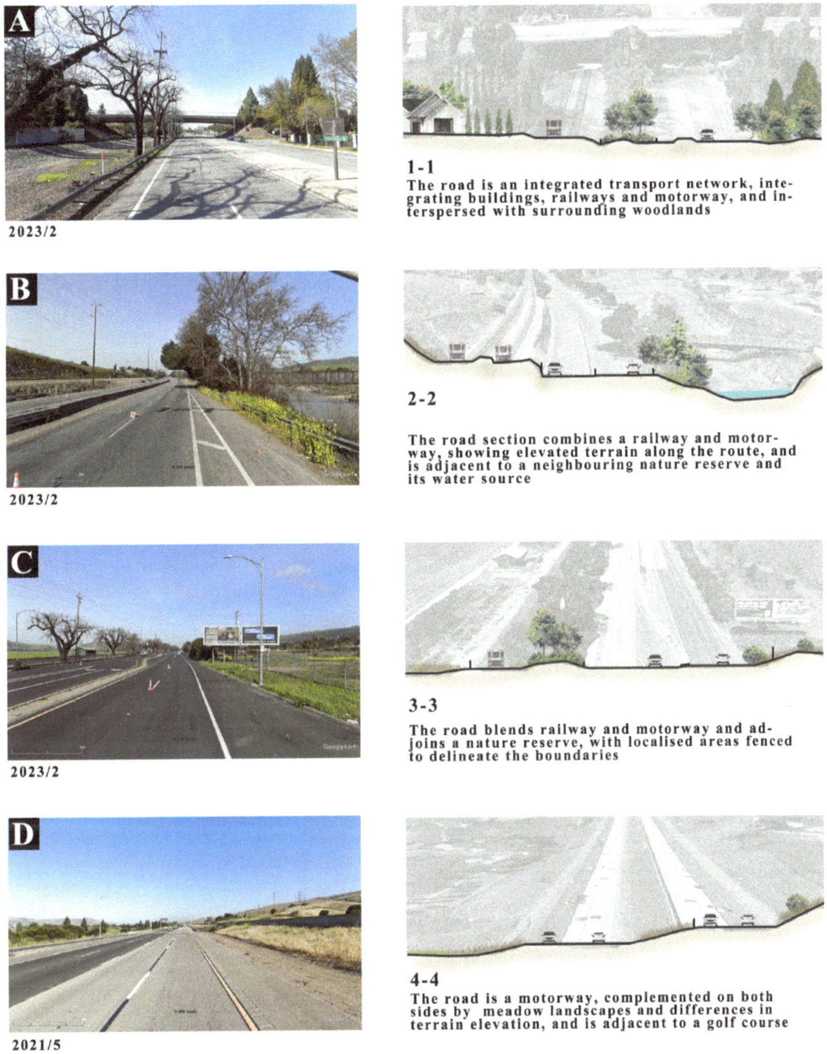

Figure 4. Representative Road Sections with Roadkill Records (**A**–**D**) are representative road sections that we selected for detailed analysis (Source: Google Earth, 2023).

Section C is characterized by its integration of rail and car lanes, adjacent to a nature reserve (Section 3-3 in Figure 4). The fences here somewhat delineate the boundaries between wildlife and human activities. However, bobcats and other wildlife adapt their behaviors to avoid human activities and habitat fragmentation [99]. The partial fencing along one side of Section C may not prevent bobcats from crossing to the other side of the road. Studies indicate a higher incidence of roadkill at fence ends than in fenced or unfenced sections, suggesting fences should be continuous or sufficiently long to encourage the use of crossing structures rather than movement around fence ends [100]. Besides barriers, wildlife underpasses or highway crossings should be added to ensure bobcat safety. In addition to passages and fencing, the habitat variables related to forests, crucial for bobcats as emphasized by Woolf et al. [73], should be considered. When designing and constructing such ecological infrastructure, efforts should be made to simulate and preserve key features of the bobcat's natural habitat, like vegetation type and density, and prey availability.

Overpasses and underpasses should be surrounded by ample vegetation resembling the adjacent habitat, avoiding elements that could startle or hinder wildlife [101]. In Section C, considering the impact of large man-made structures like billboards and streetlights on bobcat use of crossing facilities should be included in the design of ecological infrastructure. Moreover, given the proximity to a nature reserve, competition and interactions with other feline species may affect bobcat distribution [102]. In planning reserves and wildlife corridors, this competition should be accounted for by planning resources such as food, water, and shelter to minimize direct competition between species for the same resources. Section C is characterized by its blend of railway and motorway and its proximity to the nature reserve. Fencing here demarcates the boundary between wildlife and human activity to some extent; however, bobcats and other wildlife adjust their behaviors to avoid human activity and habitat fragmentation [99]. Fencing around Section C that only partially encloses an area on one side may not prevent bobcats from abandoning their crossings to the other side of the road. Further measures, such as wildlife crossings or bridges, may be needed subsequently to ensure the safety of bobcats.

Section D, with its expansive meadow landscapes and significant topographical elevation differences (Section 4-4 in Figure 4), provides bobcats with open vistas and abundant hunting grounds. This might also make them more prone to approaching human activity areas, such as nearby golf courses. While some studies suggest golf courses in urban landscapes can serve as refuges for wildlife, offering various habitats [103,104], associated factors like roads fragmenting habitats, extensive grasslands, and human activity presence may increase road mortality risks for species like bobcats. Previous research indicates that bobcat collision areas are characterized by smaller, fewer habitat patches, and larger, more isolated grassland patches [105]. In a simulation study on Deer–Vehicle Collisions (DVCs), researchers found that reducing speed limits and roadside clearings are powerful mitigation tools to decrease DVC numbers [106]. Although there is no similar simulation study on bobcat roadkill probability and vehicle speed, setting specific road signs and locally reducing speed on fast cross-city highways could be an effective and cost-efficient solution to mitigate roadkill incidents.

In conclusion, to reduce roadkill incidents, establishing crossing zones, such as wildlife underpasses or overpasses, along with barriers to prevent animal crossings, is a feasible approach [107]. Furthermore, studies have shown that carnivores prefer large, open overpasses [36,108,109]. Therefore, we recommend incorporating fences and wildlife passages into highway upgrade plans. Installing barriers along highways near reserves and other highly suitable habitats can reduce the likelihood of wildlife like bobcats entering roadways. Additionally, wildlife bridges can help bobcats cross, minimizing fragmentation of suitable habitats on either side of the highway.

In the practical implementation of these conservation measures, it is essential to tailor the design and placement of wildlife crossings and barriers based on the specific behavioral patterns and habitat requirements of bobcats. This entails conducting detailed studies of bobcat movement patterns, preferred habitats, and road-crossing behaviors. Such data can inform the strategic placement of wildlife overpasses and underpasses, ensuring these structures are located at key points where bobcats are most likely to cross roads. Additionally, the design of these crossings should mimic the natural environment to encourage usage by bobcats [110] incorporating elements like native vegetation and ensuring an appropriate scale and layout. Barrier installations along roadways should be carefully planned to minimize habitat fragmentation while effectively deterring bobcats from entering high-risk road areas [111]. Collaborative efforts with local authorities and communities are also vital in implementing traffic calming measures, such as speed limit reductions in areas with high wildlife activity, to further mitigate the risk of roadkill [35]. In summary, a comprehensive and data-driven approach, considering the unique ecological characteristics of bobcats, is crucial to effectively implement these measures and enhance the safety and connectivity of bobcat habitats.

6. Limitations

While our study presents a comprehensive assessment of suitable locations for bobcat crossings in the San Jose region, several inherent limitations should be acknowledged. First, our study focuses on the last 5 years of data for San Jose, which means that direct extrapolations to other areas may not be entirely applicable due to differences in ecological, topographical, temporal, and urban characteristics. The reliance on observational data and historical records, although valuable, may not fully capture the nuances of recent urban developments or transient bobcat populations. The criteria employed to determine crossing suitability, grounded in current knowledge, might require updates as new research emerges or bobcat behaviors evolve. Our predictive models, despite their robustness, operate under certain assumptions and may not capture the intricate web of all ecological, behavioral, and infrastructural factors. Moreover, we did not delve deeply into potential shifts in human commuting patterns, which could significantly influence the effectiveness of proposed crossings as San Jose's urban landscape evolves. Lastly, the long-term success of these proposed crossings in mitigating wildlife–vehicle collisions and bolstering bobcat genetic diversity remains to be validated. As such, while our findings shed light on urban bobcat challenges and potential interventions, they should be interpreted considering these caveats, and future research could further refine our understanding by addressing these limitations.

7. Conclusions

This study has systematically evaluated the viability of bobcat habitats in the San Jose area of California in the face of urban expansion, using established ecological and geospatial analysis methods. Our evaluation criteria are grounded in accepted habitat suitability modeling techniques, incorporating Weighted Overlay methods, Kernel Density Analysis, and Euclidean Distance Analysis [67,112,113]. These are not new inventions but rather a reintegration of proven methods, customized to address the unique challenges that bobcats face in urban environments. Our findings underscore the adverse impacts of urban growth and transportation infrastructure on habitat fragmentation, increasing the likelihood of roadkill and compromising the ecological integrity of conservation areas. The evidence points to an urgent need for implementing wildlife crossings to facilitate the coexistence of urban development and wildlife habitats.

Through the application of geospatial design and ecological modeling, we have pinpointed key potential bobcat activity zones and high-risk locations for roadkill incidents. This provides urban planners and wildlife managers with a strategic methodological framework to effectively integrate wildlife conservation and road safety in urban fringe regions, emphasizing the importance of preserving biodiversity and maintaining genetic connectivity in rapidly urbanizing landscapes. Crucially, our study brings to light the profound influence of human activities on bobcat habitat selection and suitability. By considering ecological factors, geospatial data, and human activities in tandem, we gain a deeper understanding of wildlife habitat needs, laying a scientific foundation for formulating effective conservation strategies.

The bobcat, as an indicator species, plays a pivotal role in mirroring the health of the ecosystem. Its sensitivity to habitat alterations and fragmentation serves as a critical indicator of the impacts of urbanization on wildlife. By focusing conservation efforts on the bobcat, our work not only tends to their survival but also supports biodiversity at large, in line with global initiatives like the Convention on Biological Diversity (CBD) and the United Nations' Sustainable Development Goals pertaining to life on land (SDG 15) [114,115]. This study, therefore, not only addresses local ecological challenges but also contributes to global biodiversity conservation efforts, marking the interconnectivity of urban ecology, transportation planning, and wildlife management. By addressing the needs of bobcats, a key indicator species, our study acts as a barometer for the success of urban planning and conservation initiatives, ensuring a balanced ecosystem in the San Jose region.

Author Contributions: Conceptualization, Y.Z. and Y.W. (Yuxi Wang); methodology, Y.W. (Yuxi Wang) and S.G.; data collection, Y.Z.; data curation, Y.W. (Yuxi Wang); formal analysis, Y.Z. and Y.W. (Yuxi Wang); visualization, Y.W. (Yuxi Wang) and X.W.; project administration, Y.W. (Yuxi Wang); supervision, Y.W. (Yuxi Wang); software, Y.W. (Yuxi Wang) and X.W.; validation, Y.W. (Yuxi Wang); writing—original draft, Y.W. (Yuxi Wang), X.W. and Y.W. (Yuhan Wen); writing—review and editing, Y.W. (Yuxi Wang), X.W., S.G., Y.W. (Yuhan Wen) and X.W. All authors have read and agreed to the published version of the manuscript.

Funding: This research received no external funding.

Data Availability Statement: The raw data supporting the conclusions of this article will be made available by the authors on request.

Conflicts of Interest: The authors declare no conflicts of interest.

References

1. Rytwinski, T.; Soanes, K.; Jaeger, J.A.G.; Fahrig, L.; Findlay, C.S.; Houlahan, J.; van der Ree, R.; van der Grift, E.A. How Effective Is Road Mitigation at Reducing Road-Kill? *A Meta-Analysis. PLoS ONE* **2016**, *11*, e0166941. [CrossRef]
2. Shepherd, B.J.; Houck, J.; Lyon, C. On the Road Again: A Study Valuing Wildlife Crossings for Wetland Mitigation on State Road 40 in Volusia County, Florida. *Ecosphere* **2023**, *14*, e4566. [CrossRef]
3. Kaczensky, P.; Knauer, F.; Krze, B.; Jonozovic, M.; Adamic, M.; Gossow, H. The Impact of High Speed, High Volume Traffic Axes on Brown Bears in Slovenia. *Biol. Conserv.* **2003**, *111*, 191–204. [CrossRef]
4. Bruschi, D.; Astiaso Garcia, D.; Gugliermetti, F.; Cumo, F. Characterizing the Fragmentation Level of Italian's National Parks Due to Transportation Infrastructures. *Transp. Res. Part D Transp. Environ.* **2015**, *36*, 18–28. [CrossRef]
5. Blanton, P. The Distribution and Impact of Roads and Railroads on the River Landscapes of the Coterminous United States. Ph.D. Thesis, University of Oregon, Eugene, OR, USA, 2010. Available online: https://www.proquest.com/openview/addc5907d5aa2fcf4559e21e9c48af62/1?pq-origsite=gscholar&cbl=18750 (accessed on 13 January 2024).
6. Tarabon, S.; Godet, C.; Coskun, T.; Clauzel, C. Coupling Spatial Modeling with Expert Opinion Approaches to Restore Multispecies Connectivity of Major Transportation Infrastructure. *Landsc. Urban Plan.* **2022**, *221*, 104371. [CrossRef]
7. Tatewaki, T.; Koike, F. An Overview of Roadkill Records on Japanese Municipalities Revealed by a Questionnaire Survey. *Wildl. Hum. Soc.* **2016**, *3*, 15–28. [CrossRef]
8. Vickers, T.W.; Sanchez, J.N.; Johnson, C.K.; Morrison, S.A.; Botta, R.; Smith, T.; Cohen, B.S.; Huber, P.R.; Ernest, H.B.; Boyce, W.M. Survival and Mortality of Pumas (*Puma concolor*) in a Fragmented, Urbanizing Landscape. *PLoS ONE* **2015**, *10*, e0131490. [CrossRef]
9. Carroll, C.; Hartl, B.; Goldman, G.T.; Rohlf, D.J.; Treves, A.; Kerr, J.T.; Ritchie, E.G.; Kingsford, R.T.; Gibbs, K.E.; Maron, M.; et al. Defending the Scientific Integrity of Conservation-Policy Processes. *Conserv. Biol.* **2017**, *31*, 967–975. [CrossRef]
10. Larrucea, E.S.; Serra, G.; Jaeger, M.M.; Barrett, R.H. Censusing Bobcats Using Remote Cameras. *West. N. Am. Nat.* **2007**, *67*, 538–548. [CrossRef]
11. Reed, G.C.; Litvaitis, J.A.; Callahan, C.; Carroll, R.P.; Litvaitis, M.K.; Broman, D.J.A. Modeling Landscape Connectivity for Bobcats Using Expert-Opinion and Empirically Derived Models: How Well Do They Work? *Anim. Conserv.* **2017**, *20*, 308–320. [CrossRef]
12. Riley, S.P.D. Spatial Ecology of Bobcats and Gray Foxes in Urban and Rural Zones of a National Park. *J. Wildl. Manag.* **2006**, *70*, 1425–1435. [CrossRef]
13. Riley, S.; Sauvajot, R.; Fuller, T.; York, E.; Kamradt, D.; Bromley, C.; Wayne, R. Effects of Urbanization and Habitat Fragmentation on Bobcats and Coyotes in Southern California. *Conserv. Biol.* **2003**, *17*, 566–576. [CrossRef]
14. Mimet, A.; Clauzel, C.; Foltête, J.-C. Locating Wildlife Crossings for Multispecies Connectivity across Linear Infrastructures. *Landsc. Ecol.* **2016**, *31*, 1955–1973. [CrossRef]
15. Zeller, K.A.; Wattles, D.W.; Conlee, L.; Destefano, S. Response of Female Black Bears to a High-Density Road Network and Identification of Long-Term Road Mitigation Sites. *Anim. Conserv.* **2021**, *24*, 167–180. [CrossRef]
16. Forman, R.T.T.; Alexander, L.E. Roads and Their Major Ecological Effects. *Annu. Rev. Ecol. Syst.* **1998**, *29*, 207–231. [CrossRef]
17. Serieys, L.E.K.; Rogan, M.S.; Matsushima, S.S.; Wilmers, C.C. Road-Crossings, Vegetative Cover, Land Use and Poisons Interact to Influence Corridor Effectiveness. *Biol. Conserv.* **2021**, *253*, 108930. [CrossRef]
18. Torres, A.; Jaeger, J.A.G.; Alonso, J.C. Assessing Large-Scale Wildlife Responses to Human Infrastructure Development. *Proc. Natl. Acad. Sci. USA* **2016**, *113*, 8472–8477. [CrossRef]
19. Lewis, J.S.; Logan, K.A.; Alldredge, M.W.; Bailey, L.L.; VandeWoude, S.; Crooks, K.R. The Effects of Urbanization on Population Density, Occupancy, and Detection Probability of Wild Felids. *Ecol. Appl.* **2015**, *25*, 1880–1895. [CrossRef]
20. Sergio, F.; Newton, I.; Marchesi, L. Conservation: Top Predators and Biodiversity. *Nature* **2005**, *436*, 192. [CrossRef] [PubMed]
21. Newbury, R.K.; Hodges, K.E. Regional Differences in Winter Diets of Bobcats in Their Northern Range. *Ecol. Evol.* **2018**, *8*, 11100–11110. [CrossRef]
22. Jones, L.R.; Johnson, S.A.; Hudson, C.M.; Zollner, P.A.; Swihart, R.K. Habitat Selection in a Recovering Bobcat (*Lynx rufus*) Population. *PLoS ONE* **2022**, *17*, e0269258. [CrossRef]

23. Fischer, J.D.; Cleeton, S.H.; Lyons, T.P.; Miller, J.R. Urbanization and the Predation Paradox: The Role of Trophic Dynamics in Structuring Vertebrate Communities. *BioScience* **2012**, *62*, 809–818. [CrossRef]
24. Leighton, G.R.M.; Froneman, W.; Serieys, L.E.K.; Bishop, J.M. Trophic Downgrading of an Adaptable Carnivore in an Urbanising Landscape. *Sci. Rep.* **2023**, *13*, 21582. [CrossRef] [PubMed]
25. Rubalcava-Castillo, F.A.; Sosa-Ramírez, J.; Luna-Ruíz, J.J.; Valdivia-Flores, A.G.; Díaz-Núñez, V.; Íñiguez-Dávalos, L.I. Endozoochorous Dispersal of Forest Seeds by Carnivorous Mammals in Sierra Fría, Aguascalientes, Mexico. *Ecol. Evol.* **2020**, *10*, 2991–3003. [CrossRef]
26. Rubalcava-Castillo, F.A.; Sosa-Ramírez, J.; de Jesús Luna-Ruíz, J.; Valdivia-Flores, A.G.; Íñiguez-Dávalos, L.I. Seed Dispersal by Carnivores in Temperate and Tropical Dry Forests. *Ecol. Evol.* **2021**, *11*, 3794–3807. [CrossRef] [PubMed]
27. Millions, D.G.; Swanson, B.J. Impact of Natural and Artificial Barriers to Dispersal on the Population Structure of Bobcats. *J. Wildl. Manag.* **2007**, *71*, 96–102. [CrossRef]
28. Ruell, E.W.; Riley, S.P.D.; Douglas, M.R.; Antolin, M.F.; Pollinger, J.R.; Tracey, J.A.; Lyren, L.M.; Boydston, E.E.; Fisher, R.N.; Crooks, K.R. Urban Habitat Fragmentation and Genetic Population Structure of Bobcats in Coastal Southern California. *Am. Midl. Nat.* **2012**, *168*, 265–280. [CrossRef]
29. Reding, D.M.; Bronikowski, A.M.; Johnson, W.E.; Clark, W.R. Pleistocene and Ecological Effects on Continental-Scale Genetic Differentiation in the Bobcat (*Lynx rufus*). *Mol. Ecol.* **2012**, *21*, 3078–3093. [CrossRef] [PubMed]
30. Serieys, L.E.K.; Lea, A.; Pollinger, J.P.; Riley, S.P.D.; Wayne, R.K. Disease and Freeways Drive Genetic Change in Urban Bobcat Populations. *Evol. Appl.* **2015**, *8*, 75–92. [CrossRef]
31. Nielsen, C.K.; Woolf, A. Survival of Unexploited Bobcats in Southern Illinois. *J. Wildl. Manag.* **2002**, *66*, 833–838. [CrossRef]
32. França Balbino Da Silva, A.C.; Fernando Saraiva De Menezes, J.; Rodrigues Oliveira Santos, L.G. Roadkill Risk for Capybaras in an Urban Environment. *Landsc. Urban Plan.* **2022**, *222*, 104398. [CrossRef]
33. Sprem, N.; Duduković, D.; Keros, T.; Konjevic, D. Wildlife-Vehicle Collisions in Croatia—A Hazard for Humans and Animals. *Coll. Antropol.* **2013**, *37*, 531–535. Available online: https://www.researchgate.net/publication/255910368_Wildlife-vehicle_collisions_in_Croatia_-_A_hazard_for_humans_and_animals (accessed on 13 January 2024). [PubMed]
34. Rose, N.; Prange, S.; Landry, S. Extirpated, Immigrated, Genetically Stratified—First Demographic Assessment of a Recovering Bobcat (*Lynx rufus*) Population after a Century of Extinction. *Mammal Res.* **2020**, *65*, 423–434. [CrossRef]
35. Bencin, H.L.; Prange, S.; Rose, C.; Popescu, V.D. Roadkill and Space Use Data Predict Vehicle-Strike Hotspots and Mortality Rates in a Recovering Bobcat (*Lynx rufus*) Population. *Sci. Rep.* **2019**, *9*, 15391. [CrossRef] [PubMed]
36. Clevenger, A.P.; Waltho, N. Performance Indices to Identify Attributes of Highway Crossing Structures Facilitating Movement of Large Mammals. *Biol. Conserv.* **2005**, *121*, 453–464. [CrossRef]
37. Ng, S.J.; Dole, J.W.; Sauvajot, R.M.; Riley, S.P.D.; Valone, T.J. Use of Highway Undercrossings by Wildlife in Southern California. *Biol. Conserv.* **2004**, *115*, 499–507. [CrossRef]
38. Caldwell, M.R.; Klip, J.M.K. Wildlife Interactions within Highway Underpasses. *J. Wildl. Manag.* **2019**, *84*, 227–236. [CrossRef]
39. Caldwell, M.R.; Klip, J.M.K. Mule Deer Migrations and Highway Underpass Usage in California, USA. *J. Wildl. Manag.* **2021**, *85*, 880–886. [CrossRef]
40. Lewis, J.S.; Rachlow, J.L.; Horne, J.S.; Garton, E.O.; Wakkinen, W.L.; Hayden, J.; Zager, P. Identifying Habitat Characteristics to Predict Highway Crossing Areas for Black Bears within a Human-Modified Landscape. *Landsc. Urban Plan.* **2011**, *101*, 99–107. [CrossRef]
41. Wierucka, K.; Hatten, C.E.; Murphy, D.; Allcock, J.A.; Andersson, A.A.; Bojan, J.W.; Kong, T.C.; Kwok, J.K.; Lam, J.Y.; Ma, C.H.; et al. Human-Wildlife Interactions in Urban Asia. *Glob. Ecol. Conserv.* **2023**, *46*, e02596. [CrossRef]
42. Sperling, D.; Bissonette, J.A.; Clevenger, A.P. *Road Ecology: Science and Solutions*; Island Press: Washington, DC, USA, 2003.
43. Kay, C.A.M.; Rohnke, A.T.; Sander, H.A.; Stankowich, T.; Fidino, M.; Murray, M.H.; Lewis, J.S.; Taves, I.; Lehrer, E.W.; Zellmer, A.J.; et al. Barriers to Building Wildlife-inclusive Cities: Insights from the Deliberations of Urban Ecologists, Urban Planners and Landscape Designers. *People Nat.* **2022**, *4*, 62–70. [CrossRef]
44. Paquet, P.C.; Darimont, C.T. Wildlife Conservation and Animal Welfare: Two Sides of the Same Coin? *Anim. Welf.* **2010**, *19*, 177–190. [CrossRef]
45. Blair, M.E.; Le, M.D.; Sethi, G.; Thach, H.M.; Nguyen, V.T.H.; Amato, G.; Birchette, M.; Sterling, E.J. Importance of an Interdisciplinary Research Approach to Inform Wildlife Trade Management in Southeast Asia. *BioScience* **2017**, *67*, 995–1003. [CrossRef]
46. Rice, A.H. Interdisciplinary Collaboration in Health Care: Education, Practice, and Research. *Natl. Acad. Pract. Forum Issues Interdiscip. Care* **2000**, *2*, 59–73.
47. Kirk, H.; Soanes, K.; Amati, M.; Bekessy, S.; Harrison, L.; Parris, K.; Ramalho, C.; van de Ree, R.; Threlfall, C. Ecological Connectivity as a Planning Tool for the Conservation of Wildlife in Cities. *ScienceDirect* **2013**, *10*, 101989. [CrossRef] [PubMed]
48. Hirzel, A.H.; Le Lay, G. Habitat Suitability Modelling and Niche Theory. *J. Appl. Ecol.* **2008**, *45*, 1372–1381. [CrossRef]
49. Shen, X.; Handel, S.N.; Kirkwood, N.G.; Huang, Y.; Padua, M.G. Locating the Responsive Plants for Landscape Recovery: A Toolkit for Designers and Planners. *Ecol. Restor.* **2022**, *40*, 33–35. [CrossRef]
50. Muhammed, K.; Anandhi, A.; Chen, G. Comparing Methods for Estimating Habitat Suitability. *Land* **2022**, *11*, 1754. [CrossRef]
51. Bellamy, C.; Boughey, K.; Hawkins, C.; Reveley, S.; Spake, R.; Williams, C.; Altringham, J. A Sequential Multi-Level Framework to Improve Habitat Suitability Modelling. *Landsc. Ecol.* **2020**, *35*, 1001–1020. [CrossRef]

52. Eared-Pheasant, B. Model to Assess the Habitat Suitability for Endangered Bird Species: Brown Eared Pheasant Crossoptilon Mantchuricum Swinhoe in Xiaowutaishan Reserve, China. *Pol. J. Ecol.* **2008**, *56*, 723–729.
53. de Oliveira Averna Valente, R.; Vettorazzi, C.A. Definition of Priority Areas for Forest Conservation through the Ordered Weighted Averaging Method. *For. Ecol. Manag.* **2008**, *256*, 1408–1417. [CrossRef]
54. Osaragi, T. *Classification Methods for Spatial Data Representation*; University College London: London, UK, 2008.
55. Pettit, C.J.; Klosterman, R.E.; Delaney, P.; Whitehead, A.L.; Kujala, H.; Bromage, A.; Nino-Ruiz, M. The Online What If? Planning Support System: A Land Suitability Application in Western Australia. *Appl. Spat. Anal.* **2015**, *8*, 93–112. [CrossRef]
56. Data Classification Methods—ArcGIS Pro. Available online: https://pro.arcgis.com/en/pro-app/latest/help/mapping/layer-properties/data-classification-methods.htm (accessed on 21 November 2023).
57. Store, R.; Kangas, J. Integrating Spatial Multi-Criteria Evaluation and Expert Knowledge for GIS-Based Habitat Suitability Modelling. *Landsc. Urban Plan.* **2001**, *55*, 79–93. [CrossRef]
58. Dong, J.; Peng, J.; Liu, Y.; Qiu, S.; Han, Y. Integrating Spatial Continuous Wavelet Transform and Kernel Density Estimation to Identify Ecological Corridors in Megacities. *Landsc. Urban Plan.* **2020**, *199*, 103815. [CrossRef]
59. Delso, J.; Martin, B.; Ortega, E.; Van De Weghe, N. Integrating Pedestrian-Habitat Models and Network Kernel Density Estimations to Measure Street Pedestrian Suitability. *Sustain. Cities Soc.* **2019**, *51*, 101736. [CrossRef]
60. Kaszta, Ż.; Cushman, S.A.; Macdonald, D.W. Prioritizing Habitat Core Areas and Corridors for a Large Carnivore across Its Range. *Anim. Conserv.* **2020**, *23*, 607–616. [CrossRef]
61. Lyon, K.; Cottrell, S.P.; Siikamäki, P.; Van Marwijk, R. Biodiversity Hotspots and Visitor Flows in Oulanka National Park, Finland. *Scand. J. Hosp. Tour.* **2011**, *11*, 100–111. [CrossRef]
62. Chamling, M.; Bera, B. Likelihood of Elephant Death Risk Applying Kernel Density Estimation Model along the Railway Track within Biodiversity Hotspot of Bhutan–Bengal Himalayan Foothill. *Model. Earth Syst. Environ.* **2020**, *6*, 2565–2580. [CrossRef]
63. Cushman, S.A.; Elliot, N.B.; Bauer, D.; Kesch, K.; Bahaa-El-Din, L.; Bothwell, H.; Flyman, M.; Mtare, G.; Macdonald, D.W.; Loveridge, A.J. Prioritizing Core Areas, Corridors and Conflict Hotspots for Lion Conservation in Southern Africa. *PLoS ONE* **2018**, *13*, e0196213. [CrossRef]
64. Cantrell, D.L.; Rees, E.E.; Vanderstichel, R.; Grant, J.; Filgueira, R.; Revie, C.W. The Use of Kernel Density Estimation with a Bio-Physical Model Provides a Method to Quantify Connectivity among Salmon Farms: Spatial Planning and Management with Epidemiological Relevance. *Front. Vet. Sci.* **2018**, *5*, 269. [CrossRef]
65. How Kernel Density Works—ArcGIS Pro. Available online: https://pro.arcgis.com/en/pro-app/latest/tool-reference/spatial-analyst/how-kernel-density-works.htm (accessed on 1 December 2023).
66. Brunsdon, C. Estimating Probability Surfaces for Geographical Point Data: An Adaptive Kernel Algorithm. *Comput. Geosci.* **1995**, *21*, 877–894. [CrossRef]
67. Zhang, G.; Zhu, A.-X.; Windels, S.K.; Qin, C.-Z. Modelling Species Habitat Suitability from Presence-Only Data Using Kernel Density Estimation. *Ecol. Indic.* **2018**, *93*, 387–396. [CrossRef]
68. Bailey, D.; Schmidt-Entling, M.H.; Eberhart, P.; Herrmann, J.D.; Hofer, G.; Kormann, U.; Herzog, F. Effects of Habitat Amount and Isolation on Biodiversity in Fragmented Traditional Orchards. *J. Appl. Ecol.* **2010**, *47*, 1003–1013. [CrossRef]
69. Rüdisser, J.; Tasser, E.; Tappeiner, U. Distance to Nature—A New Biodiversity Relevant Environmental Indicator Set at the Landscape Level. *Ecol. Indic.* **2012**, *15*, 208–216. [CrossRef]
70. Ergen, B. Euclidean Distance Mapping and the Proposed Greenway Method in Malta. *J. Urban Plan. Dev.* **2014**, *140*, 04013002. [CrossRef]
71. Goldstein, R.M.; Meador, M.R. Multilevel Assessment of Fish Species Traits to Evaluate Habitat Degradation in Streams of the Upper Midwest. *N. Am. J. Fish. Manag.* **2005**, *25*, 180–194. [CrossRef]
72. Driscoll, K. Use of Camera Trapping to Determine Spatial Distribution, Habitat Use, and Environmental Factors Affecting Mesopredators on Reclaimed Mine Lands at the Wilds. *J. Am. Soc. Min. Reclam.* **2017**, *2017*, 15–33. [CrossRef]
73. Woolf, A.; Nielsen, C.K.; Weber, T.; Gibbs-Kieninger, T.J. Statewide Modeling of Bobcat, *Lynx rufus*, Habitat in Illinois, USA. *Biol. Conserv.* **2002**, *104*, 191–198. [CrossRef]
74. Young, J.K.; Golla, J.M.; Broman, D.; Blankenship, T.; Heilbrun, R. Estimating density of an elusive carnivore in urban areas: Use of spatially explicit capture-recapture models for city-dwelling bobcats. *Urban Ecosyst.* **2019**, *22*, 507–512. [CrossRef]
75. Wang, L.; Ge, M.; Chen, N.; Ding, J.; Shen, X. An Evaluation Model of Riparian Landscape: A Case in Rural Qingxi Area, Shanghai. *Land* **2022**, *11*, 1512. [CrossRef]
76. Abouelezz, H.G.; Donovan, T.M.; Mickey, R.M.; Murdoch, J.D.; Freeman, M.; Royar, K. Landscape Composition Mediates Movement and Habitat Selection in Bobcats (*Lynx rufus*): Implications for Conservation Planning. *Landsc. Ecol.* **2018**, *33*, 1301–1318. [CrossRef]
77. Young, J.K.; Golla, J.; Draper, J.P.; Broman, D.; Blankenship, T.; Heilbrun, R. Space Use and Movement of Urban Bobcats. *Animals* **2019**, *9*, 275. [CrossRef] [PubMed]
78. Shilling, F.M.; Collins, A.; Longcore, T.; Vickers, W. Understanding Behavioral Responses of Wildlife to Traffic to Improve Mitigation Planning. *UC Davis Natl. Cent. Sustain. Transp.* **2020**. [CrossRef]
79. Dyck, M.A.; Shoemaker, K.T.; Dennison, C.C.; Popescu, V.D. Simulated Effects of Roadkill and Harvest on the Viability of a Recovering Bobcat Population. *J. Wildl. Manag.* **2023**, *87*, e22460. [CrossRef]

80. Hanley, V. Bobcat Identification, Abundance, and Behavior at Road Mitigation Structures in South Texas. Ph.D. Thesis, The University of Texas Rio Grande Valley, Edinburg, TX, USA, 2022.
81. Jacobson, S.L.; Bliss-Ketchum, L.L.; de Rivera, C.E.; Smith, W.P. A Behavior-Based Framework for Assessing Barrier Effects to Wildlife from Vehicle Traffic Volume. *Ecosphere* **2016**, *7*, e01345. [CrossRef]
82. Blackburn, A.; Heffelfinger, L.J.; Veals, A.M.; Tewes, M.E.; Young, J.H. Cats, Cars, and Crossings: The Consequences of Road Networks for the Conservation of an Endangered Felid. *Glob. Ecol. Conserv.* **2021**, *27*, e01582. [CrossRef]
83. Lu, C.Y.; Gu, W.; Dai, A.H.; Wei, H.Y. Assessing Habitat Suitability Based on Geographic Information System (GIS) and Fuzzy: A Case Study of *Schisandra sphenanthera* Rehd. et Wils. in Qinling Mountains, China. *Ecol. Model.* **2012**, *242*, 105–115. [CrossRef]
84. Poessel, S.A.; Burdett, C.L.; Boydston, E.E.; Lyren, L.M.; Alonso, R.S.; Fisher, R.N.; Crooks, K.R. Roads Influence Movement and Home Ranges of a Fragmentation-Sensitive Carnivore, the Bobcat, in an Urban Landscape. *Biol. Conserv.* **2014**, *180*, 224–232. [CrossRef]
85. Rolley, R.E.; Warde, W.D. Bobcat Habitat Use in Southeastern Oklahoma. *J. Wildl. Manag.* **1985**, *49*, 913–920. [CrossRef]
86. Litvaitis, J.A.; Sherburne, J.A.; Bissonette, J.A. Bobcat Habitat Use and Home Range Size in Relation to Prey Density. *J. Wildl. Manag.* **1986**, *50*, 110–117. [CrossRef]
87. Schmidt, G.; Lewison, R.L.; Swarts, H.M. Identifying Landscape Predictors of Ocelot Road Mortality. *Landsc. Ecol.* **2020**, *35*, 1651–1666. [CrossRef]
88. Cain, A.T.; Tuovila, V.R.; Hewitt, D.G.; Tewes, M.E. Effects of a Highway and Mitigation Projects on Bobcats in Southern Texas. *Biol. Conserv.* **2003**, *114*, 189–197. [CrossRef]
89. Farrell, L.E.; Levy, D.M.; Donovan, T.; Mickey, R.; Howard, A.; Vashon, J.; Freeman, M.; Royar, K.; Kilpatrick, C.W. Landscape Connectivity for Bobcat (*Lynx rufus*) and Lynx (*Lynx Canadensis*) in the Northeastern United States. *PLoS ONE* **2018**, *13*, e0194243. [CrossRef]
90. Roads, Roadsides and Wildlife Conservation: A Review. Available online: https://trid.trb.org/view/376076 (accessed on 13 January 2024).
91. Carvalho, F.; Santos, S.M.; Mira, A.; Lourenço, R. Methods to Monitor and Mitigate Wildlife Mortality in Railways. In *Railway Ecology*; Borda-de-Água, L., Barrientos, R., Beja, P., Pereira, H.M., Eds.; Springer International Publishing: Cham, Switzerland, 2017; pp. 23–42. ISBN 978-3-319-57496-7.
92. Spanowicz, A.G.; Teixeira, F.Z.; Jaeger, J.A.G. An Adaptive Plan for Prioritizing Road Sections for Fencing to Reduce Animal Mortality. *Conserv. Biol.* **2020**, *34*, 1210–1220. [CrossRef] [PubMed]
93. St. Clair, C.C.; Backs, J.; Friesen, A.; Gangadharan, A.; Gilhooly, P.; Murray, M.; Pollock, S. Animal Learning May Contribute to Both Problems and Solutions for Wildlife–Train Collisions. *Philos. Trans. R. Soc. B Biol. Sci.* **2019**, *374*, 20180050. [CrossRef] [PubMed]
94. Trombulak, S.C.; Frissell, C.A. Review of Ecological Effects of Roads on Terrestrial and Aquatic Communities. *Conserv. Biol.* **2000**, *14*, 18–30. [CrossRef]
95. Grilo, C.; Bissonette, J.; Santos-Reis, M. Spatial–Temporal Patterns in Mediterranean Carnivore Road Casualties: Consequences for Mitigation. *Biol. Conserv.* **2009**, *142*, 301–313. [CrossRef]
96. Canal, D.; Camacho, C.; Martin, B.; de Lucas, M.; Ferrer, M. Fine-Scale Determinants of Vertebrate Roadkills across a Biodiversity Hotspot in Southern Spain. *Biodivers. Conserv.* **2019**, *28*, 3239–3256. [CrossRef]
97. Červinka, J.; Riegert, J.; Grill, S. Large-Scale Evaluation of Carnivore Road Mortality: The Effect of Landscape and Local Scale Characteristics. *Mammal Res.* **2015**, *60*, 233–243. [CrossRef]
98. Seidensticker, J. Review of the Florida Panther: Life and Death of a Vanishing Carnivore. *Wildl. Soc. Bull.* **1998**, *26*, 357–359.
99. Tigas, L.A.; Van Vuren, D.H.; Sauvajot, R.M. Behavioral Responses of Bobcats and Coyotes to Habitat Fragmentation and Corridors in an Urban Environment. *Biol. Conserv.* **2002**, *108*, 299–306. [CrossRef]
100. Plante, J.; Jaeger, J.A.G.; Desrochers, A. How Do Landscape Context and Fences Influence Roadkill Locations of Small and Medium-Sized Mammals? *J. Environ. Manag.* **2019**, *235*, 511–520. [CrossRef] [PubMed]
101. Ruediger, W.C. Management Considerations for Designing Carnivore Highway Crossings. *UC Davis Road Ecol. Cent.* 2007. Available online: https://escholarship.org/uc/item/45b5183f (accessed on 4 December 2023).
102. Sánchez-Cordero, V.; Stockwell, D.; Sarkar, S.; Liu, H.; Stephens, C.R.; Giménez, J. Competitive Interactions between Felid Species May Limit the Southern Distribution of Bobcats *Lynx rufus*. *Ecography* **2008**, *31*, 757–761. [CrossRef]
103. Tanner, R.A.; Gange, A.C. Effects of Golf Courses on Local Biodiversity. *Landsc. Urban Plan.* **2005**, *71*, 137–146. [CrossRef]
104. Hodgkison, S.; Hero, J.-M.; Warnken, J. The Efficacy of Small-Scale Conservation Efforts, as Assessed on Australian Golf Courses. *Biol. Conserv.* **2007**, *135*, 576–586. [CrossRef]
105. Kolowski, J.M.; Nielsen, C.K. Using Penrose Distance to Identify Potential Risk of Wildlife–Vehicle Collisions. *Biol. Conserv.* **2008**, *141*, 1119–1128. [CrossRef]
106. Meisingset, E.L.; Loe, L.E.; Brekkum, Ø.; Mysterud, A. Targeting Mitigation Efforts: The Role of Speed Limit and Road Edge Clearance for Deer–Vehicle Collisions. *J. Wildl. Manag.* **2014**, *78*, 679–688. [CrossRef]
107. Polak, T.; Rhodes, J.R.; Jones, D.; Possingham, H.P. Optimal Planning for Mitigating the Impacts of Roads on Wildlife. *J. Appl. Ecol.* **2014**, *51*, 726–734. [CrossRef]
108. Brennan, L.; Chow, E.; Lamb, C. Wildlife Overpass Structure Size, Distribution, Effectiveness, and Adherence to Expert Design Recommendations. *PeerJ* **2022**, *10*, e14371. [CrossRef]

109. Kusak, J.; Huber, D.; Gomerčić, T.; Schwaderer, G.; Gužvica, G. The permeability of highway in *Gorski kotar* (Croatia) for large mammals. *Eur. J. Wildl. Res.* **2008**, *55*, 7–21. [CrossRef]
110. Lombardi, J.V.; Comer, C.E.; Scognamillo, D.G.; Conway, W.C. Coyote, fox, and bobcat response to anthropogenic and natural landscape features in a small urban area. *Urban Ecosyst.* **2017**, *20*, 1239–1248. [CrossRef]
111. Alonso, R.S.; Lyren, L.M.; Boydston, E.E.; Haas, C.D.; Crooks, K.R. Evaluation of Road Expansion and Connectivity Mitigation for Wildlife in Southern California. *Southwest. Nat.* **2014**, *59*, 181–187. [CrossRef]
112. Ferretti, V.; Pomarico, S. Ecological Land Suitability Analysis through Spatial Indicators: An Application of the Analytic Network Process Technique and Ordered Weighted Average Approach. *Ecol. Indic.* **2013**, *34*, 507–519. [CrossRef]
113. Hirzel, A.H.; Arlettaz, R. Modeling Habitat Suitability for Complex Species Distributions by Environmental-Distance Geometric Mean. *Environ. Manag.* **2003**, *32*, 614–623. [CrossRef] [PubMed]
114. Convention on Biological Diversity. Available online: https://www.cbd.int/ (accessed on 10 January 2024).
115. Goal 15: Life on Land. Available online: https://globalgoals.org/goals/15-life-on-land/ (accessed on 10 January 2024).

land

MDPI

Article

Landscape Architecture Professional Knowledge Abstraction: Accessing, Applying and Disseminating

Yue Qiu [1], Zheng Cong [2], Karla Nicole Opiniano [3], Xuesong Qiao [4] and Zheng Chen [5,*]

1 School of Architecture and Urban Planning, Huazhong University of Science and Technology, Wuhan 430074, China
2 Shanghai TIANHUA Landscape & Design Co., Ltd., Shanghai 200235, China
3 UNLV School of Architecture, University of Nevada, Las Vegas, NV 89154, USA
4 Shanghai Urban Construction Design and Research Institute (Group) Co., Ltd., Shanghai 200125, China
5 Key Laboratory of Ecology and Energy-Saving Study of Dense Habitat, Department of Landscape Studies, College of Architecture and Urban Planning, Tongji University, Shanghai 200092, China
* Correspondence: zhengchen@tongji.edu.cn

Abstract: Rigorous abstract knowledge, such as academic knowledge, is vital to a profession's resilience against other modern professional competition. In the field of landscape architecture, a growing number of concerns about a lack of rigorous knowledge have been observed, which may jeopardise the jurisdiction of its professional practice. A study was conducted that collected behaviours and attitudes from various members of the American Society of Landscape Architects (ASLA) regarding how they accessed, disseminated, and applied knowledge in practice. Their responses concerning the knowledge were analysed by dividing and ranking the options according to the degree of knowledge abstraction. Knowledge abstraction refers to theories and commonplace best practices that are established within a profession through shared knowledge, experience, and research. The results showed that (1) most practitioners tended to access new knowledge through tacit experience, which is indicative of a lower level of abstraction in their practice; (2) design decisions were based less on higher and rigorously abstracted knowledge such as research findings and, in circumstances where it was deployed in the design process, such knowledge was seldom used to guide design independently; (3) the majority of practitioners rarely share knowledge through high-abstracted publications; and (4) compared with accessing relatively diverse levels of knowledge abstraction, practitioners were less resourceful in knowledge application and even less in dissemination. The knowledge acquired, used and circulated in the workplace of landscape architects—as indicated by this survey—is still not comprehensively abstracted to a rigorous level, which may provide an insight into the concerns of practitioners regarding this profession's breadth of knowledge and jurisdiction.

Keywords: landscape architecture; profession knowledge; knowledge abstraction; knowledge accessing, applying, and disseminating

Citation: Qiu, Y.; Cong, Z.; Opiniano, K.N.; Qiao, X.; Chen, Z. Landscape Architecture Professional Knowledge Abstraction: Accessing, Applying and Disseminating. *Land* **2023**, *12*, 2061. https://doi.org/10.3390/land12112061

Academic Editors: Xiwei Shen, Mary Padua, Niall Kirkwood, Bo Zhang, Yang Song and Rosalea Monacella

Received: 9 October 2023
Revised: 1 November 2023
Accepted: 8 November 2023
Published: 13 November 2023

1. Introduction

In July of 2023, the US Department of Homeland Security designated landscape architecture as a Science, Technology, Engineering, and Mathematics (STEM) profession, with the American Society of Landscape Architects as an advocate for this decision [1]. Historically, landscape architecture as a practice has relied on a broad spectrum of knowledge in which tacit knowledge plays an especially pivotal role [2]. The early practitioners in this profession, in the United States, for example, were generalists with practical skills and various interests in construction, preserving natural landscapes, and offering opportunities for urban residents to access nature easily [3]. The training of landscape architecture practitioners has traditionally focused on preparing students to practise in their profession through studio-based apprenticeships [4,5]. This hands-on approach to learning is congruent with tacit knowledge, which is evidently difficult to acquire solely through research and is, therefore, difficult to disseminate or share within the workplace [6]. Despite traditional means of training successfully establishing the profession as a key role in modern society—especially in the wake of worldwide concerns regarding climate change and rapid urbanisation—it remains questionable whether this field's daily practices warrant its new science-based designation. Modern professionals are expected to offer sound professional services and justify the uniqueness and necessity of their expertise in their services [7]. Abraham Flexner, an American educator who was a key figure in legitimising the medical field and higher education in the 20th century, states the several tenets of what defines the concept of a "profession" [8]. One of these tenants suggests, "[A profession] must add and improve our stock of theoretical and stock knowledge" [9]. This, therefore, expresses that the profession of landscape architecture has been developed largely on tacit knowledge, but in order to meet the qualifications of a profession that needs to gradually function as a scientific and scholarly discipline, knowledge established within the profession must be structured systematically in conjunction with antecedent knowledge to justify and improve performance in professional practice.

Concerns about a lack of rigorous knowledge have been a recurring subject of interest amongst practitioners and academics of landscape architecture [10–13]. This is due, in part, to having greater difficulty in efficiently deriving information from larger systems of knowledge for landscape architecture than other professions such as engineering or medicine, which possess explicit abstractions [14]. Landscape architecture, as addressed by several landscape architects, does not seem to be structured in a way that can identify research needs, support a response to those needs, or integrate this response into the practice of landscape architecture [15]. Clients' need for data in relation to a post-construction evaluation was sometimes, if not often, left unsatisfied due to a lack of academic research [16]. Fellows of the American Society of Landscape Architects (ASLA) revealed a shared concern about the profession's "losing ground" due to an overly broad scope and knowledge base that is not academically rigorous enough. They suggest that landscape architects need 'better knowledge' in order to be regarded as part of a more sophisticated and holistic profession. It is a broad concern that is defined in the following three ways: a need for better theoretical and/or technical expertise; a need for research; and a need for greater academic rigour [17] (p. 68). Albert Fein also commented in his report on a survey of the profession in the 1970s that tacit abstracted artistic ability, or design creativity, was only one factor by which individual practitioners differentiate their solutions; however, all solutions, in order to be valid, must satisfy certain scientific needs and criteria with explicit abstraction [18]. Additionally, as Andrew Abbott explained in his award-winning book *The System of Professions* [7], the survival of a modern profession from competition with other related professions relies on a profession's control of knowledge or its knowledge abstraction and logical consistency. A higher level of abstract explicit knowledge allows practitioners to expand their jurisdiction to adjacent jurisdictions. Compared with explicit high abstraction, tacit low abstraction is often more vulnerable in terms of logical consistency and scientific rationality, which could leave a profession's jurisdiction vulnerable when demonstrating its cultural legitimacy to the public [7].

As a response to the concerns about knowledge and jurisdiction above, some disciplines or professions have begun to consider the abstraction of professional knowledge in their respective disciplines. The civil engineering profession enhanced its body of knowledge, defining the knowledge, skills, and attitudes necessary to enter the practice of civil engineering at a professional level in the 21st century [19]. The audit profession further examined a scenario where jurisdictions and boundaries between professional groups are ambiguous [20]. Recent developments in the profession of landscape architecture have gradually led to a more rigorous research tradition. Longitudinal questionnaires on landscape architecture educators indicate that there has been an increasing pursuit of solid knowledge and evidence-based practice in landscape architecture over the past few decades.

University administrators increasingly apply teaching services and research as evaluation indicators for landscape architecture faculty, with research outputs, particularly publications, generally recognised as the most essential [21]. The role of academics has shifted from a focus on professional education to one that includes contributing to the research and development of the discipline [4]. At the same time, professional institutions and firms have begun to take steps to contribute to disciplinary research. Through a professional lens, a landscape architecture firm that has been pushing the boundaries of research and applying design to the field is the OLIN studio [22]. With two offices located in Philadelphia, Pennsylvania, and Los Angeles, the California OLIN studio possesses a dedicated lab focused on landscape architecture research [23]. The labs' purpose and mission state that "OLIN Labs acts as a conduit between academia and practice, recognizing the value of open dialogue and knowledge sharing among individuals with different expertise and methods but with common purpose" [24], maximising innovations by approaching their research through a multi-disciplinary lens. In December of 2022, two OLIN lab representatives published an essay discussing "New Modes and Models for Purpose-Driven Design", further supporting the notion that professionals of the landscape architecture field are acknowledging the professions' shared concerns and developing methods of advancement in the industry as a solution [25].

Despite the growing abundance of rigorous abstract knowledge in landscape architecture, the gap between explicit academic knowledge and tacit professional knowledge remains a concern. In fact, the existing literature generates contradictory results when talking about what research is or what attitudes the practitioners may have about research. Although landscape architects certainly agree, research is important to landscape architecture, and current research seems unsatisfactory for broad application [26]. For example, dedicated research facilities, such as OLIN labs in association with landscape architectural design, are unique to large-scale firms that perform international services, while smaller, local firms still struggle to implement quantitative research and structure into their design practice. As a result of this stagnation, research questions such as the following have rarely appeared in the literature, which relates to the professional application of the scientific knowledge of landscape architecture created by academics: What types of knowledge do landscape architecture practitioners use in practice to support their decisions? How abstract is such practical knowledge? How much of it comes from rigorous academic research, either directly or indirectly? Where do practitioners acquire new knowledge? How many practitioners keep records of academic updates? When practitioners obtain new knowledge, how do they disseminate it? How efficient is such knowledge dissemination on a professional level? What levels of knowledge abstraction may we observe from landscape architecture? In order to resolve these issues, it is important to understand how professionals access, use, and disseminate knowledge in practice and the knowledge abstraction gained from them.

This study collected first-hand empirical data about landscape architecture practitioners' research from three aspects: accessing, disseminating, and applying knowledge in practice. Through a survey given to members of the American Society of Landscape Architects (ASLA) regarding knowledge use and perceptions, we sought to understand how abstract knowledge is implemented in practice. Specifically, how much abstraction of knowledge is involved in accessing, supporting decision making, and disseminating in professional practices?

2. Methods Section

2.1. *The Classification of Knowledge*

Abbott (1988) argued that modern professions are a type of occupation control through the legitimisation of expertise, as well as structural guarantees [7] (p. 54). In order to legitimise this jurisdiction, some design-based professions have attempted to classify research knowledge of the profession. For example, civil engineering proposed six levels of categorisation to systematically differentiate outcome characteristics, including knowledge, comprehension, application, analysis, synthesis, and evaluation. The aerospace profession describes knowledge-generating activities into seven types: transfer from science, invention, theoretical engineering research, experimental engineering research, design practice, production, and a direct trial [27]. Landscape architecture is a young discipline relative to these fields, so it needs to have clearly defined knowledge domains recognised by other disciplines and the public [28]. Some researchers have proposed a "research circle", which locates design at the beginning of the research process as the incentives or topics of research and locates research at the end of the circle as an application of the insights and understandings acquired from the results. This circle can include research into design, research as a response to design, or research leading to or informing design [4]. A study used four sources of forms of evidence to identify ten types of core landscape architecture knowledge domains, none of which were unique to landscape architecture [28]. However, it is unclear which research domains should form the core of the future landscape architecture research agenda [29].

Today's landscape architecture research, as in many oversimplified categorisations, has much of its action and interest centred in the borderlands between categories. Past categorisations that only indicate implicit and explicit knowledge may not be sufficient to describe the many types of research work conducted and published in this field. For example, recent research suggests that technology has evolved into an important area of the landscape architecture profession and that this discipline has shifted from a binary art–science narrative to a tripartite core of emerging 21st-century concepts comprised art, science, and technology [30]. Therefore, this study divides the categories of professional practice into the following four types of knowledge: tacit knowledge, operational knowledge, conceptual knowledge, and systematic knowledge. This classification draws heavily on the research of Nanaka and Takeuchi et al. [31] and reorganises the order of the four types of knowledge according to Abbott's understanding. The proposed method of classification has logical similarities with the 'research circle' [4], other scholars' assertions of knowledge systems, and the classification proposed by the civil engineering profession [19].

2.2. The Levels of Knowledge Abstraction

In this study, tacit knowledge and systematic knowledge are set as two datum points of the degree of knowledge abstraction. Tacit knowledge is considered to be the lowest level of abstraction, which embodies the nature of landscape architecture as an anthropocentric design profession that inherently relies upon human experience [10,32]. It includes knowledge that can be gained through common sense, intuition, individual experience, etc., which can be gained without the practitioner undergoing a systematic and disciplined university education or professional training. Operational knowledge is considered to be a lower level of abstraction, which consistently influences decision making in the design process [33] and, additionally, may influence performance [34]. Fundamentally, operational knowledge is "knowing how thing work". Conceptual knowledge is considered to be a higher level of abstraction, which makes tacit knowledge explicit and codifies it as principles and protocols [35]. This type of knowledge requires practitioners to have a basic knowledge of scientific research and to be able to generalise tacit and practical knowledge. For the purpose of this study, the highest degree of knowledge abstraction is systematic knowledge, which often advances system knowledge building in modern professions through rigorous research. Such advancements are "knowing propositions of a factual nature" [34], such as the discovery of the truth and facts, which require practitioners to have had a systematic university education and research experience, and the ability to explore research and generalise theories. The explanations generated from systematic knowledge often define the prestige of a modern profession. This leads to a gradual increase in the degree of knowledge abstraction from tacit knowledge to systematic knowledge.

2.3. Constructs and Measurements

This study measured the level of knowledge abstraction in practice from three perspectives: access to new knowledge, knowledge dissemination, and knowledge that supports decision making. Questions were designed based on previous peer-reviewed published research on this subject, like Fein's survey [17] of a similar topic in landscape architecture and Milburn and Brown's study [4] on academic contributions in landscape architecture, and these options have been adapted to the literature in recent years in the areas of design practice, research themes [36,37], and methods of investigation (Supplementary Materials).

The perception of where and how landscape architects obtain new knowledge in this profession was gauged by ASLA members' answers to the following two questions: one question asks how ASLA members obtain new information in this profession and how frequently they interact with other professionals in complementary fields that exhibit higher levels of knowledge abstraction (e.g., engineers). This study used a four-degree scale to measure the first question: rarely, occasionally, often, very often, and unsure. For the second question, a five-degree Likert scale was utilised: not at all, not too much, a fair degree, great extent, and very great extent.

Regarding where ASLA members disseminate their research findings, this study considered the following four modes—writing, teaching, presentations, and others—using the question, "How often do you use the following media to disseminate your research findings?"—measured by self-reporting frequencies (rarely, occasionally, often, very often, and not sure).

The degree of knowledge abstraction that supports decision making in practice was measured by self-reporting the frequency of types of thinking or sources of knowledge that are considered during the decision-making process in practice. This study also considers ASLA members' responses to finishing the statement, which types of knowledge is utilised in the practice of landscape architecture today. This question was measured through self-reporting on a scale of not at all, not too much, a fair degree, great extent, very great extent, and or not sure.

2.4. Data Collection and Sampling

As this study is primarily concerned with the level of knowledge abstraction in practice, landscape architecture practitioners are the best sources of information. Therefore, members of the American Society of Landscape Architecture (ASLA), the largest professional association for landscape architecture in North America (15,155 members), were chosen for this study [38]. In order to keep the survey more manageable, this study only randomly sampled a small portion, about 5%, of the selected population, who were all full and honorary members with valid email addresses listed in the ASLA online member directory. In addition, considering the advantages of the Internet and email today, such as collecting and organising a large amount of data at a relatively low cost [39], quick access [40,41], less time-intensive and more participant-controlled surveys [40,41], etc., this research conducted online surveys for data collection via the software SurveyMonkey®, which was used to manage the questionnaire design, distribution, and maintenance.

An invitation email was delivered to all the 769 individuals who were able to successfully receive an email from the sample with a link to the questionnaire website. Conclusively, 239 complete responses were collected. The response rate (adjusted response rate = 31%) is around the average of other web-based surveys in recent years [42]. The distribution of age, gender, educational degree, serving organisation, and job function suggested that the sample was not heavily skewed on any of the background variables.

2.5. Data Analysis

This study used the following three analysis techniques: descriptive statistics, dimension analysis, and cross-analysis.

2.5.1. Descriptive Statistics

Descriptive statistics were used to describe the general research use and perceptions of research by survey participants. The data collected were concluded from multiple-choice, measured on a four or five-degree Likert scale and were treated as quantitative data for most statistical analyses [43].

In multiple-choice questions, options were coded as numbers. This study coded options for each question; the four-degree scale options 'rarely, occasionally, often, very often' measured by frequency were assigned from 1 to 4, and the option 'not sure' was assigned 0. The five-degree Likert scale options 'not at all, not too much, fair degree, great extent, and very great extent' were assigned from 0 to 4.

Standard data in the descriptive statistics are reported by frequencies or by proportion, and unexpected patterns are highlighted. All statistical analyses were conducted using IBM SPSS statistic software (version 26).

2.5.2. Dimensional Analysis

Dimensional analysis breaks down the research objects chosen into different dimensions and then assesses each of these different dimensions. It was used to help classify items into meaningful groups that are easier to understand.

Questions that utilised a scoring system were categorised into two kinds of eigenvalue: the options with scores 0–2 were assigned a value of 1, indicating that the participants considered the option to be less critical or less frequent in the given situation, and the options with scores 3–4 were assigned a value of 2, which expresses the fact that they are considered to be more critical or more frequent.

To represent the gradient change in the degree of abstraction of knowledge more clearly, this study reclassifies the options based on their features and definitions of the four types of knowledge, ranking them according to their degree of abstraction. The classification is as follows: 'accessing to new knowledge' into seven categories (daily life, Internet, consult, conference, design materials, professional printed media, academic materials), 'knowledge dissemination' into five categories (teaching co-workers, Internet, conference, teaching in university, academic materials), and 'knowledge that supports decision-making' into six five categories (rational and intuitive thinking, consult, individual experiences, professional materials, technical standard, academic materials). The specific classification results are summarised in the following table (Table 1).

2.5.3. Cross-Analysis

Cross-analysis is a data-basic analysis method used to describe the interrelationship between variables; this method crosses two or more indicators to find the relationship between variables and discover more features of data [44]. Researchers have proposed that the connections between knowledge domains, rather than the domains themselves, might more appropriately define the core domains of landscape architecture [28]. In this study, to find out the co-occurrence relationship between different variables, multidimensional cross-analysis was performed on options of the same category and options in different categories to count the frequency of the co-occurrence of variables.

Among them, quantitative data are statistical in the same category represented by string diagrams, which are mainly used to show relationships between multiple variables. The line connecting any two segments on the circle is called a string, which represents the relationship between two variables. This diagram is adequately suited for analysing the relationships of complex data, especially bi-directional relationships and the flow of data. On the other hand, quantitative data are represented in different categories utilising a Sankey diagram; this is a specific type of flowchart where data flow from the left option to the right option, and the width of the extended line corresponds to the quantity of the data flow, the same line width is consistent back and forth, and the total number stays the same regardless of how the data flows. These two types of charts were created via an online software called dycharts (https://dycharts.com, accessed on 1 November 2023), which is used to create various types of charts online.

Table 1. Categorisation of options completed according to the definition of the four types of knowledge. (The background colour is consistent with the legend in the results, the darker the colour the higher the abstraction level).

	Access to New Knowledge	Knowledge Dissemination	Supports Decision Making
Tacit knowledge	Travelling; Everyday life; Internet;	Common sense Logic and reasoning; Intuition;	Teaching in co-workers
	Clients; Related professionals; Architects and planners; Engineers; Behavioural scientists; Natural scientists; Humanistic academicians; Applied artists; Liberal artists; Systems analysts and computer specialists;	Client expressed desires; Other specialists;	Publishing on website
	Professional conferences; Short courses and workshops;	Your professional experience; Your professional education;	
Operational Knowledge	Other landscape architects; Design historic precedents; Design competitions;	Specialised knowledge and skills developed by its practitioners; Other specialists; Historical information; The work of other landscape architects;	Presentations and lectures other than conference; Presentations and lectures at professional conferences
Conceptual Knowledge	Professional newsletters; Professional documents and reports; Professional magazines; Professional databases	Technical standard	Teaching in university
Systematic Knowledge	Books; Refereed journals	Research findings; Scientific knowledge from natural sciences (e.g., forestry and biology); Scientific knowledge from social sciences (e.g., psychology); Abstract knowledge from humanistic disciplines (e.g., history; and art)	Writing books; Writing articles for refereed journals

3. Results

3.1. Access to New Knowledge

Practitioners primarily relied on more common and less abstract knowledge to keep updated with new knowledge (Figure 1). They expected this knowledge to guide professional actions directly, while explanations and justifications for these actions were of little concern to them. They more often supported new knowledge in the profession through tacit knowledge (50.23%), consultations (27.77%), operational knowledge (20.61%), and design materials (12.45%). These paths have a lower degree of abstraction and access threshold, making access easy and timely. They also relied on conceptual knowledge like professional magazines (7.82%), as well as documents and reports (5.68%) to obtain new information about landscape architecture, all of which often provide first-hand practice-oriented knowledge with a low degree of abstraction. Fewer practitioners accessed the highest level of abstract systematic knowledge through books (6.9%) and refereed journals (2.48%), both of which are often used to describe the discovery of the truth and facts.

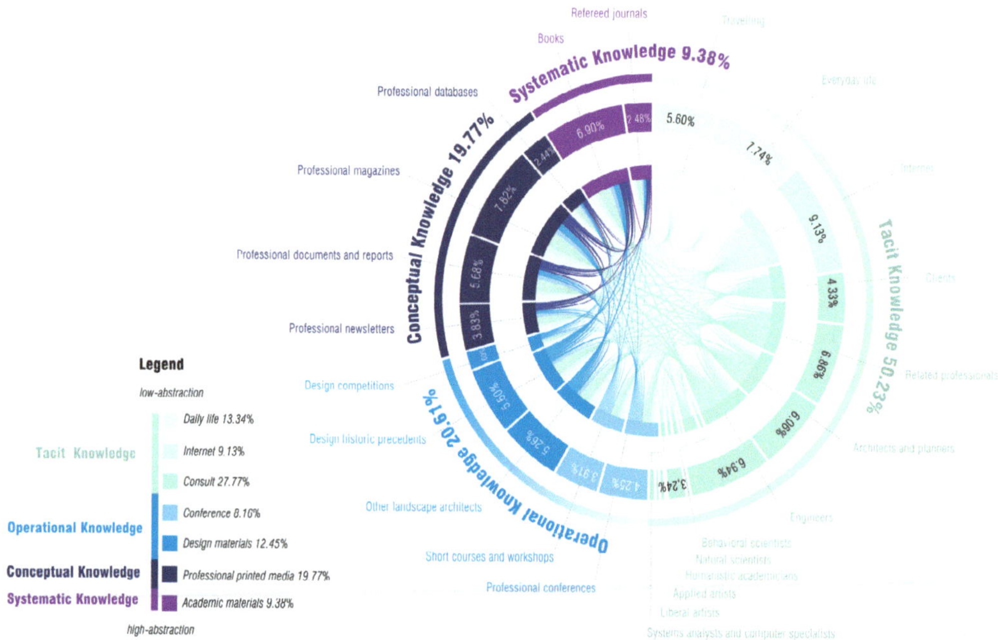

Figure 1. The proportion of practitioners accessing new knowledge at different levels of abstraction and the relationship between access to knowledge in each category, with the outer ring diagram showing the proportion of practitioners who scored 1.5 or more in this category, and the inner string diagram showing the joint selection of categories by each practitioner.

Practitioners who tended to access new knowledge in practice through tacit and operational knowledge rarely accessed new information through systematic knowledge at high levels of abstraction. Parallel to this result, among those who often accessed new professional knowledge through academic materials, only 22.16% also accessed operational knowledge (design materials = 13.66%, conference = 8.5%); another 77.84% acquired new knowledge through other paths. Those who obtained new information through daily life represented 23.12% of respondents. Respondents who acquired knowledge from consult were (28.37%), and 9.06% of respondents acquired the majority of their new knowledge from the Internet. Only 2.12% of respondents acquired new knowledge from only systematic knowledge. In total, 90.38% of those who chose to obtain new knowledge about the profession within their inner professional networks did so through ways other than academic materials, with 18% within this category learning through operational knowledge alone.

3.2. Knowledge Application during Decision Making

Landscape architecture practice, as revealed by the survey, was still perceived as primarily based on tacit (39.83%) and operational (35.71%) knowledge with low abstraction and developed by its practitioners (Figure 2). Practitioners were guided most by rational and intuitive thinking (24.99%), followed by professional knowledge gained from materials (19.10%), including those that are specialised. A relatively significant proportion of practitioners guided their practice through systematic knowledge (16.42%) with the highest abstraction. However, it should be noted that research findings (6.00%) belonging to systematic knowledge are less often used than the individual experience (16.61%), and technical standards (8.04%) belong to conceptual knowledge. Comparatively, even fewer ASLA members believed that this practice is based on natural sciences (4.68%), humanistic knowledge (2.57%), or social sciences (3.17%); together with research findings, these aforementioned subjects constitute research-oriented knowledge that supports decision making in practice. Therefore, these results might imply that professional practice focuses more on providing design solutions than on justifying them through knowledge with a high degree of abstraction.

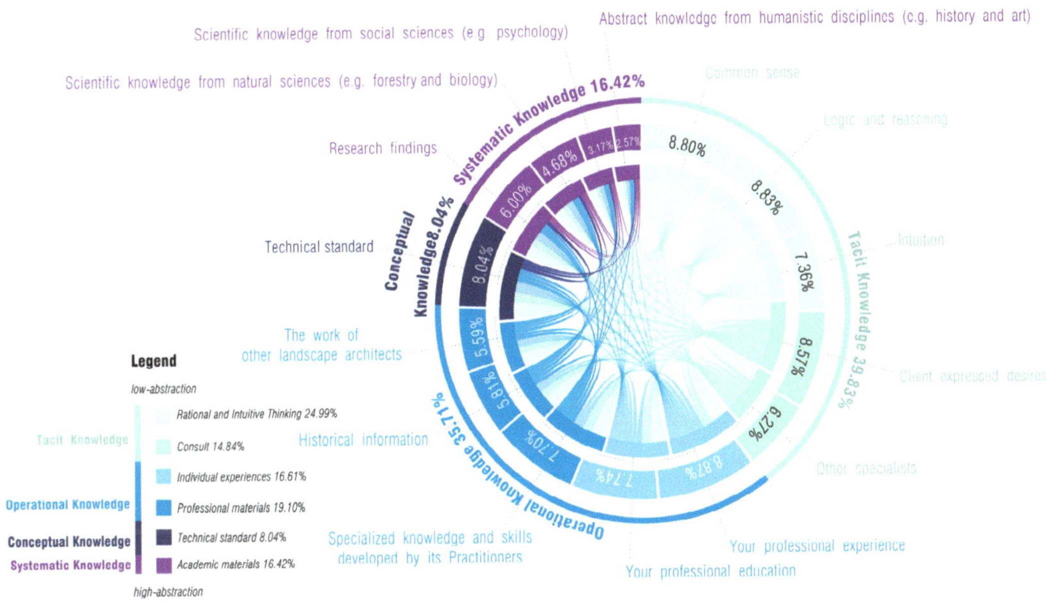

Figure 2. The proportion of practitioners applying new knowledge at different levels of abstraction supports decision making and the relationship between applying knowledge in each category.

Similarly, for obtaining and applying knowledge, about 20.04% of practitioners who applied the highest level of abstraction and used systematic knowledge to guide their practice referred to operational knowledge as well, 64.31% referred to both systematic and other knowledge, and only a small number referred to systematic knowledge only (15.65%). In practice, rational and intuitive thinking, individual experience, professional materials, and technical standards were considered in combination, and systematic knowledge also played an important role but was considered not sufficiently independent to guide practice. Less abstract operational and tacit knowledge still dominate in guiding decision making.

In addition, among the ASLA members who chose systematic knowledge to guide their practice, significantly fewer members thought that landscape architecture practice was based on humanities knowledge (3.41%) or social sciences (3.32%) than members who considered this practice to be based on natural sciences (4.58%). Due to their difficulties in justifying empirically, judgmental design knowledge and human systems knowledge were perceived as less vital, and practitioners could be less motivated to seek out and learn from research studies that fall under this category.

3.3. Knowledge Dissemination

ASLA members often disseminated tacit knowledge via co-workers in working environments (48.61%), which is usually a very small circle, possibly influenced by the traditional design studio pattern of designers taking on apprentices and the confidentiality required for some professional outputs (Figure 3). The dissemination and sharing of operational knowledge through conferences and oral presentations (26.39%) were also one of the significant ways of dissemination, which can reach both researchers and practitioners, as well as business collaborators, government officers, institution organisers, and so on, and which can help them to expand their social network to access more information and resources. These two pathways of knowledge dissemination via oral communication could be modified using the expression of the expresser and, together with other modes of expression, to effectively reduce the abstraction of knowledge. A few ASLA members disseminated knowledge to a larger audience via other sources, such as tacit knowledge via the Internet (5.56%), conceptual knowledge via the university curriculum (10.07%), and fewer members disseminated knowledge systematically via academic materials (professional magazines = 4.86%, refereed journal articles = 2.43%, and books = 2.08%) to disseminate. These pathways required disseminators to have the ability to discover facts and truths and to summarise them into research knowledge at a high level of abstraction.

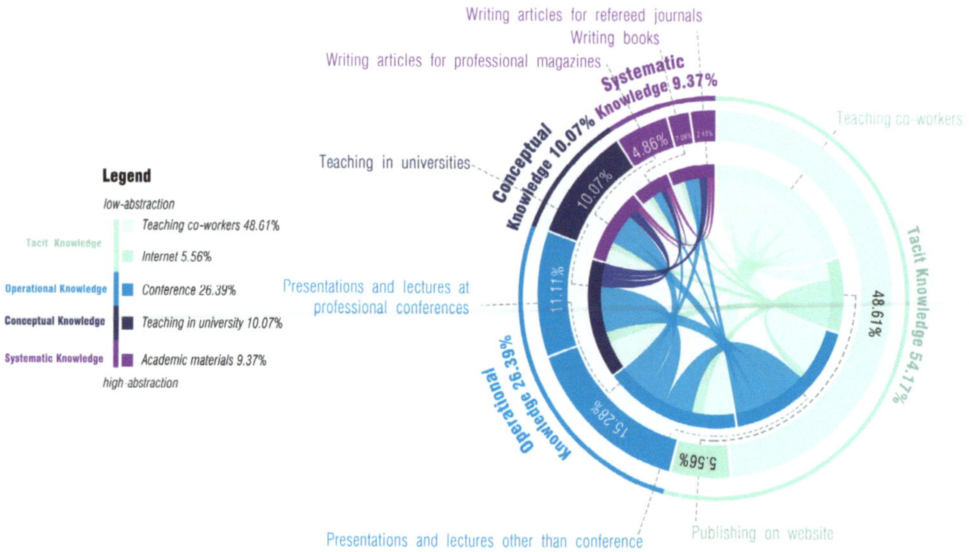

Figure 3. The proportion of practitioners disseminating new knowledge at different levels of abstraction and the relationship between disseminating knowledge in each category.

Most practitioners choose to disseminate tacit and operational knowledge by teaching co-workers and presenting at conferences, respectively, with fewer members disseminating both high-abstraction knowledge and other knowledge as the level of knowledge abstraction increases. Fewer members disseminated only high abstraction knowledge. About half of the practitioners who disseminated by teaching co-workers also did so through presenting at conferences (51.82%), and less than a quarter did so through university teaching (18.18%) and academic materials (18.18%), respectively. When looking into subcategories, it can be seen that from dissemination through presentations at professional conferences (18.10%) to teaching at the university (14.74%), professional magazines (9.33%), peer-reviewed journals (5.97%), etc., the percentage of members flowing into these categories decreases as the level of knowledge abstraction gradually increases. For members who disseminated their knowledge systematically through academic materials with high abstraction, 14.95% disseminated only through the various subcategories of academic materials and more disseminated through other ways, with 38.31% disseminating through conferences and oral presentations as their main additional ways of dissemination.

3.4. Relationship between Knowledge Acquisition, Knowledge Application during Decision Making, and Knowledge Dissemination at Different Levels of Abstraction

Practitioners who tended to access knowledge at various levels of abstraction disseminated tacit knowledge through the less abstract way of 'Teaching co-workers' (54.58%). The probability was that the level of abstraction of knowledge accessed and disseminated by the same practitioner was inconsistent, similar to those who accessed tacit knowledge with the most disseminated systematic knowledge and the highest abstraction. In total, practitioners tended to gain tacit knowledge through the lowest level of abstraction, 'daily life, internet, and consult', most of them disseminated their tacit knowledge through co-workers (55.03%), and less disseminated their systematic knowledge through the pathway of academic materials (6.7%). Only 8.01% of the practitioners who obtained operational information disseminated academically, while the other 51.57% disseminated through the teaching of co-workers. In total, 7.14% of the practitioners acquired systematic knowledge through academic materials and then disseminated them academically, while the other 55.10% disseminated their tacit knowledge by teaching co-workers. Those who tended to acquire systematic knowledge at a high level of abstraction might not have the capacity or preference to disseminate systematic knowledge (Figure 4).

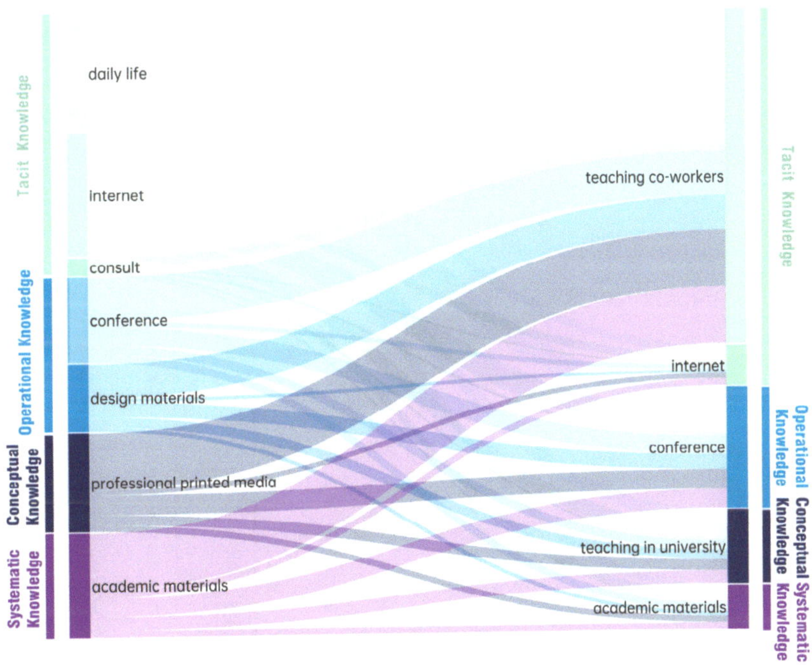

Figure 4. Relationship between knowledge acquisition and dissemination at different levels of abstraction. (The categories shown contain subcategories detailed in the Section 2).

Most of those who accessed knowledge with various levels of abstraction guided their practice by transforming it into tacit and operational knowledge, which are of a lower level of abstraction. In total, 77.10% of the practitioners accessed various knowledge, and guided their practice through other less abstract tacit and operational knowledge, with tacit knowledge (rational and intuitive thinking and consult) guiding the practice the most (40.15%), followed by operational knowledge (individual experience and professional materials) (36.95%). Only a few members who acquired systematic knowledge through academic materials also guided their application through academic knowledge (4.27%), while those who guided their practice through rational and intuitive thinking (20.16%), consultation (20.28%) and individual experience (20.52%) still predominated, and the same tendency was applied to those practitioners who acquired tacit knowledge through daily life; this meant that for both those who acquired systematic knowledge from a high level of abstraction and those who acquired tacit knowledge from a low level of abstraction knowledge, they ultimately tended to guide their practice through tacit and operational knowledge like rational and intuitive thinking, consulting and individual experience with lower abstract (Figure 5).

Practitioners who disseminated knowledge with low abstraction also tended to apply low abstraction knowledge to guide practice, whereas those who disseminated knowledge with high abstraction were less likely to apply high-abstract knowledge to guide practice; furthermore, practitioners who applied high-abstract knowledge to guide practice might not be able or willing to disseminate high-abstract knowledge (Figure 6). Those who disseminated tacit knowledge by teaching co-workers guided their practice mainly through tacit knowledge (40.05%), followed by operational knowledge (36.82%), which is in line with the basic pattern of traditional landscape studios, where experiential knowledge is disseminated mainly through a tiny circle, actions are orientated by task requirements, and there is a need to keep their projects secret. Only 3.95% of those who disseminated systematic knowledge through academic materials applied systematic knowledge to make practical decisions. In general, those who disseminated their systematic knowledge in an academic way had fewer people guiding their decisions through all kinds of knowledge, which might mean that fewer members who tended to disseminate this way were likely to practise. Only 5.36% of practitioners who applied academic knowledge to guide their practice disseminated systematic knowledge in an academic way, implying that those who were able to apply high abstraction knowledge to guide their practice might not possess the ability to condense their knowledge into high abstraction to disseminate, or may not be willing to do so.

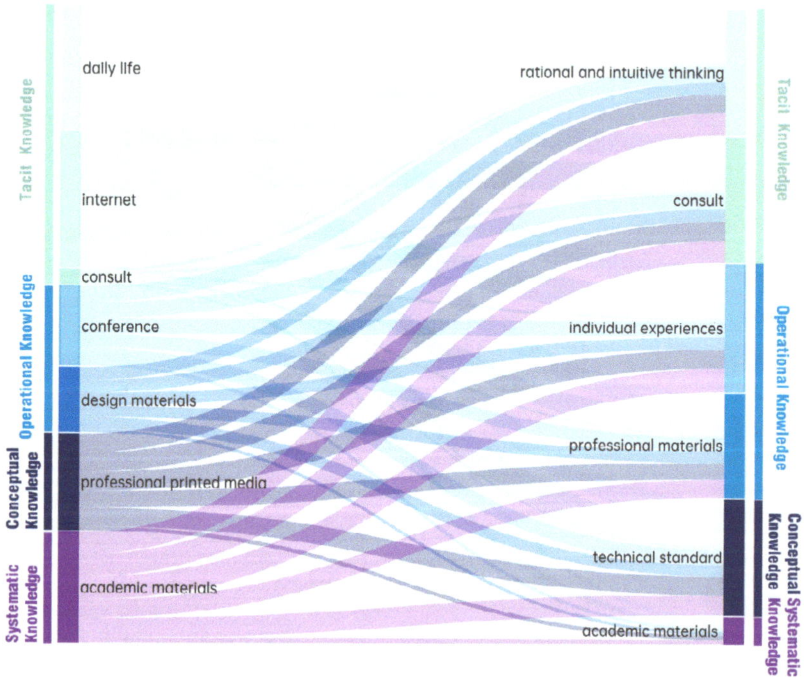

Figure 5. Relationship between knowledge acquisition and knowledge application during decision making at different levels of abstraction.

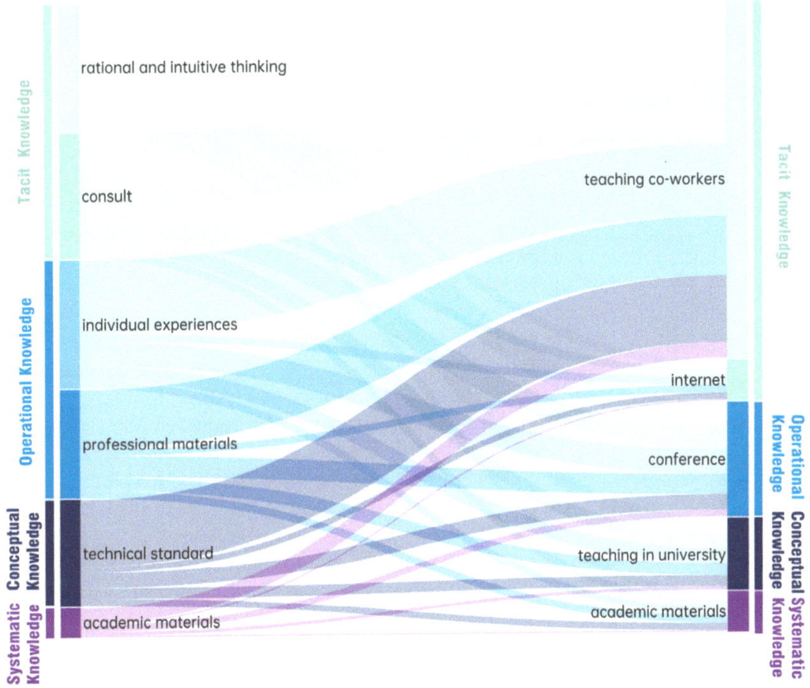

Figure 6. Relationship between knowledge application and knowledge dissemination at different levels of abstraction.

4. Discussion and Conclusions

The ability to demonstrate expertise through rigorous knowledge and research with high abstraction is vital for the survival of landscape architecture as a modern profession. Therefore, modern practitioners are expected not only to perform professional actions successfully but also to be able to back up their ideas with facts and offer rational explanations for their actions. To justify their design actions, the profession expects research to build a solid system of knowledge on which to base justifications. However, as found in a literature review, most independent landscape architecture research is not connected by any subject matter to each other, which may mean that the profession does not yet have an interconnected system [45]. This study's results also confirmed that knowledge used and disseminated in landscape architecture practice still remains at a lower abstraction level compared to other professional fields.

This study investigated the degree of knowledge abstraction in 'accessing', 'disseminating', and 'supporting decision making' aspects of landscape architecture practice. This study aimed to provide researchers and practitioners with information on the state of the profession's knowledge system building and what factors and segments may prevent the profession from enhancing its knowledge through research, which is intended to contribute to removing barriers and eventually facilitating the construction of its body of knowledge. The following section discusses the main findings of this study in the order of access, support for decision making, the dissemination and relationship between them, and the study's main limitations and improvements.

4.1. Major Findings and Suggestions

Professional decisions were made largely based on tacit knowledge (39.83%) and operational knowledge (35.71%). The results showed that design decisions were made more on professional materials (19.10%) and logical and reasoning (8.83%) than research findings (6.00%). Professional knowledge and skills are less abstract, developed, and mastered by practitioners. Research findings are not considered sufficient to guide practice independently, which might not indicate the sufficient contextualisation of academic knowledge and its application. Instead, research findings are more often used in combination with common sense, individual experience, professional materials, and technical standards. This result could be because, in the empirical approach, practitioners often approach design solutions through trial and error [46]. Practitioners gradually improve their solutions and, meanwhile, gain knowledge by learning from their own practice and practice from others. This process does not necessarily involve rational explanations for why specific actions should be taken. Suppose this profession is expected to build a research-oriented practice. In that case, it would need to change its perceptions about research and advance its knowledge through the empirical approach instead of the rational approach. More post-occupancy evaluations can be performed systematically to evaluate new or renovated environments and to identify any mismatch between the original design intent and the actual delivered environment so that evaluators can summarise knowledge from the design solutions derived from their repeated experiments and rectifications can be made (if feasible) to obtain a better match in the future [47].

The knowledge accessed by landscape architecture is primarily less abstract, like tacit knowledge. The results showed that most practitioners rely more on tacit knowledge with the lowest abstraction level, such as consultations (27.77%), daily life (13.34%), and the Internet (9.13%), but less on systematic knowledge with the highest abstraction such as books (6.90%) and refereed journals (2.48%). It may indicate that high-abstraction knowledge is not sufficient to support the broad scope of landscape architecture practice or not able to generate practical design knowledge efficiently [48]; therefore, professionals still have to largely rely on less abstract forms of knowledge acquisition. However, while low abstraction may be sufficient to guide professional actions, knowledge of high abstraction is more important to demonstrate or justify professional actions such as diagnoses, inferences, and treatments, which may impair public trust in the jurisdiction of the landscape architecture profession. Meanwhile, practitioner flows between academics and practice still appear to be limited since practitioners who tend to acquire new knowledge from operational knowledge often do not tend to acquire new knowledge from systematic sources and vice versa. This is consistent with the results of previous literature reviews, and researchers may be inclined to conduct studies independently [45]. Therefore, in order to strengthen the body of knowledge in the landscape architecture profession, practitioners and researchers should work more closely to share the responsibilities regarding knowledge growth. If the competitive advantage of a profession is a concern, they should consider the advantage only in the short term but also in the long term, which requires their investment in knowledge and research. In addition, in practice, they could use research as a prerequisite goal and choose design as a method of realisation in 'research through designing' (RTD) or actively employ the task of designing within the research process [49].

Practitioners mostly share knowledge through low-abstracted tacit knowledge, like teaching co-workers in the workplace (48.61%), similar to traditional apprentice-based studio forms, which have been the designers' traditional forte [50]. In the knowledge dissemination stage, the most chosen 'colleague knowledge-sharing model' is often used in sharing design knowledge [6]. This knowledge is often advanced through an approach in which designers reflect on their previous practical experience, and the tacit nature of this knowledge cannot offer solid justifications for design actions. This form of knowledge sharing is not efficient enough for modern professions in a fast-changing society, and the practitioners in landscape architecture are also not very motivated to conduct or share research. Both may impede the advancement of knowledge in landscape architecture in a fast-changing world and prevent the profession from entering the research world and enhancing academic knowledge. Therefore, practitioners need to externalise this knowledge for more efficient knowledge sharing [51], and the ASLA members need to be more aware of their responsibility to advance knowledge and actively expand how it is disseminated.

Based on the analysis of these three problems, this study also explored the level of abstraction of each practitioner's knowledge choices in the different stages. The results showed that practitioners who tend to access new knowledge through various levels of abstraction and coincidentally chose to disseminate knowledge through tacit knowledge with the lowest level of abstraction, like teaching colleagues (54.58%) and guiding their practice using tacit knowledge like common sense, logical thinking, and intuition; those who disseminate and guide their practice through low abstract knowledge at the same time overlapped, for the most part. In addition, practitioners may not prefer to use knowledge with similar levels of abstraction in all three stages of acquisition, dissemination, and supported decision making. They showed a bias toward selecting knowledge at different levels of abstraction depending on his or her needs at that stage. Understandably, they would be conservative about expanding the knowledge of landscape architecture into a deep holistic understanding. Every profession prior to modernisation experienced a gradual increase in knowledge abstraction and implemented holistic methods of practice. For example, the civil engineering profession has gone through similar changes in response to globalisation, sustainability requirements, and emerging technology; the steps they took were to reform the educational manner to prepare civil engineers before entering the practice field, developing a detailed and rigorous set of barriers to entry that "...supports the attainment of a body of knowledge (BOK) for entry into the practice of civil engineering at the professional level" [19] (p. 2). The aerospace profession has gone through similar changes, and they have taken the approach of extending theoretical tools to make scientific knowledge useful for engineers in practice [27]. The origins of landscape architecture were fundamentally developed from tacit and operational knowledge, which then manifested into the explorations of conceptual knowledge in academia. Evidently, the landscape architecture practitioners of today have found themselves at the precipice of an evolution in the field. The future development of this discipline should make reference to the successful experience of other related professions and establish a solid knowledge system as soon as possible so that the discipline can gain an advantage in professional competition. In the profession, practitioners and researchers should collaborate closely together, fostering a cultural consensus at the early stage of professional education to participate in academic discourses via venues and tools that appeal to practitioners. As a profession, we should also work together performing more research that is directly relevant to practise to bridge the gap between the academy and practice. In recent years, a number of professional firms and institutions have explored some instructive solutions. As previously mentioned, to response growing societal concerns regarding worldwide climate change and rapid urbanisation, many design firms (i.e., OLIN labs, Sasaki, SWA) and organisations (i.e., Landscape Architecture Foundation's Landscape Performance Division) have taken proactive measures like labs to progress in the field of landscape architecture to higher levels of knowledge abstraction and jurisdiction [52].

4.2. *Limitations and Improvements*

This study obtained empirical data on accessing, supporting decision making, and disseminating knowledge in practice, as well as the perceptions of knowledge system-building in the landscape architecture profession. Based on these data, suggestions were provided to enhance knowledge advancement and meet the increasing knowledge and needs of the profession. However, this study is limited and could be improved in the following aspects.

4.2.1. Data Classification

This study divides the answers to questions by the degree of knowledge abstraction and four types of knowledge to help understand the construction of the knowledge system in the three stages of access, dissemination, and support of decision making. This is the basis for categorising. However, the classification of the four knowledge types is not sufficiently granular, although it is able to cover some knowledge outside the boundaries of these two classifications.

4.2.2. Data Verification

Although online questionnaires can target a larger sample size, they do not allow verifications or articulations of data once they are collected. For example, different practitioners may interpret the same option differently, and the cognitive bias involved cannot be verified. Moreover, the forced-choice questionnaire also limits the depth of inquiry. For example, the results showed that many people chose to disseminate knowledge by teaching their colleagues, but the reasons for this result were not confirmed.

4.2.3. Research Object

Considering the significance of the landscape architecture profession's entry into STEM for the development of the discipline, we currently only administered the questionnaire to members of the American Society of Landscape Architecture (ASLA), but this may have limited the results to the developments, and attitudes of practitioners in the United States.

4.2.4. Future Research Considerations

This study provides suggestions on how the profession can be improved through advancing high-abstraction research; however, the validity of these recommendations and their acceptance among practitioners has yet to be discovered, and the next step could be to raise these issues in the current sampling population to ask these questions and also solicit their suggestions. Additionally, this study obtained the trend of knowledge abstraction in each practical stage; however, the reasons for making these choices have yet to be discovered, and a subsequent study may be able to explore the reasons behind these options further. Furthermore, in order to capture the progress in the construction of the academic system in the discipline more holistically, future research could consider more requirements of academic knowledge when designing and consider more dimensions when dividing data. Finally, the conclusions drawn from this study are not yet demonstrative of the profession at a global scale. Further developments in this research could employ the perspective of those in firms and organisations abroad.

Supplementary Materials: The following supporting information can be downloaded at https://www.mdpi.com/article/10.3390/land12112061/s1, Table S1: Questionnaires.

Author Contributions: Conceptualisation, Z.C. (Zheng Chen); methodology, Z.C. (Zheng Chen); formal analysis, Y.Q., Z.C. (Zheng Cong) and X.Q.; investigation, Z.C. (Zheng Chen); data curation, Z.C. (Zheng Chen) and Y.Q.; writing—original draft, Y.Q. and Z.C. (Zheng Chen); writing—review and editing, Z.C. (Zheng Chen), Y.Q., K.N.O., Z.C. (Zheng Cong) and X.Q.; visualisation, Y.Q.; supervision, Z.C. (Zheng Chen). All authors have read and agreed to the published version of the manuscript.

Funding: The work is funded by Shanghai Tongji Urban Planning & Design Institute Co., Ltd. & China Intelligent Urbanization Innovation Center for High Density Region (No. KY-2023-YB-A02).

Data Availability Statement: No new data were created or analyzed in this study. Data sharing is not applicable to this article.

Acknowledgments: Thanks are given to Patrick A. Miller, Terry L. Clements, and Mintai Kim of Virginia Technical University for their help in the question design, data collection and analysis, and results interpretation. In addition, special thanks are given to Patrick A. Miller for his detailed advice and dedication to the writing process.

Conflicts of Interest: The authors declare no conflict of interest. The article does not involve any intellectual property rights of the companies to which the co-authors belong, and they have contributed only the fundamental views and classifications.

References

1. Hardaway, L. Dept. of Homeland Security Designates Landscape Architecture a STEM Discipline. The Dirt (ASLA). 2023. Available online: https://dirt.asla.org/2023/07/12/dept-of-homeland-security-designates-landscape-architecture-a-stem-discipline/ (accessed on 26 October 2023).
2. ASLA; CSLA; CELA; CLARB; LAAB. *Landscape Architecture Body of Knowledge Study Report (LABOK)*; American Society of Landscape Architects: Washington, DC, USA, 2004.
3. Eliot, C.W. A letter. *Landsc. Archit. Mag.* **1910**, *1*, 40.
4. Milburn, L.-A.S.; Brown, R.D.; Mulley, S.J.; Hilts, S.G. Assessing academic contributions in landscape architecture. *Landsc. Urban Plan.* **2003**, *64*, 119–129. [CrossRef]
5. Powers, M.N. A Study of Self-Regulated Learning in Landscape Architecture Design Studios. Ph.D. Thesis, Virginia Polytechnic Institute and State University, Blacksburg, VA, USA, 2006.
6. Schön, D.A. *The Reflective Practitioner: How Professionals Think in Action*; Basic Books: New York, NY, USA, 1983.
7. Abbott, A.; Abbott, A.A.; Abbott, J. The System of Professions. An Essay on the Division of Expert Labour. *Am. J. Sociol.* **1988**, *35*, 141–144.
8. Flexner, A. Is Social Work A Profession? Social Welfare History Project. 1915. Available online: https://socialwelfare.library.vcu.edu/social-work/is-social-work-a-profession-1915/ (accessed on 1 November 2023).
9. Duffy, T.P. The Flexner Report—100 years later. *Yale J. Biol. Med.* **2011**, *84*, 269–276.
10. Rosenberg, A.M. An emerging paradigm for landscape architecture. *Landsc. J.* **1986**, *5*, 75–82. [CrossRef]
11. Vroom, M.J. Landscape architecture and landscape planning in Europe: Developments in education and the needs for a theoretical basis. *Landsc. Urban Plan.* **1994**, *30*, 113–120. [CrossRef]
12. Rodiek, J.E.; Steiner, F.R. Special issue: Landscape architecture research and education. *Landsc. Urban Plan.* **1998**, *42*, 73–74.
13. Taylor, P.D. RE SEARCH: The Generation of Knowledge in Landscape Architecture. *Landsc. J.* **2000**, *19*, 102. [CrossRef]
14. Zube, E.H. Research and design: Prospects for the 1980s. In *Conference Proceedings: Selected Papers Presented at the Annual Meeting of the Council of Educators in Landscape Architecture*; Alanen, A.R., Ed.; Council of Educators in Landscape Architecture and Department of Landscape Architecture, University of Wisconsin: Madison, WI, USA, 1981; pp. 1–11.
15. Riley, R.B.; Brown, B.J. Most important questions. *Landsc. J.* **1992**, *11*, 160–181.
16. Jost, D.; Lamba, B.S. Make research matter: How can we bridge the gap between researchers and designers? A panel at the ASLA Annual Meeting discussed the issue. *Landsc. Archit. Mag.* **2010**, *100*, 26–69.
17. Miller, P. A profession in peril? *Landsc. Archit.* **1997**, *87*, 66–71.
18. Fein, A. *A Study of the Profession of Landscape Architecture*; The Gallup Organization, Inc.: Princeton, NJ, USA, 1972.
19. ASCE. *Civil Engineering Body of Knowledge for the 21st Century: Preparing the Civil Engineer for the Future*; American Society of Civil Engineers: Reston, VA, USA, 2008. [CrossRef]
20. Evans, L.; Honold, K. The division of expert labour in the European audit market: The case of Germany. *Crit. Perspect. Account.* **2007**, *18*, 61–88. [CrossRef]
21. Milburn, L.-A.S.; Brown, R.D. Research productivity and utilisation in landscape architecture. *Landsc. Urban Plan.* **2016**, *147*, 71–77. [CrossRef]
22. OLIN. OLIN Studio. Available online: https://www.theolinstudio.com/ (accessed on 1 November 2023).
23. OLIN. OLIN Studio: Who We Are. Available online: https://www.theolinstudio.com/philosophy (accessed on 1 November 2023).
24. OLIN Labs. OLIN Labs: Labs as Conduit. Available online: https://olinlabs.com/ (accessed on 1 November 2023).
25. Popowsky, R.; Williams, S. Excerpt: 'The Landscape Project' | Weitzman. Stuart Weitzman School of Design. 9 December 2022. Available online: https://www.design.upenn.edu/landscape-architecture/post/excerpt-landscape-project (accessed on 1 November 2023).
26. Steinitz, C. On Landscape Architecture Education and Professional Practice and Their Future Challenges. *Land* **2020**, *9*, 228. [CrossRef]
27. Vincenti, W.G. *What Engineers Know and How They Know It: Analytical Studies from Aeronautical History*; Johns Hopkins University Press: Baltimore, MD, USA, 1990.
28. Langley, W.; Corry, R.; Brown, R. Core Knowledge Domains of Landscape Architecture. *Landsc. J.* **2018**, *37*, 9–21. [CrossRef]
29. Meijering, J.V.; Tobi, H.; van den Brink, A.; Morris, F.; Bruns, D. Exploring research priorities in landscape architecture: An international Delphi study. *Landsc. Urban Plan.* **2015**, *137*, 85–94. [CrossRef]
30. Shen, X. Identifying the role of technology within the discipline of 21st century landscape architecture. *Des. J.* **2023**, *26*, 351–361. [CrossRef]
31. Nanaka, L.; Takeuchi, H. *The Knowledge Creating Company: How Japanese Companies Create the Dynamics of Innovation*; Oxford University Press: New York, NY, USA, 1995.
32. Thompson, I.H. Environmental ethics and the development of landscape architectural theory. *Landsc. Res.* **1998**, *23*, 175–194. [CrossRef]
33. Ryle, G. Knowing how and knowing that: The Presidential Address. *Proc. Aristot. Soc.* **1945**, *46*, 1–16. [CrossRef]
34. Roland, J. On "Knowing How" and "Knowing That". *Philos. Rev.* **1958**, *67*, 379–388. [CrossRef]

35. Deming, M.E.; Swaffield, S. *Landscape Architecture Research: Inquiry, Strategy, Design*; John Wiley & Sons, Ltd.: New York, NY, USA, 2011.
36. Cushing, D.; Renata, A. Themes in Landscape Architecture Publishing: Past Trends, Future Needs. *Landsc. J.* **2015**, *34*, 15–36. [CrossRef]
37. Newman, G.; Li, D.; Tao, Z.; Zhu, R. Recent Trends in LA-Based Research: A Topic Analysis of CELA Abstract Content. *Landsc. J.* **2021**, *39*, 51–73. [CrossRef]
38. Chen, Z. *The Role of Research in Landscape Architecture Practice*; Virginia Polytechnic Institute and State University: Blacksburg, VA, USA, 2013.
39. Witt, K.J. Best Practices in Interviewing Via the Internet. In Proceedings of the Sawtooth Software Conference, Washington, DC, USA, August 1997.
40. Dillman, D.A.; Smyth, J.D.; Christian, L.M. *Internet, Mail, and Mixed-Mode Surveys: The Tailored Design Method*, 3rd ed.; Wiley & Sons: Hoboken, NJ, USA, 2009.
41. Dillman, D.A.; Tortora, R.D.; Bowker, D. *Principles for Constructing Web Surveys: An Initial Statement*; Technical Report 98-50; Social and Economic Sciences Research Center, Washington State University: Pullman, WA, USA, 1998.
42. Cook, C.; Heath, F.; Thompson, R.L. A Meta-Analysis of Response Rates in Web- or Internet-Based Surveys. *Educ. Psychol. Meas.* **2000**, *60*, 821–836. [CrossRef]
43. Babbie, E.R. *The Practice of Social Research*, 10th ed.; Wadsworth Publishing Co.: Belmont, CA, USA, 2004.
44. Lee, S.Y.; Hwang, D.I.; Kim, M.J.; Koh, I.G.; Kim, S.Y. Cross-correlations in volume space: Differences between buy and sell volumes. *Phys. A Stat. Mech. Its Appl.* **2011**, *390*, 837–846. [CrossRef]
45. Burley, J.B.; Singhal, V.B.P.; Burley, C.J.; Fasser, D.; Churchward, C.; Hellekson, D.; Raharizafy, I. Citation Analysis of Transportation Research Literature: A Multi-Dimensional Map of the Roadside Universe. *Landsc. Res.* **2009**, *34*, 481–495. [CrossRef]
46. Alexander, C. *Notes on the Synthesis of Form*; Harvard University Press: Cambridge, UK, 1964.
47. Pati, D.; Pati, S. Methodological Issues in Conducting Post-Occupancy Evaluations to Support Design Decisions. *HERD Health Environ. Res. Des. J.* **2013**, *6*, 157–163. [CrossRef] [PubMed]
48. Powers, M.N.; Walker, J.B. Twenty-five years of Landscape Journal: An analysis of authorship and article content. *Landsc. J.* **2009**, *28*, 97–110. [CrossRef]
49. Lenzholzer, S.; Duchhart, I.; Koh, J. 'Research through designing' in landscape architecture. *Landsc. Urban Plan.* **2013**, *113*, 120–127. [CrossRef]
50. Selman, P. Landscape design as research: An emerging debate. *Landsc. Res.* **1998**, *23*, 195–196. [CrossRef]
51. Nonaka, I.; Byosiere, P.; Borucki, C.C.; Konno, N. Organisational knowledge creation theory: A first comprehensive test. *Int. Bus. Rev.* **1994**, *3*, 337–351. [CrossRef]
52. Landscape Performance: About. Landscape Architecture Foundation. Available online: https://www.lafoundation.org/what-we-do/research/landscape-performance (accessed on 1 November 2023).

MDPI AG
Grosspeteranlage 5
4052 Basel
Switzerland
Tel.: +41 61 683 77 34

Land Editorial Office
E-mail: land@mdpi.com
www.mdpi.com/journal/land

www.ingramcontent.com/pod-product-compliance
Lightning Source LLC
Chambersburg PA
CBHW041357210926
51850CB00013B/3109